EXS 43:
Experientia Supplementum
Vol. 43

Birkhäuser Verlag
Basel · Boston · Stuttgart

New Trends in Research and Utilization of Solar Energy through Biological Systems

Edited by

Hans Mislin
Reinhard Bachofen

1982

Birkhäuser Verlag
Basel · Boston · Stuttgart

This review has been published in two parts in the journal
EXPERIENTIA, Vol. 38, Fasc. 1, pp. 3–66, 1982 and
EXPERIENTIA, Vol. 38, Fasc. 2, pp. 145–228, 1982.

Library of Congress Cataloging in Publication Data

Main entry under title:

New trends in research and utilization of
solar energy through biological systems.

(Experientia supplementum ; v. 43)
1. Solar energy -- Addresses, essays,
lectures. 2. Biomass energy -- Addresses, essays,
lectures. I. Mislin, Hans, 1907–
II. Series.
TJ810.N44 1982 662'.8
ISBN 3-7643-1335-8

CIP-Kurztitelaufnahme der Deutschen Bibliothek

New trends in research and utilization of solar energy through biological systems / ed. by Hans Mislin ; Reinhard Bachofen. - Basel ; Boston ; Stuttgart : Birkhäuser, 1982.
(Experientia : suppl. ; Vol. 43)
ISBN 3-7643-1335-8
NE: Mislin, Hans [Hrsg.]; Experientia / Supplementum

Printed in Switzerland by Birkhäuser AG, Graphisches Unternehmen, Basel
ISBN 3-7643-1335-8

Contents

Editorial commentary 7
Preface 8

Hall, D.O.
Solar energy through biology: fuels from biomass 9

Higher plants as energy converters

Wittwer, S.H.
Solar energy and agriculture 16
Lipinsky, E.S. and Kresovich, S.
Sugar crops as a solar energy converter 19
Calvin, M., Nemethy, E.K., Redenbaugh, K. and Otvos, J.W.
Plants as a direct source of fuel 24
Schwarzenbach, F.H. and Hegetschweiler, Th.
Wood as biomass for energy: results of a problem analysis 28
Pirie, N.W.
Leaf protein as a food source 34

Algae and water plants as energy converters

Wilcox, H.A.
The ocean as a supplier of food and energy 37
Aaronson, S. and Dubinski, Z.
Mass production of microalgae 42
Santillan, C.
Mass production of *Spirulina* 46
Tornabene, T.G.
Microorganisms as hydrocarbon producers 49
Bachofen, R.
The production of hydrocarbons by *Botryococcus braunii* 53
Ben-Amotz, A., Sussman, I. and Avron, M.
Glycerol production by *Dunaliella* 55

Biological photoproduction of hydrogen and ammonia

Guerrero, M.G., Ramos, J.L. and Losada, M.
Photosynthetic production of ammonia 59
Bothe, H.
Hydrogen production by algae 65
Zürrer, H.
Hydrogen production by photosynthetic bacteria 70

Conversion of biomass to fuel and chemical raw material

Stout, B. A.
Agricultural biomass for fuel 73
Wiegel, J.
Ethanol from cellulose 79
Eriksson, K.-E.
Degradation of cellulose 84
Higuchi, T.
Biodegradation of lignin: Biochemistry and potential applications 87
Kaiser, J.-P. and Hanselmann, K. W.
Aromatic chemicals through anaerobic microbial conversion of lignin monomers 95
Hanselmann, K. W.
Lignochemicals 104

The formation of methane from biomass – ecology, biochemistry and applications

Hungate, R. E.
Methane formation and cellulose digestion – biochemical ecology and microbiology of the rumen ecosystem 117
Wuhrmann, K.
Ecology of methanogenic systems in nature 121
Wolfe, R. S.
Biochemistry of methanogenesis 126
Pfeffer, J. T.
Engineering, operation and economics of methane gas production 129
Hobson, P. N.
Biogas production from agricultural wastes 134
Gandolla, M., Grabner, E. and Leoni, R.
Possibilities of gas utilization with special emphasis on small sanitary landfills 137

Future systems

Ghisalba, O. and Heinzer, F.
Methanol from methane – a hypothetical microbial conversion compared with the chemical process 146
Cuendet, P. and Grätzel, M.
Artificial photosynthetic systems 151

Concluding remarks 156

"We are on the threshold of a new solar age."
Wernher von Braun

Editorial Commentary

The idea for Experientia's review, 'New trends in research and utilization of solar energy through biological systems' evolved from my membership in the Swiss Federal Commission for a Total Energy Conception (die Eidgenössische Kommission für die Gesamtenergiekonzeption) as well as from my work in the area of comparative studies in international energy planning. Emerging from these activities has been the realization that if man is to survive, he must seek a new rapport with nature. Establishing this rapport will entail the shift in reliance on ever-diminishing conventional fuels (oil, coal, natural gas, uranium) to the single regenerable and environmentally safe energy source we have – solar energy. And in the effort to efficiently utilize this energy, biological systems are bound to play an increasingly important role. For this reason we are pleased the coordinator of this review, Professor R. Bachofen, has succeeded in putting together such a comprehensive study on the energy potential of these systems. True, there are those who see little point in expending vast efforts to develop solar energy programs when 'the simplest and most efficient answer to all energy supply problems lies in nuclear fusion'. We disagree. Ecological and economic considerations not only justify, they make mandatory intensified research in the field of solar energy – it is this work which will force practical breakthroughs in the generation and harnessing of power for the future.
The articles published here first appeared as a two-part multi-author review in Experientia, 1982 (Vol. 38, issues 1 and 2). The Publishers Birkhäuser have decided to make the entire review available in a special omnibus volume in order that these reports may reach as many specialists and as wide an interested and concerned public as possible.

Professor Hans Mislin
Experientia

Preface

In 1970, when the report by the Club of Rome, 'Limits of Growth', was published and discussed widely, it became obvious to everyone that the conventional forms of energy we use for our daily requirements, as well as many other important resources, are not inexhaustible. The energy shortage in 1973 demonstrated the necessity for discussing and evaluating alternatives to oil. The development of nuclear power is quite advanced in many countries and helps to satisfy a substantial part of the energy demand. But even power from nuclear fission is not unlimited. Moreover, heightened popular opposition to this form of energy and rapidly increasing costs compared to those incurred by earlier hydroelectric power plants may limit severely its further distribution. Therefore, the oldest from of energy used by mankind, solar energy, is becoming fashionable again. Actually, solar energy is not merely an alternative energy source; it is the only form of energy which is able to sustain life by giving food to all living organisms through the process of photosynthesis. Plants are able to convert sunlight into a chemical form of energy which serves as the energy source for all heterotrophs.

The amount of energy from the sun that reaches the surface of the earth is by several orders of magnitude higher than the world's energy requirement (approximately 10^{24} J versus 10^{20} J annually). However, the energy density of the solar radiation on earth is rather low (ca. 1 kW/m^2 at the vertical position of the sun). Therefore, a technical utilization of solar energy requires large absorbing areas. Such light absorbing areas are provided by nature in the form of plants covering the surface of the earth. The complex reactions of photosynthesis have converted sunlight into chemical energy for billions of years. Engineers constantly deny the energetic future of bioenergy and biomass based on the rather low efficiency of energy conversion of photosynthesis of about 0.1% worldwide. In fact, the conversion efficiency for fields and forests is in the range of 1–5%; the low figure is due to the fact that the light is usually not the limiting factor for the biomass yield and that large areas on the earth are not suited for plant production. Furthermore, it is often forgotten that biomass, the product of the photosynthetic energy conversion, is very stable and can be stored for a long period of time in contrast to heat that is gained from sun collectors or electricity from sun cells.

In the present review, we have tried to bring together diverse ideas of many experts in order to give the reader a concrete impression of the potentials in biological energy conversion. Many projects are still in the state of basic research, of laboratory size, and may never be realized on a large scale. Other projects, while not necessarily significant for the energy balance of the world as a whole, may prove to be important locally. Still others, one hopes, may be of universal importance; these include systems for producing biogas from wastes. The first half of this review deals with how solar energy is converted into biomass by higher plants and microorganisms (the latter forming especially interesting products which might, for example, be used directly as a fuel.) The final half is concerned with biological procedures for converting the rather unhandy biomass into convenient forms of energy, with a large section devoted to anaerobic digestion yielding methane.

Professor Reinhard Bachofen
Institute of Plant Physiology
University of Zürich

Solar energy through biology: fuels from biomass

by D.O. Hall

King's College, London SE24 9JF (England)

1. *Introduction*[1-6]

Hardly a day goes by without there being a news item warning us of the impending shortage of oil and what it is going to cost us – assuming we can get it! The belated realization that non-renewable liquid fuels are going to increase in price, and possibly even be rationed, is one of the main reasons why biomass is being looked into so seriously by so many of the developed countries. For the developing countries, the energy problem is as acute – if not more so. The 'woodfuel crisis' is revealing the long-term detrimental agricultural, social and economic consequences of deforestation.

The majority of the people in the world live by raising plants and by processing their products, but now, their governments – and most particularly the governments of developing nations – are confronted with the critical problem of maintaining and possibly increasing consumption without harming ecological systems. A more efficient use of existing biomass and energy alternatives upon which technology is solar- and wind-based is absolutely essential if the present trend of excessive biomass use is to be reversed. The biomass that provides a source of energy now can continue to do so in the future – but to what extent it will be able to contribute to the overall provision of energy will very much depend on the special economic and geographic circumstances within any given country, and the extent to which each country is capable of realistically assessing and planning for its energy requirements in the future.

The oil and energy problem of the last 8 years has already made a clear impact on the use and development of biomass energy. First, in the developing countries there has been an accelerated use of biomass as oil products have become too expensive or even unavailable. Second, in a number of developed countries large research and development programs have been instituted to establish the potential and costs of energy from biomass. Estimated current expenditure is over $100 million per annum in North America and Europe. While this work is still in its early stages, results look far more promising than was thought possible even three years ago. Finally, already in Brazil for example, (a country which currently spends over half of its foreign currency on oil imports), large scale biomass energy schemes are being implemented as rapidly as possible – the current investment is over one billion dollars per annum.

It is well-known that all our fossil carbon reserves are products of past photosynthesis. Photosynthesis is *the* key process in life and, as performed by plants, can be simply represented as

$$H_2O + CO_2 \xrightarrow[\text{solar energy}]{\text{plants}} \text{organic materials} + O_2$$

In addition to C, H and O, the plants also incorporate nitrogen and sulphur into the organic material via light-dependent reactions – this latter function is often not appreciated.

Where, in the past, photosynthesis has given us coal, oil and gas, fuelwood, food, fiber and chemicals, it now seems necessary to look at how photosynthesis fits into the biosphere and explore in what new ways solar energy conversion can become a source of raw materials in the future.

Most people are not aware of the magnitude of present photosynthesis: it produces an amount of stored energy in the form of biomass which is roughly 10 times the amount of energy which the world uses annually. Table 1 shows that the total amount of proven fossil fuel reserves below the earth is only equal to the present standing biomass (mostly trees) on the earth's surface while the fossil fuel resources are probably only 10 times this amount. This massive-scale capture of solar energy and conversion into a stored product occurs with only a low overall efficiency of about 0.1% on a world-wide basis, but because of the adaptability of plants it takes place and can be used over most of the earth.

It is not widely appreciated that 15% of the world's annual fuel supplies are biomass (equivalent to 20 million barrels of oil a day – the same as the USA consumption rate) and that about half of all the trees cut down are used for cooking and heating. This use is confined mostly to developing countries where biomass in rural areas supplies more than 85% of the energy. Total biomass energy in these countries is approximately 4×10^{10} GJ annually. In the non-OPEC developing countries, which contain over 40% of the world's population, non-commercial fuel (including wood, dung and agricultural wastes) accounts for up to 90% of the total energy used. Total wood-fuel consumption is probably 3 times that usually shown in statistics; supply statistics of non-commercial energy can be out by factors of 10 or even 100.

In the present paper, I would like to expand on this evidence that fuels produced by solar energy conversion are a very important source of energy now and will continue to be increasingly so in the forseeable future. However, with today's increased population and standard of living, we cannot rely only on old technology, but must develop new means of utilizing present-day photosynthetic systems more efficiently. Solar biological systems could be realized to varying degrees over the short and long term and some systems using, for example, wood, biological and agricultural wastes, and energy farming, could be put into practice immediately. Photobiological systems can be tailored to suit an individual country according to the energy available, local food and fiber production, ecological aspects, climate and land use. In all cases the total energy input (other than sunlight) into any biological system must be weighed against the energy output and also against the energy consumed in the construction and operation of any other competing energy producing system.

Table 1. Fossil fuel reserves and resources, biomass production and CO_2 balances (Hall[1])

Proven reserves	Tonnes coal equivalent
Coal	5×10^{11}
Oil	2×10^{11}
Gas	1×10^{11}
	8×10^{11} t $= 25\times 10^{21}$ J
Estimated resources	
Coal	85×10^{11}
Oil	5×10^{11}
Gas	3×10^{11}
Unconventional gas and oil	20×10^{11}
	113×10^{11} t $= 300\times 10^{21}$ J
Fossil fuels used so far (1976)	2×10^{11} t carbon $= 6\times 10^{21}$ J
World's annual energy use	3×10^{20} J (5×10^{9} t carbon from fossil fuels)
Annual photosynthesis	
a) Net primary production	8×10^{10} t carbon (2×10^{11} t organic matter) $= 3\times 10^{21}$ J
b) Cultivated land only	0.4×10^{10} t carbon
Stored in biomass	
a) Total (90% in trees)	8×10^{11} t carbon $= 20\times 10^{21}$ J
b) Cultivated land only (standing mass)	0.06×10^{11} t carbon
Atmospheric CO_2	7×10^{11} t carbon
CO_2 in ocean surface layers	6×10^{11} t carbon
Soil organic matter	$10\text{–}30\times 10^{11}$ t carbon
Ocean organic matter	17×10^{11} t carbon

These data, although imprecise, show that a) the world's annual use of energy is only $\frac{1}{10}$ the annual photosynthetic energy storage, b) stored biomass on the earth's surface at present is equivalent to the proven fossil fuel reserves, c) the total stored as fossil fuel carbon only represents about 100 years of net photosynthesis, and d) the amount of carbon stored in biomass is approximately the same as the atmospheric carbon (CO_2) and the carbon as CO_2 in the ocean surface layers.

Solar energy is a very attractive source of energy for the future but it does have disadvantages. It is diffuse and intermittent on a daily and seasonal basis, thus collection and storage costs can be high. However, as plants are designed to capture diffuse radiation and store it for future use, very serious thought (and money) is being given to ideas promoting biomass as a source of, for example, liquid fuels and also for power generation (table 7). I am aware of biomass programs in the UK, Ireland, France, Germany, Denmark, Sweden, USA, Canada, Mexico, Brazil, Sudan, Kenya, Zimbabwe, Australia, New Zealand, India, Indonesia, Philippines, Thailand, Israel, South Korea and China. The greatest obstacle in implementing them seems to lie in the simplicity of the idea – the solution is too simple for such a complex problem! Fortunately for us, plants are very adaptable and exist in great diversity – they could continue indefinitely to supply us with renewable quantities of food, fiber, fuel and chemicals. If the serious liquid fuel problem which is predicted to befall us within the next 10–15 years comes about, we may turn to plant products sooner than we axpect. Let us be prepared!
What I am definitely not suggesting is that any one country should or will ever be able to derive all its energy requirements from biomass; this is highly unlikely except under especially favorable circumstances. What each country (or even region), should do, however, is to look closely at the advantages of and problems with biomass energy systems (table 2). The long term advantages are considerable but implementation of significant programs will take time and require important economic and political commitments. The programs will vary in their emphasis and thus most of the research and development should be done locally. Such R & D is an ideal opportunity to encourage the work of local scientists, engineers and administrators in one field of energy supply. Even if biomass systems do not become significant suppliers of energy in a specific country in the future, the spin off in terms of benefits to agriculture, forestry, land use patterns and bioconversion technology can, I think, be significant.

2. *Efficiency of photosynthesis*[5]

Plants use radiation between 400 and 700 nm, the so-called photosynthetically active radiation (PAR). This PAR comprises about 50% of the total sunlight which (total) on the earth's surface has an average normal-to-sun daytime intensity of about 800–1000 W/m^2.
The overall practical maximum efficiency of photosynthetic energy conversion is approximately 5–6% (table 3) and is derived from the process of CO_2 fixation and the physiological and physical losses involved. Fixed CO_2 in the form of carbohydrate has an energy content of 0.47 MJ/mol of CO_2 and the

Table 2. Some advantages and problems foreseen in biomass for energy schemes (Hall[2])

Advantages	Problems
1 Stores energy	1 Land use competition
2 Renewable	2 Land areas required
3 Versatile conversion and products; some products with high energy content	3 Supply uncertainty in initial phase
4 Dependent on technology already available with minimum capital input; available to all income levels	4 Costs often uncertain
5 Can be developed with present manpower and material resources	5 Fertilizer, soil and water requirements
6 Large biological and engineering development potential	6 Existing agricultural, forestry and social practices
7 Creates employment and develops skills	7 Bulky resource; transport and storage can be a problem
8 Reasonably priced in many instances	8 Subject to climatic variability
9 Ecologically inoffensive and safe	
10 Does not increase atmospheric CO_2	

energy of a mole quantum of red light at 680 nm (the least energetic light able to perform photosynthesis efficiently) is 0.176 MJ. Thus the minimum number of mole quanta of red light required to fix 1 mole of CO_2 is $0.47/0.176=2.7$. However, since at least 8 quanta of light are required to transfer the 4 electrons from water to fix 1 CO_2, the theoretical CO_2 fixation efficiency of light is $2.7/8=33\%$. This is for red light, and obviously for white light it will be correspondingly less. Under the most optimal field conditions values of 3% conversion can be achieved by plants; however, often these values are for short-term growth periods, and when averaged over the whole year they fall to between 1 and 3%.

In practice, photosynthetic conversion efficiencies in temperate areas are typically between 0.5 and 1.3% of the total radiation when averaged over the whole year, while values for sub-tropical crops are between 0.5 and 2.5%. The yields which can be expected under various sunlight intensities at different photosynthetic efficiencies can be easily calculated from graphical data.

Table 3. Photosynthetic efficiency and energy losses[5]

	Available light energy
At sea level	100%
50% loss as a result of 400–700 nm light being the photosynthetically usable wavelengths	50%
20% loss due to reflection, inactive absorption and transmission by leaves	40%
77% loss representing quantum efficiency requirements for CO_2 fixation in 680 nm light (assuming 10 quanta/CO_2)*, and remembering that the energy content of 575 nm red light is the radiation peak of visible light	9.2%
40% loss due to respiration	5.5%
	(Overall photosynthetic efficiency)

* If the minimum quantum requirement is 8 quanta/CO_2, then this loss factor becomes 72% instead of 77%, giving the final photosynthetic efficiency of 6.7% instead of 5.5%.

3. *Implementation of biomass energy schemes*[2, 7–9]

The main factors which will determine whether a biomass scheme can be implemented in a given country are a) the biomass resource, b) the available technology and infrastructure for conversion, distribution and marketing, and c) the political will combined with social acceptance and economic viability. These points are now considered in turn.

a) *The resource base*

The total annual production of biomass (net primary production), the amount of wood produced (including natural forest and managed plantations), and the harvested weight of the major starch and sugar crops are shown in table 4. In addition, there is a worldwide availability of crop residues and other organic wastes (table 5). Although the amount of such wastes has been calculated in some detail for the USA, Canada and certain European countries, where they have been identified as the major short term biomass-resource, such figures are not generally available for the developing countries. Such data that are available are often questionable and cannot, at present, form a basis for any energy planning discussions. In addition to established sources of wood and food, a wide range of other land and aquatic cultivation systems have been proposed for the future. Both established and future options are summarized in table 5. Two usually neglected resources must be mentioned, viz. aquatic plants and algae and also arid land plants.

b) *Technology for conversion*

Biomass as it stands in the field or is collected as wastes is often an unsuitable fuel since it has a high moisture content, a low physical and energy density

Table 4. Annual biomass production in tonnes[7]

Net primary production (organic matter)	2×10^{11}
Forest production (dry matter)	9×10^{10}
Cereals (as harvested)	1.5×10^{9}
as starch	1×10^{9}
Root crops (as harvested)	5.7×10^{8}
as starch	2.2×10^{8}
Sugar crops (as harvested)	1×10^{9}
as sugar	9×10^{7}

and is incompatible with present demands that a fuel be used in internal combustion engines, the main power source for transport and agriculture in most countries. Established conversion technology can be divided into the biological and the thermal (table 6). The great versatility of biomass energy systems is one of their most attractive features – there is a range of conversion technologies already available (and being improved) yielding a diversity of products, especially liquid fuels to which the world seems addicted and upon which most world economies have recently been based.

Plant materials may be degraded biologically by anaerobic digestion processes or by fermentation, the useful products being methane, ethanol and possibly other alcohols, acids and esters. At present the established technologies are the anaerobic digestion of cellulosic wastes to form methane or the fermentation of simple sugars to form ethanol. The most suitable feedstocks for anaerobic digestion are manures, sewage, food wastes, water plants and algae.

The most suitable materials for thermal conversion are those with a low water content and high in lignocellulose, for example wood chips, straw, husks, shells of nuts, etc. The most likely processes to be adopted will use part of the material as fuel for the production of the required mixture of carbon monoxide and hydrogen (synthesis gas) for the subsequent catalytic formation of alcohols and hydrocarbons. During gasification oxygen or steam may be introduced in order to enhance the degree of conversion to synthesis gas and to increase its purity.

Two basic routes of catalytic conversion, of synthesis gas to further products can be recognized. The gas may be converted directly to hydrocarbons via the Fisher-Tropsch synthesis, or may be used for the formation of methanol. Both routes are well established in connection with use of gas produced from coal with plants operating in countries such as South Africa and Germany. Some plants using sorted domestic rubbish are operating and considerable research is being carried out on gasification of wood. On a smaller scale commercial wood-fueled gasification plants have been available for some time, the gas produced being suitable for use in stationary engines.

Table 5. Sources of biomass for conversion to fuels[7]

Wastes	*Land crops*	*Aquatic plants*
Manures	*Ligno-cellulose crops*	*Algae*
Slurry	Trees: Eucalyptus	Uni-cellular:
Domestic rubbish	Poplar	*Chlorella*
Food wastes	Firs, pines	*Scenedesmus*
Sewage	*Leuceana,*	*Navicula*
	Casuarina	Multi-cellular:
Residues	*Starch crops*	Kelp
Wood residues	Maize	
Cane tops	Cassava	*Water plants*
Straw		Water hyacinth
Husks	*Sugar crops*	Water reeds/rushes
Citrus peel	Cane	
Bagasse	Beet	
Molasses		

c) *Energy ratios and economics*

In an ideal world the main factors to be considered in adopting a specific biomass route would relate to the energy gain and the economics. The benefit to be derived by converting plant material to ethanol for example can be expressed in terms of the net energy ratio (NER) which is obtained by dividing the final yield of energy in useful products by the total energy inputs derived from sources other than the biomass itself. In computing the inputs, in addition to fuel, fertilizer and irrigation, a value has to be assigned to the farm and process machinery and to ongoing maintenance. In general a net energy gain is seen where the fermentation and distillation is powered by the burning of crop residues, as in the case of sugar cane, or by burning of wood obtained from close by – as for a cassava alcohol-distillery powered using Eucalyptus wood. Reported NER values for such systems vary from about 2.4 to over 7. For most starch

Table 6. Solar energy for fuels: conversion process and products[7]

Resource	Process	Products	Users
Dry biomass (e.g. wood, residues)	Combustion	Heat, electricity	Industry, domestic
	Gasification	Gaseous fuels → methanol, hydrogen, ammonia	Industry, transport, chemicals
	Pyrolysis	Oil, char, gas	Industry, transport
	Hydrolysis and distillation	Ethanol	Transport, chemicals
Wet biomass (e.g. sewage, aquatics)	Anaerobic digestion	Methane	Industry, domestic
Sugars (from juices & cellulose)	Fermentation and distillation	Ethanol	Transport, chemicals
Water	Photochemical/photobiological catalysis	Hydrogen	Industry, chemical, transport

Simplified table: numerous cross-links exist. Agriculture included in industry. Many important final products not listed.

crops and sugar beet the values are close to or below 1, i.e., more energy is used than is produced. However, this may still be worthwhile if the fuel source is for instance cheap coal of poor quality, wood or residues, etc., which are in effect converted to a high quality fuel.
For the thermal conversion routes an efficiency can be calculated as the ratio of energy in the end product as a fraction of the energy content of the starting material. Since part of the feed is completely combusted to power the conversion, this value must be less than 1. Here the justification is again related to the production of a high quality, higher energy density liquid fuel, from a bulky wet biomass source. At present the efficiency of methanol production from wood is probably about 25%, however efficiencies of around 60% are theoretically feasible.

The estimates of the cost of producing alcohol by fermentation of biomass vary enormously from US 10c/ l to over 60c/l. However, many of these estimates are based on paper studies. Realistic figures from Brazil (1979) are as follows: 30.5c/l for sugar cane alcohol and 31.7c for cassava-derived alcohol as compared to gasoline at an ex-refinery selling price of 23c/l and a retail price of 39.6c/l. The alcohol prices are FOB distillery selling prices calculated for alcohol produced by autonomous distilleries computed to yield the investor a 15% annual return on investment calculated according to the discounted cash flow method and on Proalcool funding of 80% of the fixed investment. Ethanol production, from farm crops, in the USA is profitable at present due to the tax structure. The Federal Government has passed an exemption of gasoline tax on Gasohol (a 10% ethanol:gasoline blend) equivalent to $0.4 per gallon. Various states offer further tax incentives: in Iowa, for instance, the combined subsidies work out at over $1 per gallon. The justification for this lies in the fact that in order to maintain corn prices the government subsidizes each bushel of corn *not* produced with 1 dollar. A bushel of corn can produce 2.5 gallons (US) of ethanol to be used in 25 gallons of gasohol.

Most paper studies indicate that methanol produced by gasification of wood and catalytic resynthesis will be considerably cheaper than ethanol produced by fermentation. The only problem is that no production plants are operating at present. A detailed analysis for a methanol plant in New Zealand can be summarized as follows. At an efficiency of 50% for a 2500 oven-dry tons per day plant at 1977 prices using NZ national cost benefit economics (10% on capital, DCF over 30 years, no tax or depreciation) the product price was $214 per ton using wood at $55 a ton or $146 a ton for wood at $25 a ton. These values are equivalent to product costs of between 17 and 19c a liter, comparable with those summarized recently by the US Solar Energy Research Institute (SERI) where methanol costs from wastes or fuel crops varied from 11c to 35c/l at raw material costs from a negative value for waste to about $50 a ton, with assumed efficiencies of methanol production of between 25 and 50%.

d) *Implementation*

I have already mentioned that the assessment and implementation of biomass energy programs in individual countries is an excellent opportunity for a country to develop its own research, and to demonstrate its capabilities in this area. The types of biomass available for conversion to energy are very much region-dependent, e.g., sugar cane and cassava in hotter climates, cellulose in temperate areas and hydrocarbon shrubs in arid zones. No one country has a monopoly on biomass-for-energy expertise. Indeed the expertise is widespread – note the ethanol program in Brazil, the biogas plants in China and India, gasifiers in Germany, straw burners in Denmark, agro-forestry in East Africa, village woodlots in Korea and parts of India, and so on. There is also the opportunity to encourage collaboration among scientists, engineers, foresters, agronomists, sociologists, economists, and administrators within regions within a country, and among countries. Biomass conversion to energy is an 'old-but-new' and rapidly developing area which interests many young scientists and engineers because it has both immediate practical and also longer-term basic research and development features.
It is imperative that in individual countries accurate energy assessments be made of energy flows, priority needs and available resources; at the same time, the limitations of the data at hand must be recognized. Proposals to implement biomass energy systems must be specific. But even then, policy changes can be effected only if the energy programs have the full support of the decision makers and the people themselves who are convinced of the practical importance of these systems. Otherwise, as experience shows, nothing can be accomplished.

4. *Status of existing biomass projects*

In tables 7 and 8 a short summary of biomass and energy costs of some schemes around the world are listed. Further details are given for a number of these schemes in the references cited[6,10,11]. At present, I will briefly refer only to some European studies.
In Europe a number of countries and the EEC are conducting extensive feasibility studies on the potential which biomass may have for supplying a source of energy and fuels in the future. Trial plantings of alder, willows, poplars, etc., are being undertaken in addition to assessing energy yields from agricultural residues, urban wastes, techniques of conversion, waste land and forest potentials, and algal systems. Biological and thermal conversion equipment is available

Table 7. Estimated biomass and energy product costs (various countries)[1]

Country	Product and source	Cost
1. Brazil (1977)	Ethanol from sugar cane	US$ 16.7/10^6 BTU
	(ex distillery)	US$ 0.33/l
	Gasohol (retail)	US$ 13.18/10^6 BTU
2. Australia (1975)	Ethanol from Cassava	AU$ 250/t
	Industrial ethanol	AU$ 275/t
3. Canada (1975 and 1978)	Methanol from wood	CAN$ 0.35–0.70/gallon
4. New Zealand (1976)	Ethanol from pine trees	NZ$ 260/t
	(500 t/day capacity; credits	(13% return on
	from byproducts)	capital)
5. New Zealand (1977)	Biogas from plants	NZ$ 3.45–5.57/GJ
	Natural gas production cost	NZ$ 1.09/GJ
	Coal gas production cost	NZ$ 6.33/GJ
6. Upper Volta (1976)	Fuelwood from plantation	US$ 0.09/kWh (t)
	Kerosene (retail)	US$ 0.13/kWh (t)
	Butane gas (retail)	US$ 0.11/kWh (t)
	Electricity	US$ 0.19/kWh (t)
7. Philippines (1977)	Electricity from *Leucaena* fuelwood-fired generating station	
	(same cost as oil-fired station)	US$0.014–0.018/kWh
8. Tanzania (1976)	Biogas from dung (for cooking and lighting)/Electricity	US$0.012/kWh
	Casvarina fuelwood to replace coalfired electricity generating station	US$0.113/kWh
9. India (Tamil Nadu) (1978)	(Competitive with coal: 15–30 year payback)	US$ 12/t (dry)

Table 8. Biomass energy product costs compared to conventional – USA (1981 estimate)[15]

Product	Cost from biomass ($/$10^6$ BTU)	Conventional cost ($/$10^6$ BTU)	Biomass: Conventional
Methanol	8.4–15.9	8.4	1.0–1.9
Ethanol	15.0–36.3	19.6	0.8–1.9
Medium BTU gas	4.7–7.4	3.0–5.0	0.9–2.5
Substitute natural gas	4.8–7.3	3.0–5.0	1.0–2.4
Ammonia	5.8–11.4	7.4	0.8–1.5
Fuel oil	3.6–7.9	3.2	1.1–2.5
Electricity	0.03–0.14 ($/kWh)	0.03–0.06 ($/kWh)	0.5–4.5

and is in great demand for use in Europe and for implementing overseas programs.

A recent study by the European Commission (Brussels) shows that in the 9 EEC countries biomass could provide 4% of their total energy requirements in the year 2000 (equal to 50 million tons of oil equivalent or 1 million barrels of oil per day) – this is the same as the agricultural sector's energy requirement and could be achieved with the use of residues and wastes and by utilizing some marginal land with minimal disturbance to conventional agriculture. With a great effort and disturbance to agriculture and forestry the EEC countries could provide 20% of their total energy requirements from biomass, if they so wished. In France, which estimates a biomass potential of 9 million tons oil equivalent by 1990, the large Government R & D program for solar energy gives priority to biomass for energy schemes. The EEC has a substantial biomass program and the UK also is supporting serious assessment and trial projects.

5. *Future photosynthesis*[12–14]

Whole plants

One of the 'problems' with photosynthesis is that it requires a whole plant to function – and the problem with whole plant photosynthesis is that its efficiency is usually low (less than 1%) since many limiting factors of the environment and the plant itself interact to determine the final overall efficiency. Thus a task for researchers of photosynthesis in the future will be to try to select and/or manipulate plants which will give higher yields with acceptable energy output/input ratios. We need much more effort placed on studies of whole plant physiology and biochemistry and their interactions with external (environmental) factors. Fortunately, however, this type of research is already being increasingly funded by both industrial and government organisations.

Examples of the areas in which research is being, or needs to be, done are: photosynthetic mechanisms of carbon fixation; bioproductivity; genetic engineering using plant cell tissue cultures; plant selection and breeding to overcome stresses (drought, temperature or salinity); selection of plants and algae yielding useful products such as oil, glycerol, waxes, or pigments; nitrogen fixation and metabolism and its regulation by photosynthesis.

Artificial photosynthesis

Also to be seriously considered is long-term basic, directed research on artificial photobiological/chemical systems for production of fuels and chemicals (H_2, fixed C, and NH_3); sustained funding will be needed if these exciting possibilities are to be realized.

Since whole plant photosynthesis operates under the burden of so many limiting internal and external factors, would it be possible to construct artificial systems which mimic certain parts of the photosynthetic processes and so produce useful products at higher efficiencies of solar energy conversion? (A 13% maximum efficiency of solar energy conversion is considered a practical limit to produce a storable product). I think that this is definitely feasible from a technical point of view but it will take some time to discover whether it could ever be economic. Note must also be taken of other chemical and physical systems (light driven) which are presently being investigated and may come to fruition before biologically-based systems do so.

A number of proposals have been made to mimic photosynthesis in vitro or to use in vivo photosynthesis in an abbreviated form in order to overcome the inefficiencies and instability factors that seem to be inherent in whole plant (or algal) photosynthesis. The state of the art is still very rudimentary, but we already have some idea of what may be achieved in the future – the scope is enormous, but it may well take 10 years or longer to discover whether any of these systems has any practical potential for the future. Fortunately, the quality of the work being done, and the wide interest and range of disciplines involved augurs profitable results.

Plants perform at least two unique reactions upon which all life depends, viz., the splitting of water by visible light to produce oxygen and protons and the fixation of CO_2 into organic compounds. An understanding of how these two systems operate and attempts to mimic the processes with in vitro and completely synthetic systems is now the subject of active research by biologists and chemists alike.

In vitro systems which emulate the plant's ability to reduce CO_2 to the level of organic compounds are being actively investigated. Recent reports claim the formation from CO_2 of methanol, formaldehyde and formic acid; this is the first time that light has been used outside the plant to catalytically fix CO_2. There has also been one report of the photochemical reduction of nitrogen to ammonia on TiO powder using UV-light.

One recently published idea deals with the photosynthetic reduction of nitrate to ammonia using membrane particles from blue-green algae. This process seems to occur naturally by light reactions closely linked (via reduced ferredoxin) to the primary reaction of photosynthesis, i.e., not involving the CO_2 fixation process. It is an interesting way to produce ammonia! However, it may be possible to use intact blue-green algae (immobilized?) to continually fix N_2 to NH_3 – the cells would have to be genetically derepressed and this is now certainly possible with the recent advances in genetics.

The term 'biophotolysis' is an abbreviation applied to photosynthetic systems that split water to produce hydrogen gas. This applies to both living systems, such as algae, and to in vitro systems comprising various biological components such as membranes and enzymes. We also discuss the so-called 'artificial' systems which seek to mimic the photosynthetic systems by the use of synthetic catalysts. Our bias tends to lean heavily on this last approach.

The great interest in biophotolysis-type systems probably derives from the fact that they are the only energy systems currently known to have the following three attributes: a) a ubiquitous, substrate (water) b) an unlimited driving force (the sun), and c) a stable and non-polluting product (hydrogen). At present, the biological system is the only one that is able to use wide range visible light to catalytically split water to H_2 and O_2; we hope that other systems will be found soon.

Photovoltages and photocurrents

Over the last few years, there have been a number of experiments aimed at constructing electrochemical devices based on the principle of charge separation in photobiological membrane systems. A common approach is to deposit the pigmented membranes onto electrodes (pure chlorophyll, mixtures of chlorophyll and organic redox compounds, chloroplast membranes, and reaction centers of photosynthetic bacteria have been used). Another approach has been to use the biological system with lipid membranes such as liposomes or BLMs (bacteriorhodopsin, photosynthetic reaction centres, and chloroplast membrane extracts have been used); stability has been improved by polymer incorporation into the BLM or by the use of lipid-impregnated Millipore filters. The characteristics of such systems have been described.

6. *Concluding statement*

Photosynthesis is the key process in the living world and will continue to be so for the continuation of life as we know it. The development of photobiological energy conversion systems has long term implications. We might well have an alternative way of providing ourselves with food, fuel, fiber and chemicals in the next century.

Suggested timetable for biomass-for-fuel programs

Next 10 years: Fuels from residues, trees and existing crops; use of existing biofuels; demonstrations and training.

10–20 years: Increased residue and complete crop utilisation, local energy crops and plantations in use.

After 20 years: Energy farming; improved plant species; artificial photobiology and photochemistry.

Acknowledgment. Parts of this article are derived from studies done by the author for a UNESCO study on 'Fundamental World Energy Problems', a UNEP project on 'Photosynthesis in Relation to Bioproductivity', and a UN Biomass Panel paper. In addition various research contracts and reviews by the author have been incorporated.

Addendum. Since the completion of this article 4 symposia have been published which are of direct relevance to this article.

a W. Palz, P. Chartier and D.O. Hall, eds, Energy from Biomass. Applied Science Publishers, London 1981. 982 pp.
b T. Chandler and D. Spurgeon, International Cooperation in Agroforestry, ICRAF, P.O. Box 30677, Nairobi 1980. 469 pp.
c Proceedings Bio-Energy 1980 World Congress. Ed. Bio Energy Council, Washington D.C. 20056, USA. 586 pp.
d D.O. Hall, G.W. Barnard and P.A. Moss, Biomass for Energy in the Developing Countries. Pergamon, Oxford 1981. 206 pp.

1 D.O. Hall, Solar Energy *22,* 307–328 (1979); Nature *278,* 114–117 (1979); Outlook Agriculture *10,* 246–254 (1980).
2 D.O. Hall, Solar energy through biology – fuel for the future, in: Advances in food producing systems for arid and semi-arid lands, pp. 105–137. Ed. J.T. Manassah and E.J. Briskey. Academic Press, New York 1980.
3 M. Slesser and C.W. Lewis, Biological energy resources. Spon, London 1979. 196 pp.
4 V. Smil and W.E. Knowland, Energy in the developing world: the real energy crisis. Oxford Univ. Press, Oxford 1980. 300 pp.
5 Solar Energy: a UK assessment 1976. UK-ISES, 19 Albermarle Street, London Wl, UK, 1976. 375 pp.
6 Proceedings conf. C20 'Biomass for Energy'. UK-ISES, 19 Albemarle Street, London Wl, UK, 1979. 99 pp.
7 J. Coombs, Renewable sources of energy (carbohydrates), Outlook Agriculture *10,* 235–245 (1980).
8 P. Bente, Bio-Energy Directory, 1625 Eye St. N.W., Washington, D.C. 20006, USA, vol.3, 1980. 766 pp; vol.4, 1981. 1201 pp.
9 Office of Technology Assessment, US Congress, Energy from biological processes. OTA-E-124, Washington, D.C. 20510, USA, 1980. 195 pp.
10 P. Chartier and S. Meriaux, L'énergie de la biomasse. Recherche *11,* 766–777 (1980).
11 W. Palz and P. Chartier, Energy from biomass in Europe. Applied Science Publishers, London 1980. 234 pp.
12 J.R. Bolton and D.O. Hall, A. Rev. Energy *4,* 353–401 (1979).
13 M. Calvin, Energy *4,* 851–870 (1980).
14 D.O. Hall, M.W.W. Adams, P.E. Gisby and K.K. Rao, New Scient. *80,* 72–75 (1980).
15 S.J. Flaim and R.E. Witholder, Policy aspects of biomass utilization, in: Solar diversification. Ed. K.W. Boer and G.E. Franta, p. 56–60. AS-ISES Inc. Univ. Delaware, Newark, Del. 19711, USA, 1978.

Higher plants as energy converters

The first paper by S.H. Wittwer describes agriculture as the only major industry that processes solar energy. On one side this happens through photosynthesis and production of biomass. He discusses the possibilities for increasing the biomass yield for food, fiber and fuel as well as the use of sunlight for biological nitrogen fixation. On the other side solar energy can be used for many processes on the farm such as drying grain, heating livestock stables and greenhouses.

E.S. Lipinski and S. Kresovich discuss sugar crops (sugar cane, sugar beet and sweet sorghum) as important energy plants. They outline the features of each in cultivation and detail the procedures for processing and converting the crops into useful products and fuel.

M. Calvin et al. present plants with a high content of hydrocarbons, e.g. *Euphorbia* species, for energy crops on arid land. While these plants have a low requirement for water and nutrients, their production of biomass is relatively high.

F.H. Schwarzenbach and T. Hegetschweiler discuss energy conservation in trees and wood, the oldest form of biomass to be used as fuel by man. They deal primarily with the production of wood in the industrial countries of temperate climates rather than with energy farming in the tropical and subtropical zones.

Finally in the paper by N.W. Pirie, a special biomass product, leaf protein, is proposed as an alternative source of protein for food and fodder.

Solar energy and agriculture*

by Sylvan H. Wittwer

Agricultural Experiment Station, Michigan State University, East Lansing (Michigan 48824, USA)

Agriculture stands pre-eminent as the world's first and largest industry. It is our most basic enterprise, and its products are renewable as a result of 'farming the sun'. Through the production of green plants, agriculture is the only major industry that 'processes' solar energy. The greatest unexploited resource that strikes the earth is sunlight and the green plants are biological sun traps. Each day they store on earth 17 times as much energy as is presently consumed worldwide. The goal of agriculture is to adjust species and cultivars to locations, planting designs, cropping systems and cultural practices to maximize the biological harvest of sunlight by green plants to produce useful products for mankind. Many products of agriculture may be alternatively used as food, feed, fiber or energy. Conflicts over the agricultural use of land and water resources for food, feed or fuel production will arise as resource constraints tighten.

Thus, the greatest use of solar energy in agriculture is in agricultural production itself. All farm practices directed toward increased crop productivity must ultimately relate to an increased appropriation of solar energy in the plant system. And yet, while solar energy is clean, makes no noise, is widely available, requires no fuels, and is renewable, is non-polluting, and cannot be embargoed by nations, few research efforts on photosynthesis have focused on crop productivity and increased biomass production. Photosynthetic efficiency averages less than 0.1% during the entire year for the major food crops. For most crops that efficiency of solar energy utilization during the growing season does not expend 1%. Only under the best of conditions can it be 2–3% for such crops as sugarcane, maize, hybrid Napier grass and water hyacinths. Yields and photosynthesis are affected by varying environmental pressures and diversity among plants themselves. There are many opportunities in research for enhancing photosynthesis. The approaches may be biochemical, genetical or cultural. One way to most immediatly improve photosynthesis for a large number of food crops species is to genetically alter the plant architecture. Verticle positioning of the flag leaves of the rice plant above the panicles of grain rather than having them droop below is an example of recent technological achievement which is improving yields. A better light receiving system is thus created (fig.). Positioning the flag leaves above the panicles of grain in the rice plant is one approach currently being adopted in Southeast Asia and in the People's Republic of China. The current energy establishments (industries), however, tend to downplay the importance of solar energy in the photosynthetic process and for other agricultural uses. This is perhaps to be expected of organizations that would be threatened by the success of new energy systems. Nevertheless, some noteworthy developments are in progress.

New rice varieties with flag leaves positioned above the panicles of grain. This type of architecture creates an improved light receiving system for capture of solar energy in photosynthesis.

A shift from food production to fuel production is being encouraged in many parts of the world as motor fuel prices soar. Alcohol derived from grain and to be used for fuel is the latest rage. This may have serious repercussions not only for the resource economy of fuel production from grain, but also as land and water resources are siphoned away from food production. Energy crops not only compete for land and water but also for investment capital, fertilizer, farm management skills, farm to market roads, agricultural credit, and technical advisory services[1,2].

Maize is the primary raw material being considered for alcohol production in the USA. It is also widely adopted for temperate and subtropical agriculture, and has the highest genetic potential productivity of any major food crop. Maize also has many other productive uses – as a food stuff, animal feed, starch, cooking oil and as a basis for sweeteners. In Brazil, sugarcane is the key raw product, and that nation has a 2-year goal for achieving independence with respect to imported oil.

The potentials for food and feed from biomass, and energy from biomass are enormous. Cellulose is the world's most abundant organic compound, followed by lignin. The biochemistry of lignin's degradation and conversion to food and fuel is much more challenging than that of cellulose.

Vegetable oils, particularly sunflower, rapeseed and soybean oil, are not only useful as foods, but will power diesel engines by direct injection[3] and require only one third the processing energy of ethanol extraction from grain or cellulose. Photosynthesis through the green plant is a remarkable resource which produces an abundance of raw materials which may alternatively be used or converted to fuel, feed, or food. Aside from solar energy through photosynthesis there are many other existing and potential solar energy inputs for agriculture[4–6]. These include nitrogen fertilizer fixation – both biotic and abiotic, grain and crop drying, heating of poultry and livestock housing units and drying of wastes therein, irrigation pumping, water heating for low temperature food processing, biogas generation, greenhouse heating and cooling, solar stills in arid lands, and as solar collectors in protected cultivation utilizing various plastic covers and mulches. Many projects are under development in each area.

Nitrogen fertilizer fixation. Approximately one-third of the fossil energy required for agricultural production in the United States comes from the use of nitrogen fertilizer, and up to 35% of the total productive capacity of all crops is ascribed to this single input. Massive amounts of energy are required for nitrogen fixation whether chemical or biological. The limiting factor in the biological fixation of nitrogen for the *Rhizobium*-legume association is photosynthate production[7]. One of the great gaps in current agricultural research is the lack of integration of studies in photosynthesis with biological nitrogen fixation in crop production.

Biological nitrogen fixation research throughout the world has produced and published enormous amount of valuable information during the past 20 years, but this information has had little significance for practical application under the field conditions of traditional agriculture. A unique approach toward nitrogen fertilizer fixation for developing countries would be an abiotic technology utilizing renewable energy resources - solar, wind or water power. A scaled down electric arc system for atmospheric nitrogen fixation that was displaced over 50 years ago by the Haber-Bosch process is now under review[8].

Farm-size units are now being tested that have the potential of making even the poor small farmer self-sufficient in nitrogen fertilizer. This unit could be inexpensive and easy to operate and maintain. It would utilize nitrogen from the air, and modest electric current from a generator powered by wind, water, or sunlight freely available on many farms. It is now being tested on a farm in Sandpoint, Idaho, at the Solar Energy Project of the University of Nebraska's field laboratory near Mead, and in Nepal with its plentiful supply of fast-flowing water and very little nitrogen. Similarly, the unit appears well suited for adaptation in The People's Republic of China given that country's vast and as yet largely untapped hydroelectric power resources.

Grain and crop drying. Drying requires over 60% of the total energy needed for on-farm corn production in the northern corn belt of the USA as well as the humid south. Solar energy or sun drying in the open offers the only alternative in many developing countries. Considerable interest has now also emerged among industrialized countries, and many sophisticated technologies have been described which use variously designed solar collectors including inflated plastic structures. Their economic feasibility based on current prices of fossil fuels is questionable. Alternative sources of fuels such as crop residues are also under development[9].

Heating of poultry and livestock housing units. Numerous studies - starting with brooder houses to full scale production units - have been completed and other schemes are under development. Numerous tests are also in progress using solar energy for the drying of animal and poultry wastes. Water may be used as the collecting and storage medium. Three years of evaluating a solar collector for a poultry laying house in East Lansing, Michigan, showed fuel and feed savings from higher in-house temperatures capable of paying off the collector in about 5 years[10].

Irrigation pumping. 20% of the total agricultural energy production budget is used for irrigation. It reaches as high as 60% for irrigated agriculture. Energy used for irrigation alone may be twice the total energy inputs for rain fed crop production. While these figures apply to the USA, similar energy inputs exist in other nations. A major research effort should be directed toward the more efficient use of solar energy for crop irrigation. The approach should be most opportunistic, since crop water requirements are usually positively correlated with solar energy intensities.

An extensive project is currently underway in Nebraska using solar power to irrigate a 32-ha corn field. Solar generated electricity powers a center pivot irrigation system's motor and water pump and runs fans which dry the grain once it is harvested[11]. The extra electrical energy is used for the production of nitrogen fertilizer by the electric arc method[12].

Water heating for low temperature food processing. About 6% of the energy budget in the USA is used in food processing and packaging. The milk industry is one of the largest users, requiring much of its energy to create warm water. Currently, natural gas, propane, and electricity constitute the main source for heating of water in food processing plants. It has been demonstrated that solar water heating for food processing plants is now economically feasible in many locations in the USA and could become increasingly so as fossil energy prices escalate. Up to 30-40% of the electric energy demand for warm water in dairy plants may now be provided by solar energy[13].

Biogas generation. The primary raw products are human, animal and vegetable wastes with animal manures predominating. These are subjected to anaerobic bacterial digestion which results in the release of methane. This, in turn, is used for home cooking and electrical power generation. Use thus far has been confined largely to agriculturally developing countries[6]. A chief constraint in India, for example, is that the use of animal manures competes directly with their use as a dried raw product for cooking fuel. Hence in all of India it is estimated there are scarcely 30000 family size biogas generating units. This contrasts with The People's Republic of China where there are an estimated 7 million. China has 2 major biogas research stations, one in Sichuan Province, the other near Shanghi. Solar energy collectors are being utilized in development programs to provide the heat necessary to keep the units operational during the cold winter months. Another approach is to develop microorganisms that can function effectively over a range of temperatures from 5 to 45 °C.

Solar greenhouses, stills, plastic covers and mulches for controlled environment agriculture. Greenhouses - glass, fiberglass or plastic - are natural solar energy collectors. Essential components for further refinements are A-frame constructions with saline solar pond collectors and the provision for insulation to prevent heat loss at night. The possibilities of greenhouse-residences are under review[14].

Solar stills may be a means of providing fresh water for crop production in arid lands. They are semi-cylindrical plastic tunnels or ground mulches positioned over polluted or brackish water or over sandy irrigated soils. They can effect considerable water vapor condensation as well as reduce water losses by evaporation. The opportunity lies in integrated solar still-bubble greenhouse combinations for the production of high value crops[15].

The most extensive use of solar energy collectors for crop production is in the use of plastic covers and soil mulches for promoting early seedling growth and water conservation in temperate zone agriculture. Significant developments are those in China where rice crops from seedlings covered with plastic mature 2–3 weeks earlier. Especially significant results have been obtained with maize in France by mulching with clear plastic. Maturity is advanced by 3–4 weeks and the yields are doubled[16]. Indeed, this is one of the important options for the future enhancement of food production in agriculturally developing as well as for industrialized nations.

In conclusion, a model for future agricultural technology relating to resource inputs would be that which is scale neutral, non-polluting and environmentally benign, which would add to the resources of the earth, result in stable production at high levels, be sparing of capital, management and resources, and which could be alternatively labor intensive or labor saving. Future developments in the use of solar energy for the production and processing of agricultural products should well satisfy the above criteria.

* Agricultural Experiment Station publication No. 9708.

1 L.R. Brown, Food or Fuel: New Competition for the World's Cropland Worldwatch, paper 35. Worldwatch Institute, Washington, DC, 1980.
2 Food and Agriculture Organization of the United Nations. FAO Expert Consultation on Energy Cropping Versus Food, Production. Rome, 1980.
3 J.M. Cruz, A.S. Ogunlowo, W.J. Chancellor and J.R. Goss, Biomass based fuels for diesel engines. Paper presented at Pacific Regional Annual Meetings, American Society of Agricultural Engineers. Hilo, Hawaii, March 18–20, 1980.
4 United States Department of Agriculture. Solar Energy and Non-fossil Fuel Research, USDA/SEA Miscellaneous Publication 1378. Washington, DC, 1979.
5 B.A. Stout, Jr, J.A. Clark, P. Maycock and J. Asmussen, Overview of Solar Energy Technologies – Heating, Photovoltaics, Wind, Biomass. Paper presented at the National Society of Professional Engineers Energy Seminar, Detroit, Michigan, July 23, 1980.
6 R. Revelle, Science *209*, 164 (1980).
7 R.W.F. Hardy and U.D. Halvelka, Science *188*, 633 (1975).
8 B. Treharne, M.R. Moles and C.B. McKibben, A Nitrogen Fertilizer Generator for Farm Use. Technical Note 1. Charles F. Kettering Foundation Research Laboratories, Yellow Springs 1978.
9 F. Payne, Potential of Biomass Conversion Processes for Grain Drying in Kentucky. University of Kentucky AEES-4. Lexington, USA, 1978.
10 F. Hall, M. Esmay, Gardjito, R. Haney and C. Flegal, Solar Heating of Poultry Laying Houses in the Northern States. Michigan State University, E. Lansing, MI, 1980.
11 N.W. Sullivan, T.L. Thompson, P.E. Fischbach and R.F. Hopkinson, Management of solar power for irrigation. Paper presented at the 1978 Winter meeting, American Society of Agricultural Engineers. Chicago, Ill. December 18–20, 1978.
12 B.K. Rein, N.W. Sullivan and P.E. Fischbach, Nitrogen fertilizer production by the electric arc method. Paper presented at the 1980 Mid-Central meeting, American Society of Agricultural Engineers. St. Joseph, Mo. March 21–22, 1980.
13 F.W. Bakker-Arkema and A.L. Rippen, Solar Energy Applications in Food Processing. US Department of Energy-Michigan State University Cooperative Research Project, Contract No. 12147001901, 1980.
14 J.O. Newman and L.G. Godbey, Greenhouse Residence – a New Place in the Sun. Agric. Res. *29*, 4 (1980).
15 A.L. Kamal, Application of 'Bubble' and 'Solarstill' technology. Proc. Int. Conf. Recent Advances in Food Producing Systems for Arid and Semi-Arid Lands. Kuwait Institute for Scientific Research, April 19–24, Kuwait City 1980.
16 S.H. Wittwer, Advances in protected environments for plant growth. Proc. Int. Conf. Recent Advances in Food Producing Systems for Arid and Semi-Arid Lands. Kuwait Institute for Scientific Research, April 19–24. Kuwait City 1980.

Sugar crops as a solar energy converter*

by E.S. Lipinsky and S. Kresovich

Resource Management and Economic Analysis Department, Battelle's Columbus Laboratories, 505 King Avenue, Columbus (Ohio 43201, USA)

Biomass crops are renewable resources with multiple uses that can benefit mankind[1]. Current attention centers on replacement of petroleum as a principal source of energy for transportation applications. Fuels for electric power generation and household heating and/or cooking also are needed. Rising global demand for chemicals, food, construction materials, and paper products increase resource requirements even further.

Sugar crops are of special interest as solar energy converters because effective use of these renewable resources can make available the multiple products that will be required to ameliorate the crises affecting the availability of materials and fuels. Sugar crops achieve high yields, grow in many countries, and can be converted into desirable fuels, chemicals, and other products by application of relatively simple technology. Although sugar crops lack some of the glamour of hydrocarbon crops or aquatic crops, rapid changes are occurring in the technology for growing, processing, and converting these crops into useful products.

1. Sugar crop agriculture

The 3 major sugar crops are sugarcane (*Saccharum officianarum* L.), the sugar beet (*Beta vulgaris* L.), and sweet sorghum (*Sorghum bicolor* L. Moench). These sugar crops are compared with respect to their botanical characteristics, production requirements, current and possible yields, and compositions in the table. This comparison is necessarily oversimplified because special cultivars, production methods (e.g., drip irrigation), or environmental constraints could drastically change the tabular entries for specific geographical areas. For additional in-depth information, the interested reader is referred to some recent reviews on sugarcane[2,3], sugar beets[4], and sweet sorghum[5–7].

Sugarcane is a very high yielding crop that grows well in the wet tropics. Because it can be 'ratooned' (regrown from the plant material left in the ground following harvest), frequent investment in expensive planting operations is not required. The sugar juice is rich in sucrose and is readily crystallized. Therefore, this crop which is already a favorite of tropical less developed countries (LDCs) as a means of earning foreign exchange through sale of raw sugar offers excellent prospects as a source of fermentable sugars for production of fuels and chemicals[8].

The sugar beet presents quite a different picture. It grows best in temperate climates which includes many of the industrialized nations but few LDCs. Each crop requires a separate planting operation. The destruction of sugar beets by pests and diseases necessitates rotation with other crops[4], a distinct disadvantage compared with either sugarcane or sweet sorghum. The low fiber content of sugar beet is a disadvantage because not enough fibrous residues are available to generate the process steam required for ethanol production. However, the sugar beet fiber is a highly attractive cattle feed which is a source of cash revenue. The root of the sugar beet is a sugar-containing organ which is much more storable than is the stalk of either sugarcane or sweet sorghum.

Sweet sorghum resembles sugarcane in that the stalk storage organ contains appreciable sugar and ample fiber to generate process steam[9]. However, sweet sorghum differs markedly from sugarcane in that it is planted from seed and grows to maturity in from 3 to 6 months. The short growing season allows this tropically adapted plant to be grown in geographical areas with temperate climate during their warm season. In this respect, sweet sorghum resembles maize. Although sweet sorghum has been grown for more than 100 years on relatively small land holdings, the status of the crop development is far behind sugar beet and sugarcane. This crop is highly attractive for future development because yield improvement through conventional breeding and genetic engineering can be expected to parallel the achievements with maize.

Sweet-stemmed grain sorghum is a close relative of sweet sorghum in which the sugars that are synthesized are stored partly in the stalk as simple sugars and partly in the grain head as starch[6,7]. This dual storage of photosynthate permits part of the crop to be processed immediately and another part to be stored as readily as is maize or conventional grain sorghum.

2. Sugar crop processing and conversion

As shown in the figure, sugar crops are renewable resources that are created from land, labor, capital, and technology. The millable stalks or beets are transported to a facility whereby juice is obtained for processing into raw sugar or into ethanol or other chemicals. The fibrous beet residues are immediately useful as cattle feed. Sugarcane or sweet sorghum fibrous residues, known as bagasse, can be converted to steam or electricity to perform milling operations, juice evaporation, distillation of fermented sugar solutions, or drying of fiber products. Only about 50% of the fiber is needed to achieve energy self-sufficiency at the processing facility, if modern steam and electricity generation technology is employed. Fiber remains for conversion into construction materials such as plywood substitutes or pulp for paper making or synthesis gas for energy or chemical markets.

Both juice processing and fiber processing facilities are capital intensive and involve economies of scale. Therefore, not all of the technologies shown in the figure necessarily can be carried out at any given location. Where the intent is to ferment sugar crop juice rather than prepare food grade raw sugar, there are opportunities to reduce investment in milling and other juice extraction equipment. This approach is exemplified by the Envirogenics process[10] and in the Ex-Ferm process[11]. Alternatively, the Tilby process[12,13] involves capital equipment investment roughly equal to conventional technology but aims at higher quality plywood substitutes to compensate for the investment.

Interest in sugar crops as solar energy converters has focused on production of ethanol by fermentation of the dilute sugar solutions obtained from sugar crops or from the molasses by-product that occurs in raw sugar production. Technologically, this conversion route is quite appropriate because very little of the energy content of a sugar crop juice is lost when the fermentable sugar is converted into ethanol and carbon dioxide[14]. The effect is to convert a virtually noncombustible material into an effective liquid fuel with a high octane rating. Whether the conversion of sugar crop juice into ethanol is desirable economically as well as technologically depends on the price of raw sugar and molasses and on the willingness of customers to accept larger quantities of these products. When sugar sells for $0.60/kg, a liter of ethanol with a selling price of less than $0.50 per liter consumes enough sugar to make $1.00 worth of raw sucrose. Therefore, it may be more desirable to develop chemical products or fuels from bagasse while selling the simple sugars in food and feed markets. In many LDCs, charcoal is a major cooking fuel in rural areas. Pelletized bagasse could be a strong competitor as a fuel material because charcoal production involves the loss of much of the energy contained in the biomass resource[15].

Major trends in fermentation technologies that are relevant to sugar crops are aimed at cost reduction at every stage of the conversion process. For example, Kirby and Mardon have found that ethanol can be produced in high yields from coarsely ground sugar beets without the necessity for diffusion or other juice extraction processes[16]. This 'solid state' fermentation is said to yield a more concentrated ethanol solution. Similar experiments have been carried out on the pith

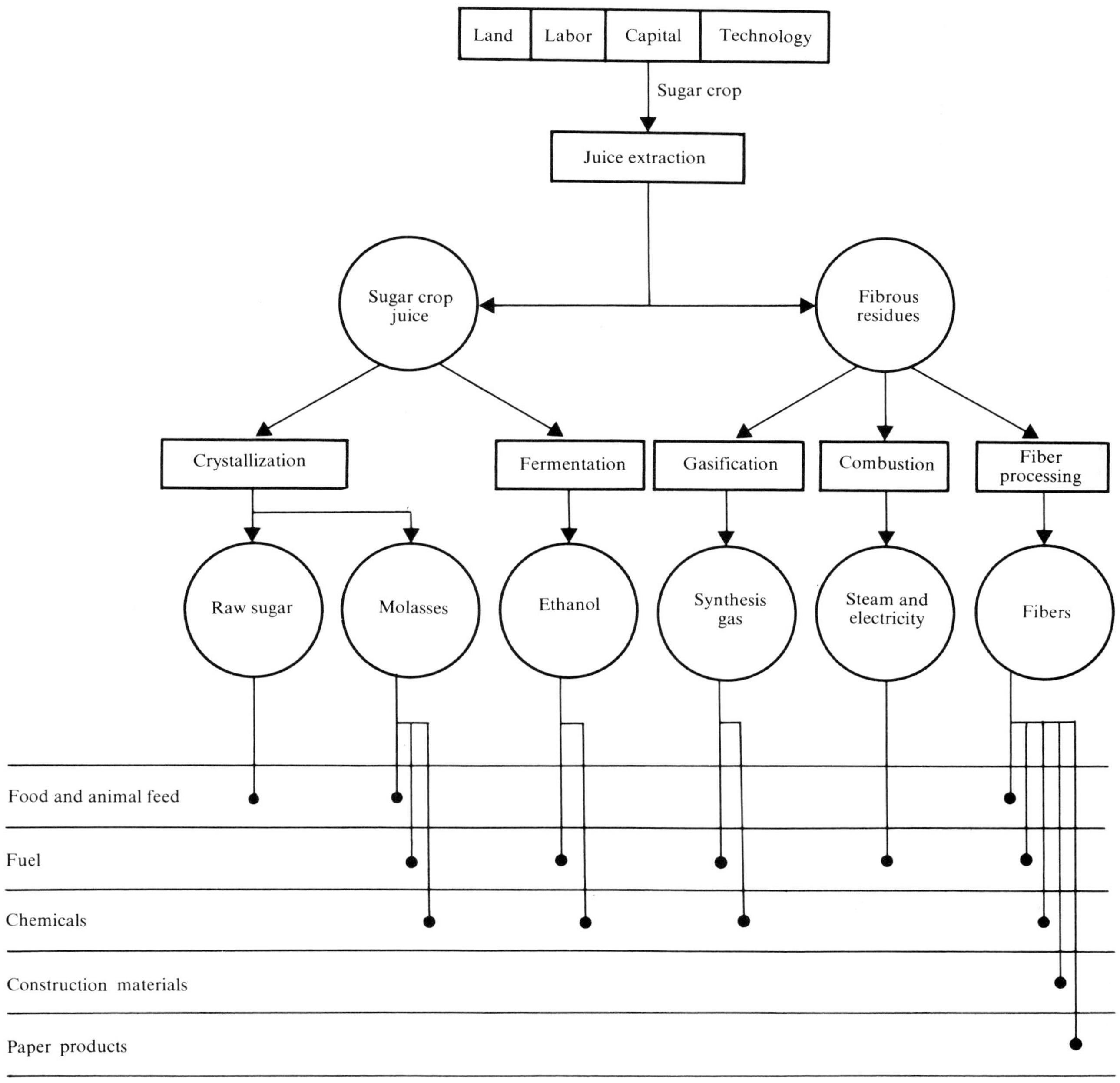

Overview of sugar crop processing systems.

of sweet sorghum. These approaches are related to the Envirogenics[10] and Ex-Ferm[11] processes mentioned previously.

The tower fermenter converts soluble sugars into ethanol in a fixed bed of yeast. The need for numerous fermenter vats is eliminated. Development of this technology requires not only the appropriate hardware but also the development of yeast with the appropriate physical properties (e.g., flocculation rate) to facilitate the process[17]. Fermentation times of 20–30% of batch fermentation times have been achieved readily with tower fermenters.

Fermentation rates are adversely affected by the toxic effects of the ethanol product on the yeast that manufactures ethanol. In recent years, several processes have been proposed and placed under development that involve removal of ethanol by vacuum distillation before it reaches a toxic concentration[18]. The initial embodiments of this concept involve continous removal of the ethanol but problems have been encountered in maintaining the appropriate vacuum when large quantities of CO_2 are evolved at the same time. More recent efforts have included staged processing in which the partially fermented broth is degassed, ethanol removed by vacuum distillation, and the broth is returned for continued fermentation[19,20].

Although yeasts have many advantages as sources of ethanol from sugar crops, other microorganisms have been found to perform this conversion much more rapidly and to function in the presence of higher concentrations of ethanol. The microorganism receiving the greatest attention at present is *Zymomonas* which is under investigation in numerous laboratories[21].

The solar energy conversion process that uses a sugar stalk crop as a collector generates at least as much lignocellulose as simple sugars. More than 70% of the lignocellulose consists of polymers of hexoses and pentoses, with the rest being lignin, waxes, and other materials[22]. If the pentoses and hexoses could be converted to ethanol, the sucrose could still be sold in high value food markets but energy products also could be made from these sugar stalk crops. Recent trends in lignocellulose research and development are greatly increasing the probability that this approach will be used for energy production before the year 2000. For example, a process is under development at MIT that involves direct fermentation of lignocellulose by *Thermocellum* microorganisms[23]. The result is production of ethanol from both cellulose and hemicellulose without need for pretreatment. Alternative approaches include a wide variety of pretreatments that involve fractionation of lignocellulose into

Comparison of sugar crops

Characteristic	Crop Sugarcane	Sugar beets	Sweet sorghum
1. Botanical family and carbon fixation pathway	Gramineae (C-4 pathway)	Chenopodiaceae (C-3 pathway)	Gramineae (C-4 pathway)
2. Sucrose storage organ	Stalk	Root	Stalk
3. Propagation method	Cutting, every 3–12 years	Fruit or seed, in rotation with other crops	Seed
4. Temperature requirements	Tropically adapted (21–40 °C), high freeze damage risk	Temperately adapted (16–28 °C)	Tropically adapted (18–40 °C)
5. Moisture requirements	3 cm rainfall yields 1 ton millable stalks (150+ cm for good yields) wide range	More than 50 cm of rainfall	More than 45 cm rainfall
6. Soil requirements	Sandy loams to heavy clay pH 4.5–8.0	Loams to heavy soils preferred ideal pH = 6.0–7.5. Poor performance in acid soil	Loams and sandy loams, relatively salt tolerant
7. Length of growing season	8–24 months	6–10 months (biennial)	3.5–6 months
8. a) Average yields (wet) b) Average yields (dry)	56–100 ton/ha · yr 15–27 ton/ha · yr	30 ton/ha · yr 6.6 ton/ha · yr	33–44 ton/ha · season 10–13 ton/ha · season
9. Possible yields (wet)	160 ton/ha · yr	65 ton/ha · yr	120 ton/ha · 6 month crop season
10. Typical composition of millable sugar storage organ	Water = 73% Soluble solids = 13% Fiber = 14%	Water = 78% Soluble solids = 16% Fiber = 6%	Water = 70% Soluble solids = 12% Fiber = 18%
11. Typical composition of soluble solids	Sucrose = 85% Glucose = 3.5% Fructose = 3.5% Nonfermentables = 8%	Sucrose = 75% Glucose = 4% Fructose + 4% Nonfermentables = 17%	Simple sugars = 70% (a) Nonfermentables = 30%

a) Sweet sorghum cultivars range widely in content of sucrose versus invert sugars (glucose and fructose) due to development of 'sugar' types and 'syrup' types and environmental conditions during sugar biosynthesis and storage.

lignin, cellulose, and hemicellulose. Steam explosion[24,25], destruction of the lignocellulose complex with lignin solvents[26], and destruction of the lignocellulose complex with cellulose solvents[27] are the 3 major thrusts in pretreatment. It is still too early to determine which approach is likely to be commercially viable.

These developments in the production of ethanol from lignocellulose do not render obsolete the production of ethanol from sucrose and other simple carbohydrates elaborated by sugar crops. Rather, they provide the opportunity for an evolution in which facilities are first constructed to make ethanol from the simple sugars and gradually converted to make use of sugar derived from the lignocellulose. The sucrose content then can be employed in food applications.

Most research and development concerning sugar crops are concerned with increasing yields, improving the energy balance, and reducing milling costs. However, the large scale manufacture of ethanol from sugar crops poses an ecological threat that is requiring increasing research attention. When ethanol is obtained by distillation, approximately 10–15 l of liquid effluent of high pollutant characteristics ('stillage') is obtained for each l of ethanol. Brazil's experience[28] with this stillage indicates that it represents a serious pollution problem. Research is under way in many laboratories but no fully satisfactory answers have been found yet. Typical methods for stillage disposal include: evaporation to dryness for use as fertilizer or animal feed[29], incineration to obtain fertilizer salts[29], spray irrigation, and anaerobic digestion to produce methane[30].

3. Prognosis

Sugar crops appear quite attractive as renewable resources for fuels, foods, chemicals, and other products. The attractiveness is based on firm agronomic foundations involving high yields of simple carbohydrates with valuable properties. These valuable properties include the nutritional and food functional values of sucrose and the fermentability of sucrose/invert sugars. The sugar crops with the best prospects as multiuse resources also contain large quantities of lignocellulose that provide both energy self-sufficiency for processing facilities and a raw material for numerous products based on the fiber, chemical, or energy content of this lignocellulose. Sugar crops will not displace trees or maize as major renewable resources but they will stand as full partners.

The reason for constructing sugar crop production and processing complexes is likely to remain the production of raw sugar to satisfy the steadily rising demand for this basic foodstuff throughout the world. Energy uses of this multiple-use resource will be sources of additional income rather than the reason for engaging in sugar crop production. In other words, sugar crops will not be energy crops but will be food crops with energy by-products.

Ethanol production from sugar crops is likely to grow, especially where stimulated by national energy self-sufficiency programs. Ethanol production can be employed to combat periodic raw sugar surpluses and to consume molasses in areas remote from cash markets for this cattle feed. In countries seeking to achieve self-sufficiency in ethylene-based chemicals, ethanol from sugar crops could provide the basis for an indigenous small-scale-ethylene industry. As with ethanol for motor fuel, national policies and subsidy programs would determine whether such an approach which is technically feasible is economically desirable. Any lignocellulose not needed for achievement of energy self-sufficiency for the ethanol facility could be used as a replacement for charcoal in countries where this cooking fuel is a major expenditure for low income families. Recent trends in fermentation research on lignocellulose indicate that some of the substantial lignocellulose by-product of raw sugar production may be of a source of ethanol.

The sugar crops provide an enormous challenge to those who support and conduct research and development on this planet. These crops and their products could become renewable resources for many countries that otherwise have few resources. The development of sugar crop systems to provide foods, fuels, chemicals, and other products from both the simple sugar constituents and the lignocellulose constituents should have high priority, especially for LDCs. Here, the issue is not food versus fuel – the system generates food, fuel, and cash to promote development.

*Research for this publication was supported in part by the cooperative US Department of Agriculture Forest Service/US Agency for International Development Office of Energy, Biomass for Energy Project. Support of our sweet sorghum research by DOE's Biomass Energy Systems Division also is gratefully acknowledged. The authors thank B.R. Allen and A. Bose of Battelle-Columbus for helpful discussion of fermentation technology.

1 E.S. Lipinsky, Science *199*, 644 (1978).
2 J.E. Irvine, in: Cane Sugar Handbook, 10th edn, p. 1. Ed. G.P. Meade and J.C.P. Chen. Wiley-Interscience, New York 1977.
3 R.P. Humbert, The Growing of Sugar Cane. Elsevier Publ., New York 1968.
4 J.H. Martin, W.H. Leonard and D.L. Stamp, Principles of Field Crop Production, 3rd edn. Macmillan Publ., New York 1976.
5 K.C. Freeman, D.M. Broadhead and Natale Zummo, Agriculture Handbook No. 441. US Dept. of Agriculture 1973.
6 S. Kresovich, in: Handbook of Biosolar Resources, vol. 2. Ed. O. Zaborsky, E.S. Lipinsky and T.A. McClure. CRC Press, Inc., Boca Raton 1981.
7 F.R. Miller and R.A. Creelman, Proceedings of the American Seed Trade Association Annual Meeting – Corn and Sorghum Industry Research Conference, Chicago, Illinois, December 9–11, 1980.

8 Anon., Sugar y Azucar *75* (11), 22 (1980).
9 E.S. Lipinsky, H.S. Birkett, J.A. Polack, J.E. Atchison, S. Kresovich, T.A. McClure and W.T. Lawhon, in: Sugar Crops as a Source of Fuels, vol. II, p. 82; Report No. NTIS TID-29400/2, National Technical Information Service, Springfield, Virginia, 1978.
10 H. Bruschke, G.F. Tusel and A.H. Ballweg, Proc. IVth Int. Symposium on Alcohol Fuels Technology, Brazil 1980; vol. 1, p. 153.
11 C. Rolz, S. De Cabrera and R. Garcia, Biotech. Bioengng *21*, 2347 (1979).
12 S.E. Tilby, US Patent No. 3,567,511.
13 S.E. Tilby, US Patent No. 3,976,499.
14 M. Calvin, Energy Res. *1*, 299 (1977).
15 S.M. Kohan and P.M. Barkhordar, in: Mission Analysis for the Federal Fuels from Biomass Program, 1979; vol. IV, p. 105.
16 K.D. Kirby and C.J. Mardon, Biotech. Bioengng *22*, 2425 (1980).
17 I.G. Prince and D.J. McCann, The Continuous Fermentation of Starches and Sugars to Ethyl Alcohol. Paper presented at Conf. on Alcohol Fuels, Sydney, Australia, August 9–11, 1978.
18 G.R. Cysewski and C.R. Wilke, Biotech. Bioengng *20*, 1421 (1978).
19 B. Maiorella, H.W. Blanch and C.R. Wilke, Rapid Ethanol Production Via Fermentation. Paper presented at American Institute of Chemical Engineers National Meeting, San Francisco, California, November 29, 1979.
20 C.R. Wilke and H.W. Blanch, Process Development Studies on the Bioconversion of Cellulose and Production of Ethanol. Prepared for US Dep. of Energy; Report No. NTIS LBL-9909, National Technical Information Service, Springfield, Virginia 1979.
21 Anon., Chem. Wkly *127*, 19 (December 17, 1980).
22 J.M. Paturau, in: By-Products of the Cane Sugar Industry, p. 26. Elsevier Publ., New York 1968.
23 G.C. Avgerinos, H.Y. Fang, I. Biocic and D.I.C. Wang, A Novel Single Step Microorganism Conversion of Cellulose Biomass to a Liquid Fuels Ethanol. 6th Int. Fermentation Symposium, Toronto 1980; abstracts, p. 80.
24 R.H. Marchessault and J. St-Pierre, in: Future Sources of Organic Raw Materials – Chemrawn. Ed. L.E. St-Pierre and G.R. Brown. Pergamon Press, New York 1980.
25 M. Wayman, J.H. Lora and E. Gulbinas, in: Chemistry for Energy. Ed. M. Tomlinson, ACS Symposium Series 90. American Chemical Society, Washington 1979.
26 R. Katzen, R. Frederickson and B.F. Bruch, Chem. Engng Prog. *76*, 62 (1980).
27 G.T. Tsao, M. Ladisch, C. Ladisch, D.A. Hsu, D. Dale and D. Chou, in: Annual Reports on Fermentation Processes, vol. 2, p. 1. Ed. D. Perlman and G.T. Tsao. Academic Press, New York 1978.
28 E.A. Jackman, Chem. Engr 239 (1977).
29 P. Kujala, R. Hull, F. Engstrom and E. Jackman, Sugar y Azucar *71* (4), 54 (1976).
30 B.P. Sen and T.R. Bhaskaran, J. Wat. Pollut. Control Fed. *34*, 1015 (1962).

Plants as a direct source of fuel*

by Melvin Calvin, Esther K. Nemethy, Keith Redenbaugh and John W. Otvos

Lawrence Berkeley Laboratory and Department of Chemistry, University of California, Berkeley (California 94720, USA)

Summary. Euphorbia lathyris, a plant which has been proposed as an 'energy farm' candidate yields a total of 35% of its dry weight as simple organic extractables. Chemical analyses of the extracts show that 5% of the dry weight is a mixture of reduced terpenoids, in the form of triterpenoids, and 20% of the dry weight is simple sugars in the form of hexoses. The terpenoids can be converted to a gasoline-like substance and the sugars can be fermented to alcohol. Based on a biomass yield of about 25 dry tons ha^{-1} $year^{-1}$, the total energy that can be obtained from this plant in the form of liquid fuels is 48 MJ ha^{-1} $year^{-1}$, 26 MJ in the form of hydrocarbons and 22 MJ in the form of ethanol. A conceptual process study for the large scale recovery of *Euphorbia lathyris* products indicates that this crop is a net energy producer. Several lines of investigation have been started to increase the hydrocarbon yield of this plant. Tissue cultures of *E. lathyris* have been established and will be used for selection, with the aim of regenerating a superior plant. Biochemical studies have been initiated to elucidate regulation of terpenoid metabolism. Future plans include eventual genetic engineering to select the most desirable plant for hydrocarbon production.

Introduction

We have undertaken the investigation of how green plants, such as certain *Euphorbia* species, serve as renewable resources for hydrocarbon production. All higher plants fix CO_2 into carbohydrates. However, there are a number of plant species which can reduce CO_2 further to hydrocarbon-like compounds. A well known plant is *Hevea*, the rubber tree, which belongs to the family Euphorbiaceae. In this family of plants, the genus *Euphorbia* consists of approximately 2000 species, ranging from small herbs and succulents to large trees, the large majority of which produce a milky latex which is often rich in reduced isoprenoids. In general, this genus grows well in semiarid climates. Most of the Euphorbias are native to various parts of Africa but have been found elsewhere throughout the world. At least 20 species are found in South Africa, another half dozen or so in Morocco and Ethiopia and several other candidates for oil producers are indigenous to the Canary Islands. 3 of the wild species which grow in the area of Tenerife, Canary Islands, are shown in figure 1 *(E. regis jubae, E. balsamifera* and *E. canariensis).*

We have chosen *Euphorbia lathyris*, a biennial shrub that grows wild in California as a potential 'energy farm' candidate (fig. 2). *E. lathyris*, like many other

Figure 1. *Euphorbia regis jubae*, *Euphorbia balsamifera* and *Euphorbia canariensis* from Tenerife, Canary Islands.

latex-bearing plants, has never been cultivated, nor has its chemical content and the biochemistry of its abundant secondary metabolites been studied in detail. Therefore, in order to assess this plant as a hydrocarbon producer, various kinds of basic information need to be obtained: a) Crop yield and cultivation conditions; b) optimal methods for extracting the useful components and chemical characterization of the plant extracts; c) suitable methods for modifying the plant extracts into liquid fuel form; and d) various methods of increasing the 'hydrocarbon' content via plant selection, hormone treatment or by tissue culture techniques.

Crop yields and agronomic inputs

The first effort to cultivate *Euphorbia lathyris* began in 1977–1978, when test plots from wild seeds were established at the South Coast Field Station of the University of California in Santa Ana. After a 9-month growing season, the plants attained a typical dry weight of about 500 g and grew to a height of approximately 1 m. Planting density was 30 cm, centers and the plots were irrigated, receiving a total of 50 cm of water, 25 cm of which was natural rainfall. Preliminary results from these experiments indicate that a biomass yield of 25 dry tons per ha and year may be achieved with *E. lathyris*[1].

More recently, 2 important agronomic inputs were tested by Professor R. Sachs at the University of California, Davis[2]. A plot of 0.1 ha was established to determine the yield of *E. lathyris* as a function of nitrogen. The residual nitrogen in the soil (85 kg · ha^{-1}) was sufficient for this crop; there was no significant response to any added nitrogen. A larger plot (0.25 ha) was established to determine yield as a function of irrigation. The amount of applied water varied from zero at the edge of the plot to 60 cm at the center. The corresponding biomass yields after a 7-month growing season were about 10 dry tons per ha in the region of 0–12 cm of applied water, increasing to a maximum of 20 dry tons per ha at the center of the plot where 60 cm of water was applied. However, the amount of heptane extractables per unit dry weight was not a function of irrigation.

Plant extraction

Euphorbia lathyris exudes a milky latex when cut. However, this plant is not amenable to continuous tapping like some other *Euphorbia* species. In order to obtain the reduced photosynthetic material, the entire plant is extracted after drying at 70 °C to 4% moisture content. The reduced organic material is not uniformly distributed throughout the plant; the leaves contain twice as much as the stems per unit weight. Therefore,

Figure 2.

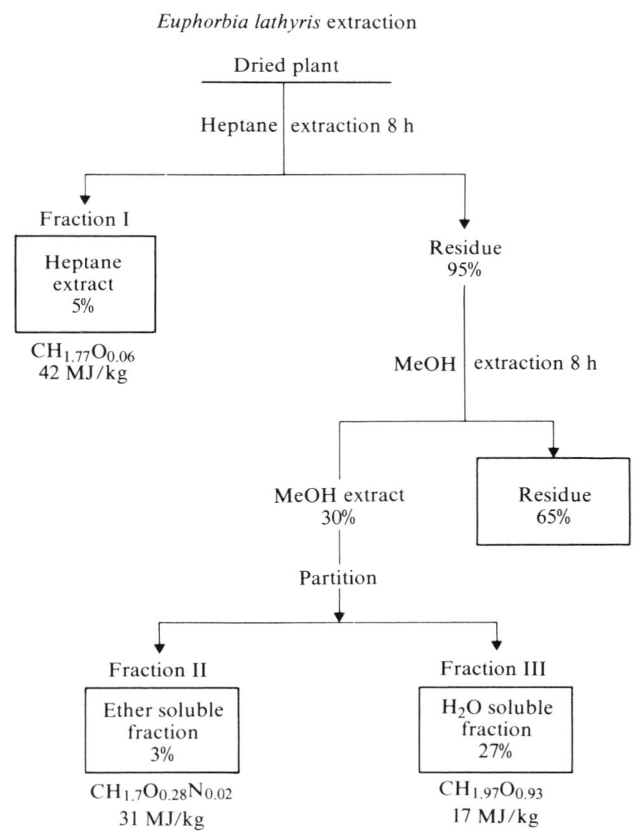

Figure 3.

for uniform sampling, the dried plant is ground in a Wiley Mill to a 2-mm particle size and subsequently thoroughly mixed. A portion of the plant material is then extracted in a Soxhlet apparatus with a boiling solvent for 8 h. Different solvents can be used to extract various plant constituents. One scheme which yields cleanly separated fractions and reproducible results is shown in figure 3.

Acetone can be used as the initial solvent instead of heptane. However, acetone brings down a variable amount of carbohydrates, which precipitate out of solution. This can be filtered off, leaving behind a pure acetone-soluble portion which is 8% of the dry weight of the plant. This is equivalent to the sum of fractions I and II of the extraction scheme (fig. 3) and has a heat value of 38 MJ/kg which corresponds to the weighted average of fractions I and II.

Chemical characterization of the plant extracts

The 2 fractions we have analyzed in detail are the heptane extract (fraction I) and the water soluble portion of the methanol extract (fraction III)[3]. The low oxygen content and high heat value of the heptane extract indicates a potential for use as fuel or chemical feedstock material; because the amount of methanol extract is substantial we investigated its chemical composition as well.

The terpenoids (heptane extract): The heptane extract is a complex mixture which can be separated into crude fractions by adsorption chromatography on silica gel. We have examined the column fractions further by gas chromatography and have obtained structural information on the major components by combined gas chromatography-mass spectroscopy (GC-MS). Molecular formulae were obtained by high resolution mass spectroscopy. The data from the GC-MS analyses indicate that over 100 individual components comprise the heptane extract. About 50 of these are major, and we have either identified or classified them. The major part of the heptane extract consists of various tetra- and pentacyclic triterpenoids functionalized as alcohols, ketones or fatty acid esters. The only non-triterpenoid components of the heptane extract are 2 long chain hydrocarbons which comprise column fraction I and a small quantity of fatty alcohols isolated from column fraction III. The 2 hydrocarbons are straight chain waxes: n-$C_{31}H_{64}$ and n-$C_{33}H_{68}$; the 3 fatty alcohols are $C_{27}H_{53}OH$, $C_{28}H_{57}OH$ and $C_{29}H_{57}OH$. These compounds, however, represent only ~ 8% of the total heptane extract, 7% of the sample is chlorophyll, so 85% of the extract is composed of only 1 class of natural products, triterpenoids[4].

The major terpenoid components of the latex itself have been identified as 5 triterpenoids. All of the latex components with the exception of euphol (a minor one) could also be detected in the whole plant extract. The plant extract, however, yields a much greater variety of triterpenoids than the latex, suggesting that terpenoid synthesis may take place in other parts of the plant as well.

The carbohydrates (methanol extract): As the data in figure 3 indicate, a substantial amount (30%) of the dried plant weight can be extracted with methanol. The empirical formula of the water soluble portion of this extract is indicative of carbohydrates. Since simple hexoses can be directly fermented to ethanol, a useful liquid fuel, we have determined the carbohydrate content of *Euphorbia lathyris* and identified the specific sugars[4]. The results of gel-permeation chromatography of fraction III (Biogel-P-2) indicated that

Comparison of liquid fuel yield for different crops

Process	Process feed, tons/ ton dry biomass	Process efficiency	Tons of liquid fuels/ ton dry biomass	Fuel value of liquid (GJ/t)	Energy in liquid fuel (GJ/t biomass)	Dry biomass yield (t/ha · year)	Water requirement (cm rainfall)	Energy in liquid fuel (GJ/ha · year · cm rainfall) related to area and water requirement	Energy in cellulose residue (10^3 GJ/ha · year)	Energy in cellulose residue related to area and water requirement (GJ/ha · year · cm rainfall)
Corn to ethanol	0.32	0.4	0.128	26.8	3.4	12.3	62.5	0.67	115 (3.4 tons)	1.84
Sugar cane to ethanol	0.2	0.4	0.08	26.8	2.15	74.1	195	0.82	813 (24 tons)	4.17
Energy cane to ethanol	0.08	0.4	0.032	26.8	0.85	112	120	1.40	2493 (73.6 tons)	20.78
Euphorbia lathyris to hydrocarbon	0.08	0.86	0.069	36.9	2.5			0.84		
						21.0	62.5		207 (6.12 tons)	3.3
Ethanol	0.2	0.4	0.08	26.8	2.15			0.72 1.56		

there are no poly- or even oligosaccharides present in this fraction. The carbohydrate-containing fractions were identified by the Molish test and were further characterized by 2-dimensional paper chromatography and high pressure liquid chromatography (HPLC). In both these systems only 4 simple sugars were detected: sucrose, glucose, galactose and fructose in a ratio of 7.4:0.5:0.5:1.0, respectively. These 4 sugars comprise 20% of the plant dry weight. The water soluble fraction (fraction III of fig. 3) also contains 3 amino acids: alanine, valine and leucine. The amino acid fraction comprises 3–4% of the dry plant weight.

This high carbohydrate content enhances the liquid fuel yield of *E. lathyris* significantly. Even though the heat value of alcohol is about two-thirds that of hydrocarbons, the great abundance of sugars in this plant doubles the energy obtained in the form of liquid fuels. In addition, no pretreatment such as acid hydrolysis or enzymatic digestion, necessary for starch producing plants, is needed to process the *E. lathyris* extract to alcohol.

Isolation and testing of the biologically active components: The latexes of several *Euphorbia* species contain chemicals which are strong skin and eye irritants, some of which exhibit tumor promoting activity. The concentration of these compounds is very low (~0.01%) compared to the triterpenoids which represent ~50% of the dry latex weight. However, nanogram levels of these compounds exhibit promoter and irritant activity. From the latex of *E. lathyris* we have isolated a fraction which is composed of several ingenol esters. Biological tests showed that this latex-derived extract contains potent compounds which act like the best known, chemically-related tumor promoter 12-tetradecanoylphorbol-13-acetate (TPA). In contrast, the heptane and methanol extracts of the dried plant show little or no activity, indicating that the drying and hot solvent extraction degrades the active components. This finding has significant bearing on future harvesting and larger scale processing plans. Our preliminary results indicate that after mechanical harvesting and drying there should be no toxicological dangers.

Conversion of plant extracts to liquid fuels

Conversion to gasoline. Euphorbia lathyris yields 5–8% of its dry weight as reduced photosynthetic material; however, if this terpenoid extract is to be used as conventional fuel, then further processing of this material is necessary. The conversion of biomass derived hydrocarbon-like materials to high grade transportation fuels has recently been demonstrated by Mobil Research Corp.[5]. Various biomaterials such as triglycerides, polyisoprenes and waxes can be upgraded to gasoline mixtures on Mobil's shape selective Zeolite catalyst. The terpenoid extract of *E. lathyris* was processed under similar conditions with this catalyst[6].

Fermentation to alcohol: The entire crude carbohydrate extract (fraction III of fig. 3) as well as the solid residue prior to methanol extraction is fermentable to ethanol. Several different fermentation conditions were tried, using various yeasts. An 80% fermentation efficiency was achieved on the crude extract; the dry residue can also be fermented directly, but with lower efficiency.

We can therefore obtain not only hydrocarbons from *Euphorbia lathyris* but a substantial quantity of ethanol as well. Per unit dry weight the sugar content of *E. lathyris* equals that of sugar cane. Our present biomass yield of 25 dry tons per ha and year yields approximately 2 tons crude extract which is converted to gasoline and 5 tons of sugars fermentable to alcohol.

Comparison with conventional biomass processes

Since *Euphorbia lathyris* and other hydrocarbon-producing crops are new species from the point of view of cultivation, their agronomic characteristics, requirements and yield potentials are not yet well known[7]. With further agronomic research and plant selection the biomass as well as the terpenoid yield is expected to increase. Nevertheless, it is interesting to compare in terms of energy yield a new crop like *E. lathyris* to other established crops such as corn or sugar cane.

The table shows the basic crop and the derived liquid fuel characteristics needed for this comparison, as well as the liquid fuel yield in terms of a very important agronomic input, the water requirement. The different fertilization requirements for these crops are not taken into account; fermentation efficiency is kept constant at 80% for the purpose of a comparison. In terms of liquid fuel production the crop which is comparable to *E. lathyris* is so-called 'energy cane' under development in Puerto Rico. Energy cane produces less sugar than normal cane (8% vs 20%). However, it yields an extraordinary 100 dry tons of biomass per ha. As the last column in the table indicates its yield of cellulose per unit of water input is significantly better than the other crops; however, its yield of liquid fuel (ethanol) is comparable to *E. lathyris.*

Increasing the biomass yield as well as the terpenoid content is of primary importance in the development *E. lathyris* as a new energy crop. In particular, raising the terpenoid yield per unit dry weight leads to large cost reduction in processing[8].

Plant tissue culture

Actively growing green cultures of *Euphorbia lathyris* have been established. The plant source materials (hypocotyl segments of seedlings and leaves of young plants) and the media and hormone requirements for producing actively growing callus cultures were determined. These cultures have been maintained for over a year. In addition, the cells have been placed in suspension culture for several months, and then brought back to callus culture on a semisolid medium. Roots have been regenerated from newly produced callus, and both roots and shoots have been produced from cultured hypocotyl segments through normal manipulation. Current efforts are directed at regenerating entire plants from callus cultures so that any yield improvements made in vitro can be tested in regenerated plants.

Triterpene biosynthesis in callus cultures: Since tissue cultures quite frequently do not synthesize the secondary metabolites characteristic of the parent plant at all, or do so in minute amounts, it was necessary to establish first whether *E. lathyris* callus tissue produces any terpenoids. By using our previously established techniques for triterpenoid isolation and characterization, we determined that *E. lathyris* callus cultures grown in the dark do produce 0.1–0.2% triterpenoids on a dry weight basis. This yield is comparable to that reported for secondary metabolism in other callus species.

Since *Euphorbia lathyris* tissue cultures do have the capability of triterpenoid biosynthesis, we now have a system which can be effectively exploited for the selection of desirable callus lines as well as for the exploration of secondary metabolic pathways.

*This work was supported by the Assistant Secretary for Conservation and Solar Energy, Office of Solar Energy, Solar Applications for Industry Division of the US Department of Energy under Contract No. W-7405-ENG-48.

1 M. Calvin, Bioscience *29*, 533 (1979); Naturwissenschaften *67*, 525 (1980).
2 R. Sachs, Agronomic Studies with *E. lathyris* – A Potential Source of Petroleum-like Products, in: Energy from Biomass and Wastes, V. Institute of Gas Technology, in press 1981.
3 E.K. Nemethy, J.W. Otvos and M. Calvin, J. Am. Oil chem. Soc. *56*, 957 (1979).
4 E.K. Nemethy, J.W. Otvos and M. Calvin, Pure appl. Chem, in press (1981).
5 P.B. Weisz, W.O. Haag and P.G. Rodewald, Science *206*, 57 (1979).
6 W.O. Haag, P.G. Rodewald and P.B. Weisz, Catalytic Production of Aromatics and Olefins from Plants Materials. Presented at ACS National Meeting, Las Vegas, Nevada, August 1980.
7 J.D. Johnson and C.W. Hinman, Science *208*, 460 (1980).
8 E.K. Nemethy, J.W. Otvos and M. Calvin, High Energy Liquid Fuels from Plants, in: Fuels from Biomass. Ed. D.L. Klass and G.H. Emert. Plenum Press, New York, in press (1981).

Wood as biomass for energy: results of a problem analysis

by Fritz Hans Schwarzenbach and Theo Hegetschweiler

Swiss Federal Institute of Forestry Science, CH-8903 Birmensdorf (Switzerland)

1. *Wood as biomass for energy*

Under the pressure of a world-wide crisis in nutrition, raw materials and energy, a new field of research has evolved which attempts, through interdisciplinary cooperation, to promote the production of economically convertible 'biomass' rapidly and effectively using modern biotechnological methods. Biomass is defined as all organic material produced by living things. Biomass has potential uses as food, as raw material for economically important products and as an energy source.

One of the main aims of this new research is to find suitable organisms producing the greatest quantity of economically convertible biomass under well-adjusted conditions. Trials on the cultivation of fast-growing woody plants in intensively managed plantation-type monocultures and the production – with the shortest possible time lapse between planting and cropping (short-rotation forestry) – of a maximum of woody biomass for energy or chemistry, are to be evaluated from this viewpoint. This cultivation method for fuel wood differs in several ways from the common practice in forestry to date. Therefore it is expedient to examine more closely the widespread suggestion that fuel wood be produced on special farms; we will do this within the framework of a general survey of the topic 'Wood as raw material and energy source'.

2. *Systems analysis approach*

Forest management is a very complex dynamic system, which can be divided into 4 closely interrelated subsystems:

a) the *ecological* sub-system deals with the relationship between woody plants and their animate and inanimate environment;

b) the *forestal* sub-system is founded on all those observations made during the course of man's efforts to guide the development of wood-producing plants towards specific aims;

c) the *timber-economic* sub-system concerns itself with the production, processing, distribution and utilization of wood in the commercial sector, and its restoration to the natural ecosystem;

d) these 3 systems are superimposed by the management system. It seeks to bring the economically-, politically- and socially-based demands on forest management into harmony with one another and with forestal realities and the ecological limits of wood production.
Such a systems analysis approach has proved especially useful for investigating the ways in which new demands of forest management affect ecology, forestry and timber economics, and for indicating possible changes in the balance of the entire system.
In the present article, the study of fuel wood production in plantations following the short rotation forestry principle serves as a basis for such an all-encompassing evaluation. The correlations within the aforementioned sub-systems are examined by addressing the following questions:

- What measures can be implemented to increase the production of woody biomass in plantations? (forestal sub-system);
- which fast-growing, highly-productive shrubs and trees are suitable for cultivation in plantations? (forestal sub-system);
- what demands do these species make on their growth sites? (ecological sub-system);
- are there ecological objections to the growing of these species in intensively cultivated monocultures? (ecological sub-system);
- how well can the plantation method compete with other methods of producing fuel wood and with other energy resources? (timber-economic sub-system);
- what are the repercussions of introducing new production methods on the collective system 'wood as raw material and energy source?' (Management system).

In answering these questions, potentials and effects of fuel wood plantations are outlined.

3. *Forestal aspects*

Short rotation forestry practice. In short-rotation forestry, the time between the establishment of a stand and wood harvesting is kept short, i.e. usually 2–10 years. In these few years the dense plantations (5000–100,000 plants per ha) produce thin stems which are mechanically cut and harvested. Work is being done in several countries on the development of suitable cropping machines, partly within the framework of a project of the International Energy Agency (IEA).
Numerous trials on the cultivation of fast-growing woody plants in short harvest cycles have been running for some years (USA, Canada, Sweden, Denmark, Federal Republic of Germany etc.)[1].
In their work on the afforestation of polders, Dutch

Figure 2. Willows grown in short rotation in Flevoland, Netherlands. 75% of the planted area (1973: 172 ha) consists of selected clones of the species *Salix viminalis* and 25% of *Salix tiandra.* A normal yield is: with a rotation of 2 years: 1700 bundles of shoots per ha; with a rotation of 3 years: 2300 bundles of shoots per ha. A bundle is after 2 years 3.5 m long, with 70 cm circumference at 30 cm from the bottom. A bundle weighs fresh 20 kg, dry 10 kg. (Photo: Ir. H.A. van der Meiden, Wageningen).

Figure 1. A cropping machine in the Netherlands is harvesting willow-rods grown in short rotation (Photo: Rijksdienst voor de Ijsselmeerpolders, Lelystad).

Figure 3. A planting machine with 6 rows is used for planting the willows. The rows are 70 cm apart, distances in the row are 50 cm. The planting stock consists of unrooted cuttings with a length of 23 cm. (Photo: Rijksdienst voor de Ijsselmeerpolders, Lelystad).

specialists have adapted the extensive experience of many years to a special form of short-rotation forestry, although not for the production of fuel wood. Cultures of a particularly fast-growing clone of *Salix* produce strong shoots within 3 years. These rods are used in large bundles to stabilize the sand as a first stage in dike construction. Propagation by short-rotation is very suitable for this special use of shoots. (Information from the Research Institute for Forestry and Landscape Planning, Wageningen, The Netherlands.)

Similar experience has been obtained with the propagation of plant material for engineering-biological constructions. The specifications for such plant material are very exact in comparison to the purely quantitative considerations in the production of wood for energy conversion.

Planting is done with cuttings, especially with bred types (e.g. poplar). The selected species should, as far as possible, display strong growth in youth as well as high sprouting capacity after several croppings.

Wood production in short rotation entails cultivation of the soil, planting in a very close arrangement, fertilization, pest control (with pesticides and insecticides), some degree of irrigation and the most efficient mechanized cropping possible[2]. In this, the technique resembles that of agriculture but differs considerably from conventional forestry methods. The term 'wood-farming' is as apt as 'short-rotation forestry'.

Potential increase in yield. The production of maximum yields requires favorable sites, care and protection. Cannell and Smith[3] compared and examined 11 reports on cultivation trials carried out under favorable growth conditions; the average yield (dry stem- and branch-wood, with bark, without leaves) did not exceed 10–15 t/ha. year for rotation times of 1–5 years.

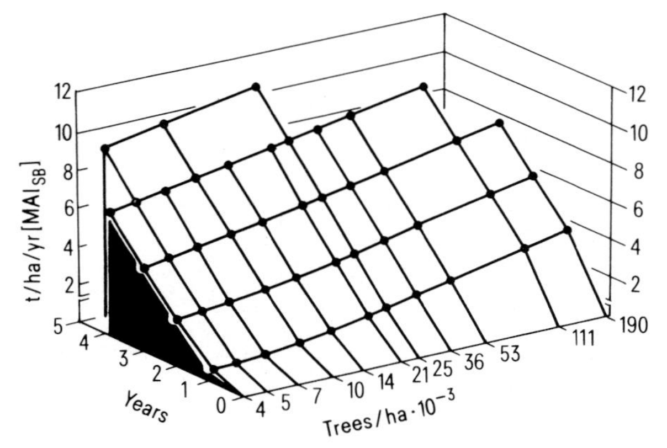

Figure 4. Dry weights of stems and branches with bark yielded by closely-planted hardwoods 1–5 years after planting.
Mean annual increments of dry stems and branches with bark (gross yield/age) (MAI_{SB}), in the 1st rotation after planting (not coppicing) with the given number of plants per ha (Figure from Cannell and Smith[3], p. 424; with permission).

The yields during the first few years after coppicing can be 10–30% greater than those after planting. The level of yields can be approximated by conventional forestry methods; agrarian cultures produce higher yields. Hence it appears doubtful whether higher yields really can be attained with extremely short rotation times than by conventional forestry methods.

In order to determine the net yield in terms of energy, an energy budget must be prepared. The energy consumed in soil cultivation, fertilization, pest control, harvesting, drying and transport must be subtracted from the gross yield. Short-rotation forestry entails an increased energy input, which correspondingly reduces the net yield.

Availability of fast-growing woody plants with high sprouting capacity. A tree species is described as fast-growing if, on good sites in central Europe, it attains a mean total increment of more than 9 solid meters by its 30th year and more than 12 solid meters by its 50th[4,5]. The requirements for sprouting capacity, fuel value, or depending on the intended use timber quality (fiber length etc.), limit the number of suitable tree species. Current projects are working mainly with the following:

- bred and selected clones of various poplar species *(Populus)*
- willow species *(Salix)*
- plane (*Platanus occidentalis* L.)
- black locust (*Robinia pseudoacacia* L.)
- alder species *(Alnus)*
- elm species *(Ulmus)*
- eucalyptus species (*Eucalyptus*)
- red oak (*Quercus rubra* D. R.)
- hornbeam (*Carpinus betulus* L.).

A more extensive survey of species suitable for fuel wood production is given by Burley[6]. The range of suitable types for any given site is, however, generally small, although Cannell and Smith[3] have shown that the choice of species influences the yield less than the density of planting and the rotation time.

4. *Ecological observations*

The growth of shrubs and trees is strongly influenced by environmental factors. Among the bioclimatic requirements in temperate zones, the temperature sum during the annual vegetation period is particularly important. Hence the growth performance of alpine trees deteriorates with increasing altitude[7,8]. A similar reduction in annual growth can be observed in local species at their northern limits of distribution in northern Europe and North America[9]. Trees in general perform less well on suboptimal sites. The implications for the cultivation of fast-growing, highly-productive species in plantations are obvious:

- the selected species must be planted on sites where it experiences optimal growth conditions.

- conditions in areas of the temperate zones with low temperatures during the growth period are unfavorable for fast-growing woody species.

The withdrawal of nutrients from the soil is higher in silvicultural energy plantations than in conventionally managed forests. It is often suggested that fertilizers be used to compensate for this loss, but this increases the load of nutrients in the ground water, as is also the case in agricultural areas.

From the ecological point of view, it must also be emphasized that silvicultural monocultures (especially where the trees are of the same age) may be swept by epidemics of plant and animal pests causing extensive and permanent damage[10]. Prolonged application of pesticides may therefore be necessary, which is disquieting from the forest ecology viewpoint. Forests represent one of the last more or less natural environments and have not only productive but also ecological functions to fulfil. These premises argue that such multiple objectives can best be achieved with natural silvicultural techniques[11], whereas a monoculture of woody coppice sprouts, cropped every few years is biased towards the productive function alone. Ecologically, short-rotation forestry can be regarded as a special form of agriculture.

5. *Timber economy considerations*

From the viewpoint of timber economy the production of fuel wood in plantations with short rotation must be compared with other possible energy supply methods. So far, under fuel wood the following has been understood:

- dry fallen wood;
- trimmings from felled trees;
- small-sized timber from thinning operations (also used as raw material for the pulp and particle board industries);
- specially-prepared classes of firewood;
- residual wood from sawmills (also used as raw material for the pulp and particle board industries);
- residual wood from wood-processing;
- unwanted wooden artifacts;
- wood resulting from the cutting and clearing of orchards, hedges, windbreaks, and shrubs and trees in parks and gardens.

The combustible end-products of industrial wood conversion (e.g. fiberboard, particle board, cardboard, paper) can also be used for energy production when they are not needed any more.

Fundamentally it can be seen that an ecologically and economically rational exploitation of wood should direct as much freshly-harvested wood as possible into commercial circulation as a raw material, and use it for energy production only as a last stage in conversion. From an overall point of view, the use of wood primarily as a raw material whether in a single form or in successive stages, and lastly, as an energy source, appears to be the most sensible and effective utilization of this natural resource.

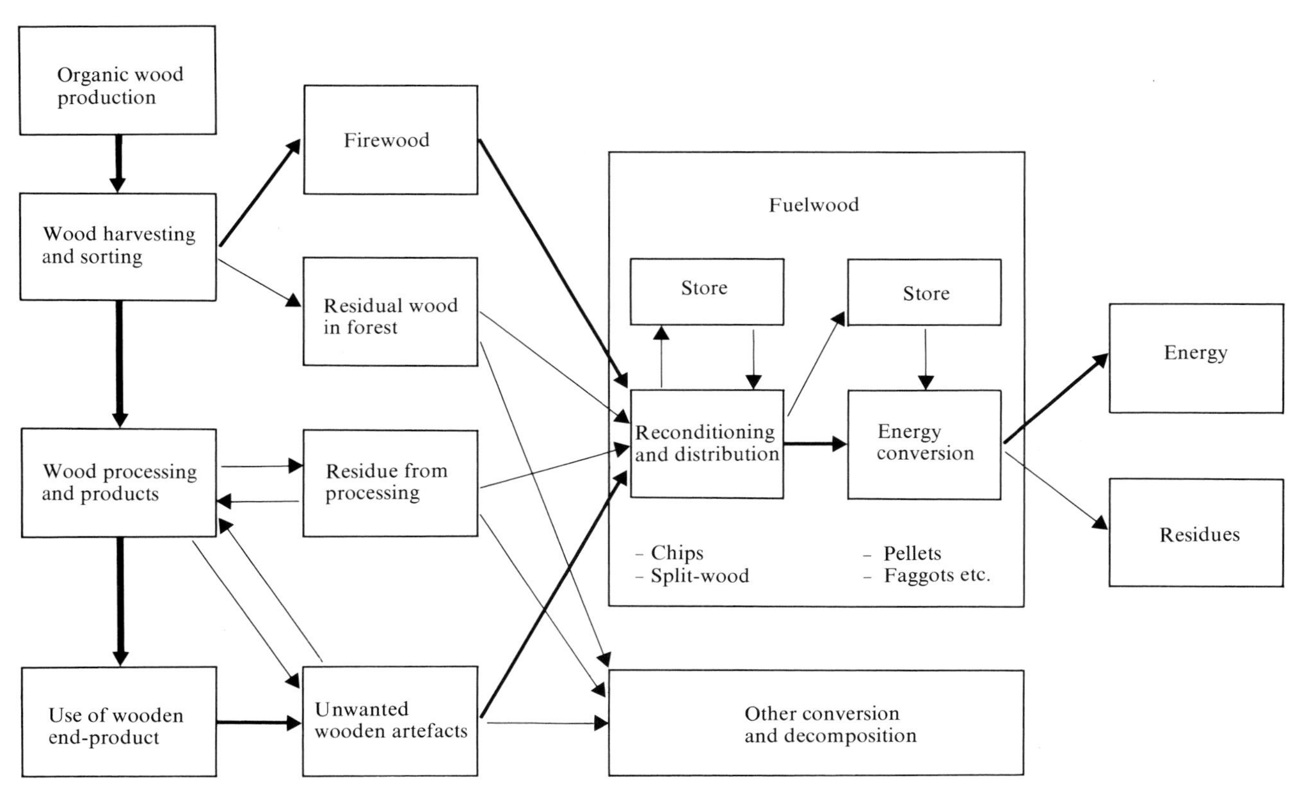

Figure 5. Schematic diagram of the sources of fuel wood in the flow of wood materials.

In the industrial lands of central Europe, the last 3 decades have seen an increasing displacement of firewood as the energy source by electricity, oil or gas in many households; some of this 'redundant' firewood is industrially converted to cellulose, paper, particle and fiber-board, while the nonconvertible residue from felling is either burned on the spot or left lying in the forest.

This contempt of wood as an energy source in the past years is in marked contrast to the attitude of earlier generations who, before the introduction of electricity and liquid fuels, were far more dependent on wood for heating and cooking.

Today there is a great interest in the utilization of wood as an energy resource for reasons of supply policy. A rapidly growing demand for firewood following the oil crisis of 1973/74 led, in Switzerland and in other countries, to a considerable price increase and the sale of some industrial wood as firewood. Consequently, efforts are being made today to increase the supply of wood for energy correspondingly. The possibilities are as follows:

- opening-up of as yet unused forest for wood production,
- intensified utilization of forest stands which, because of economic considerations, have been insufficiently exploited thus far,
- afforestation of areas abandoned by agriculture.

The production of wood in plantations can supplement these measures by replacing agriculture in mechanically cultivable areas as far as agropolitical considerations allow.

The requirements for the economical running of such wood farms on the short-rotation forestry or wood-farming principle would have to be fulfilled.

6. *Problems and conflicting aims in optimal wood production management*

Technologically, energy conversion of wood presents no particular problems. At present there is some interest in the development of plants for the production of wood gas which is a suitable fuel for engines with heatpower-coupling[12].

Many sections of the transport industry point out the possibility of producing alcohols, which are miscible with petrol, from wood.

Purely energy-economic considerations argue for the definite promotion of fuel wood in all those countries with sufficiently high wood production which depend on imported oil for a considerable portion of their energy supply. When the unusual versatility of wood as a raw material is considered, however, great objections arise over extending the energy conversion of wood at the expense of other utilization forms and thereby risking a crisis in commercial timber supply (Global 2000 report to the President, 1980)[13].

In Switzerland, for example, problems of this type would be particularly likely if the pressure towards the manufacture of fuels from wood increased. The production potential of forests in Switzerland can cover at best a small percent of the present total energy requirement[14], but a switch to energy wood plantations for the production of additional quantities of energy-convertible wood is only rational if large areas, which possess favorable conditions for wood-producing plants and are not otherwise utilized, are available.

The production of wood for energy must be assessed not only on the basis of energy-political aspects, but also from the silvicultural point of view. Forestry sees wood production as only one of the many functions a forest has to fulfil, both ecologically and socio-economically. As the history of forestry shows, over-utilization for economic reasons and biased promotion of particular commercial species sooner or later leads to progressive destabilization of the natural ecosystem. As custodians working on behalf of the public, the forestry bodies advocate the continued maintenance of the forests in their full capacity, the ensuring of ecological stability and the limiting of commercial forest exploitation in accordance with the sustained increase by natural growth.

7. *Political consequences*

In light of available information, ecological and overall economic considerations permit only the partial satisfaction of supply policy demands for a more energy-oriented timber management. At present the following positions are taken regarding the use of wood in energy production:

- wood production by the existing forestry methods is to be maintained within a framework which ensures the continued preservation and ecological stability of the forests;
- an increase in wood crop potential is primarily to be achieved through the opening-up of as yet unused forest, the improvement of forest tending and the afforestation of fallow and otherwise unused areas;
- the primary utilization of available wood as a raw material clearly takes priority over its use as an energy resource; a manifold conversion of wood over various stages and a final conversion for energy is ecologically and economically sound;
- otherwise non-convertible wood is to be used as much as possible for energy production;
- the production of fuel wood in special plantations may be considered if the potential production areas are suitable for the cultivation of fast-growing tree species and are not otherwise used. As a special form of agriculture, these plantations should not be subject to forest laws;

- in relation to energy supply policy, it is clear that no energy-production program can be convincing without a corresponding energy-saving program.

8. *Summary*

1. Wood has been used by man as a fuel since prehistoric times and therefore holds a special place among the energy-convertible forms of biomass.

2. The biotechnological possibilities for biomass production today raise the question of whether a significant increase in wood production per area and time can be achieved through fast-growing, highly-productive shrubs and trees, as compared to production by the usual forestry methods. This method has been suggested principally for the production of fuel wood for combustion, gasification and the manufacture of alcohol as a fuel additive. This wood could also be used as raw material for pulp production and wood chemistry.

3. By means of a systems analysis approach, the demand for the increased production and utilization of fuel wood is examined in terms of its ecological, forest- and overall economic effects, as is the suggestion that fuel wood be produced in plantations on the short-rotation forestry principle.

4. The increase of wood production through short-rotation forestry methods is clearly limited by the ecological requirements of suitable tree species and the slow growth of many others.

5. There are objections to the cultivation of selected, usually vegetatively-propagated, tree species in intensively-managed plantations: - susceptibility of monocultures to epidemics of plant and animal pests, - stressing of the soil through heavy fertilization or residues from pesticides and insecticides, - ecological impoverishment, concomitant with any monoculture.

6. The cultivation of woody plants in plantations requires large areas of land which are not otherwise managed, are mechanically cultivable, and which meet the ecological requirements of the species concerned.

7. Figures on operating costs and possible yields of short-rotation forestry plantations can be expected from current pilot projects within the foreseeable future.

8. The demands in various countries for an increased utilization of wood for energy create serious competition situations for consumers who need wood for a multitude of other purposes.

9. Overall economic considerations demand that all suitable wood be used as a raw material, as far as possible in successive stages, ending with a final conversion for energy.

10. Wood unsuitable for other purposes should, as much as possible, be collected and used in high-efficiency energy conversion processes.

11. Forestry management should yield to a rising demand for fuel wood only in as far as the wood cropping involved does not jeopardize the long-term silvicultural objectives:
- maintenance of the forest in such a way that it can fulfil all its functions, - safe-guarding of the permanent ecological stability of the forest, - limitation of wood harvesting in order to sustain productivity.

1 R. Bachofen, Bioenergie heute ... morgen. Vjschr. naturf. Ges. Zürich *125*, Nr. 5 (1980); Neujahrsblatt 1981.
2 J. B. Jensen, Mulighed for minirotation ved dyrkning af skovtraeer. Dansk Skovforen. Tidsskr. *63*, 1-39 (1978).
3 M. G. R. Cannell and R. I. Smith, Yields of minirotation closely spaced hardwoods in temperate regions: review and appraisal. Forest Sci. *26*, 415-428 (1980).
4 L. Dimitri, Anbau schnellwachsender Baumarten im Kurzumtrieb: Möglichkeiten und bisherige Erfahrungen. SchrReihe forstl. Fak. Uni. Göttingen *69*, 79-92 (1981).
5 E. Roehrig, Waldbauliche Aspekte beim Anbau schnellwachsender Baumarten. Forst-Holzwirt *34*, 106-112 (1979).
6 J. Burley, Selection of species for fuelwood plantations, 8 h World Forestry Congress 1978, paper FPC/3-6.
7 H. Turner, Mikroklimatographie und ihre Anwendung in der Ökologie der subalpinen Stufe. Eidg. Anst. forstl. Versuchswes., report 68, 1971.
8 H. Turner, Bergwald im Kampf gegen Natur und Mensch. Aufforstungsversuche in der oberen Gebirgswaldstufe. Eidg. Anst. forstl. Versuchswes., Report 168, 1976.
9 H. W. Windhorst, Geographie der Wald- und Forstwirtschaft. Teubner, Stuttgart 1978.
10 F. Schwerdtfeger, Die Waldkrankheiten, 3rd edn. Parey, Berlin 1970.
11 H. Leibundgut, Wirkungen des Waldes auf die Umwelt des Menschen. Rentsch, Zürich 1975.
12 J. Buchli and J. Studach, Wärmeversorgung Sent. Pilotprojekt einer möglichst autarken, wirtschaftlichen und umweltgerechten Wärmeversorgung eines Bergdorfes. IGEK, Chur 1980.
13 The Global 2000 Report to the President. Report by the Council of Environmental Quality and the Department of State. OCLC 6572993. US Government Printing Office, Washington DC 1980.
14 E. Wullschleger, Das Nutzungspotential unserer Wälder: Aufgabe und Chance für die Wald- und Holzwirtschaft. Schweiz. Z. Forstw. *130*, 1037-1046 (1979).

Leaf protein as a food source

by N.W. Pirie

Rothamsted Experimental Station, Harpenden, Herts AL5 2JQ (Great Britain)

The mixture of proteins in the juice pressed out from pulped leaves may conveniently be called 'leaf protein' (LP) although, after coagulation and pressing, it also contains lipids, starch and nucleic acid. It is sometimes called 'leaf protein concentrate' (LPC): that term, if used, should be reserved for material that has undergone some further stage of refinement. The most obvious reason for including work on LP production in a program of research on solar energy is that, because it takes protein directly from the leaf – the site where protein is synthesized – the losses that are inseparable from translocation, or conversion in an animal, are avoided. LP production has, however, a bearing on other aspects of solar energy research and on the general problem of energy conservation.
If suitable crops were used, LP could be made from the leafy by-product of an 'energy plantation', the fibrous residue from which LP has been extracted would be a convenient substrate for methane fermentation, and various chemicals that are now made by partial or total synthesis (starting from coal or oil) could be separated from the juice after protein has been removed. Furthermore, less energy would be needed for managing an irrigation system if water weeds were being removed, and used as a source of LP, so that they no longer wasted water. And, in many situations, the fibrous residue from the extra growth resulting from irrigation would produce more energy, when used as a fuel, than would be produced by the same amount of water passing through a turbine.

The process

Simple pressure does not release LP from leaf cells: rubbing and partial disintegration of the leaf structure are essential preliminaries. Equipment ideally suited for extracting leaf juice on either an industrial or an 'Appropriate Technology' scale is not at present on the market. Because roller mills and screw expellers do not apply pressure as smoothly and efficiently as their makers intended, there is enough rubbing in them to release some juice. However, such inefficient extraction wastes energy. We have made some progress[1,2] in designing a unit in which thorough rubbing precedes pressing. This unit needs about 300 W to extract juice from about 100 kg of leaf per h; the actual quantities depend on the texture of the crop. This unit can make 5–10 kg (dry weight) of LP in a day. That is the scale of production for which, at present, there seems to be the most demand, but larger units working on the same principle could be made.

Leaf juice can be coagulated by acidification to about pH 4, but LP is easier to handle if it is coagulated by heating suddenly to 60–90 °C. Sterilization and enzyme inactivation are more complete at the higher temperature. The curd is collected on a filter and pressed until it contains 60%, or less, water. When made from species with little flavor, the LP is then ready for use; LP from most species should be resuspended in water, at pH 4, filtered off, and pressed again. It can be kept, like cheese, for a few days without refrigeration; for longer storage it can be canned, frozen or preserved with acetic acid, salt or sugar[3]. Drying should be avoided if possible; if it should be essential, it must be done carefully[4] or the LP will become gritty and less digestible.

Properties

When carefully made, LP consists of 60–70% true protein (on the dry matter), about 25% lipids including pigments derived from chlorophyll, and 0.1–0.2% β carotene. The lipids are highly unsaturated which, as in other foods, raises problems in storage. Some people find the green colour unattractive. For these reasons, solvent extraction is often suggested. This would be a mistake because the lipids are nutritionally useful and the β carotene (pro-vitamin A) is as valuable a component as the protein in those countries in which vitamin A deficiency is common. Furthermore, even partial decolorisation, by any process, would make LP much more expensive and would rob it of one of its main merits – that it is a good source of protein which can be made by unskilled people from local material for local use.
The nutritive value of LP has been well established by many experiments on animals and by trials with infants in India, Nigeria and Pakistan[5]. Unless LP is being made from a species not closely related to those already used, e.g. a conifer or fern, there is no need for further nutritional trials. Although LP has a less favorable amino acid composition and is less digestible than casein, if it has not been damaged by inept handling it is as good as, or better than, the protein in cereal or legume seeds. The poisonous substances present in some leaf species are removed when the LP is washed at pH 4.

Sources of leaf

By harvesting a succession of crops from the same piece of land, the annual yield of dry, 100% protein at Rothamsted was 2 t per ha; with irrigation the yield would be larger because there is here a summer water deficit in most years. In India the yield is 3 t. Yields from perennial grasses are a little smaller. These yields – 3 or 4 times as large as those of any other protein concentrate – are possible because crops that are harvested while still green maintain a photosynthetically active cover on the ground for longer than crops that need a period of ripening or drying. Therefore, LP is potentially the most abundant of all edible protein sources. Crops destined primarily for LP production will, however, replace seed crops to only a limited extent because the need for protein concentrates is limited. Seeds will remain the main source of dietary energy, and they are more easily stored than perishable LP.

The yield of dry matter from forage crops harvested several times during a year is larger than the yield from any other type of crop. They lose part of this advantage when they are put through the inefficient process of conversion in a ruminant animal, and they contain so much water that artificial drying is an expensive prelude to their use as fuel. That is why so much sunlight is used merely as a source of heat for maturing or drying crops in the field, rather than as a source of low-entropy energy for photosynthesis.

LP cannot be satisfactorily separated from leaves with a dry, fibrous, or mucilaginous texture, nor from leaves that are acid or contain a great deal of phenolic material. In spite of these limitations, more than 100 suitable species are known that grow well in the tropics or the temperate zone. No attempt has so far been made to select varieties specifically for LP production.

Several crops, e.g. potatoes and sugar beet, are harvested while some of the leaf is still green and unwithered, and leafy material is a by-product from the cultivation of many vegetables. These leaves are at present unused or inadequately used: they are all good sources of LP. I calculate that 100,000 t of protein could be recovered annually in the UK from potato haulm (mainly from early and seed potatoes) and sugar beet tops.

Weeds growing on uncultivated ground are not likely to be useful, not because any defect characterizes all weed species, but because uncultivated ground is likely to be too rough or steep for convenient harvesting, and it is unmanured. The situation is different with water weeds. They would be easy to collect, they are often well fertilized, and much effort is already being expended on their control. LP has been extracted from a few species[5] but this potential source has been most inadequately studied.

Trees were not, in the past, thought of as probable sources of LP because it would not be easy to collect leaves from them in the conditions of conventional forestry. The recent increase in interest in coppiced trees as sources of paper pulp, and as components of 'energy plantations', changes the situation. Mechanical stripping of leaves from the uniform poles that grow in a coppice would be easy: machinery already exists in the USSR for stripping needles from conifers. Leaves collected in this way are being used as cattle fodder: I have reviewed that subject briefly elsewhere[6]. However, it would obviously be preferable to make LP from them for use as food by people and non-ruminant animals.

Few attempts have been made to extract LP from tree leaves. Casual experience at Rothamsted suggests that they tend to be dry and tough; elder *(Sambucus nigra)* is the only species from which extraction is satisfactory. It is unlikely that elder is unique. If, as seems likely, 'energy plantations' are used extensively for collecting solar energy, the possibility of extracting protein from the by-product leaves is a factor that should be borne in mind when selecting suitable species. Tree leaves could then become one of the more important sources of LP. Nitrogen-fixing leguminous trees, such as leucaena and the acacias, have obvious advantages, and there is no reason to expect poorer extraction from them than from other species.

Ethanol would be a more convenient end-product of trapped solar energy than wood. Sugar cane and cassava are the plants usually suggested as the raw materials for alcoholic fermentation. Sugar cane tops are so fibrous that it is difficult to extract what little LP they contain; LP does not extract readily from the varieties of cassava that have been studied, but other varieties may be more satisfactory[5]. Sweet sorghum and chicory have also been suggested as sources of fermentable carbohydrate. The appearance and texture of these leaves suggest that LP should extract readily from them, but experiments demonstrating this have not been published.

The by-products from LP production

By-product leaves from 'energy plantations' are important because their use will make this form of energy more economic. The same economic argument will make it essential that the pressed, fibrous, residue that remains when LP has been extracted from any type of leaf is efficiently used. Hitherto, most attention has been given to its value as ruminant fodder. Because juice has been pressed out of it, much less fuel is needed to dry it, compared with the original crop, for conserved winter feed. It is indeed possible, in most climatic conditions, to complete the drying by exposure to unheated air. Use of the fresh or conserved fiber as fodder is the most sensible course when LP is being made on a small scale for local use.

The disintegrated, fibrous residue from LP extraction has a texture that is well adapted to methane fermentation. It was primarily for use in a methane project that one of the early largescale pulpers (1 t of leaf per h), designed at Rothamsted, was sent to Uganda. So

far as I know, this use of the fiber has not been exploited further.
Vegetable matter must obviously be as dry as possible before being used as a fuel. Therefore, straw gets the most attention and, when grasses are being considered as fuels, it is assumed that they will be harvested when mature in spite of the smaller yield of dry matter compared to what can be taken with repeated harvests of young growth. The fibrous residue from juice extractors of the present type, in which there is relative movement between fiber and metal, contains 65–75% of water. An extravagant amount of power would be wasted in friction if attempts were made to get drier fiber from them. The extra juice coming out with intense pressure contains little protein because that tends to be trapped by the tightly packed fiber. The hydrated fiber from young leaves, after intense pressing, contains about 50% of water. Fiber that is to be used as fuel, or conserved for ruminant feed by drying, should therefore be subjected to a 2nd pressing in a unit of the 'horn-angle' type in which there is no differential movement between fiber and metal. About 1000 times as much energy is needed to evaporate water as to press it out. I have discussed these points more fully elsewhere[7]. When comparing these 3 ways of using the fibrous residue, it should be borne in mind that the 1–2% of nitrogen in it can be recovered for use as fertilizer when the residue is eaten or fermented, but use as fuel wastes the nitrogen.
The protein curd that separates when leaf juice is heated is easily filtered from a 'whey' which contains most of the phosphorus and potassium, and much of the nitrogen, of the leaf. The simplest way to use the 'whey' is to put it back as fertilizer on the land from which the crop was taken. However, it also contains carbohydrate on which yeasts and other microorganisms grow readily. The 'whey' from many species of leaf contains drugs, dyes and substances which could be the starting points for various syntheses. These substances are now usually made be energetically extravagant processes starting with fossil fuels. If we are taking the use of solar energy seriously, the possibility of separating them as by-products from LP production is worth thinking about even if the thought seems a little old-fashioned.

The future of leaf protein in the context of solar energy

If 'energy plantations' develop on the scale that is often envisaged, they will be an important, perhaps the most important, source of LP. Although my primary interest is in the LP, some general comments on photosynthetic methods for exploiting solar energy may be permissible. Most of them were made[8] at a UNESCO conference on 'Wind and solar energy' in 1954; they were then thought far-fetched, but have now become topical.

In 3 obvious ways photosynthesis is preferable to the use of photocells or 'solar towers'. The cost of covering an area with leaf is a tiny fraction of the cost of covering it with photocells or mirrors. Leaves keep themselves clean by continuous growth, whereas the labor involved in keeping other surfaces clean would be formidable; these methods for trapping solar energy might almost qualify as labor-intensive activities! The product of photosynthesis is there for use as and when it is needed, whereas the other methods supply energy only when the sun shines. The principal merit of photocells is that their efficiency is greater and they make electricity whereas plants merely make fuel. However, the argument that both the physical methods could be used in desert areas where plants would be incapable of growth is not quite as simple as it seems at first sight. Any process that changes the character of a large area of the earth's surface may affect the local climate.
There is general agreement that many deserts were created because of deforestation and overgrazing. A suggested mechanism for this desertification is that bare ground, when humus has been oxidised, has a greater albedo than ground covered with vegetation; convection leading to cumulous formation is therefore less over bare ground, and there is less nocturnal cooling. The matter is controversial, but it raises the possibility that, if water could be piped into a desert for a short time so as to reintroduce vegetation on an extensive area, rainfall would increase to such an extent that vegetation would survive without further attention. Although that suggested explanation is controversial, the issue may be pursued a little further. When a sandy area is covered with photocells, or mirrors angled so as to shine on a 'solar tower', there will be a diminution in albedo similar to that caused by vegetation. A hitherto cloudless area, could then become cloudy. If there is validity in this argument, the suitability of an area for photosynthesis is enhanced by using it for that purpose, whereas it is impaired if used for the physical methods. Obviously, this effect, if real, applies only to the exploitation of very large unbroken areas.
Undoubtedly, more water is consumed, in the transpiration stream, when plants rather than physical systems trap solar energy. In some circumstances the use of water by plants is more productive than the uses to which it is now put. Consider an ideal but not unrealistic set of conditions: if water passes through a turbine and is then lost because there is no irrigable land beneath a coastal cliff 100 m high, it produces only a tenth as much power as would have been produced if it had been used to increase plant growth in a semi-arid region at the top of the cliff, and if that extra crop had been burned in an engine with 20% efficiency[9].
In spite of much publicity, progress in exploiting

biological methods for trapping solar energy is slow. This suggests that the economic and other advantages of using photosynthesis have not been widely accepted. The case for photosynthesis would be strengthened if LP and other by-products were fully exploited. There is already well financed research in Berkeley and Madison (USA) on large-scale equipment suitable for this job; the most pressing need is for research on the species most suitable for it.

The future of leaf protein in the context of appropriate technology

Research on the production of edible LP on a small-scale for local use is in an almost exactly inverse position. We know many suitable sources of LP, and there is good reason to think that its practical exploitation would have a more immediate effect on the welfare of more people than an increase in the energy supply would have. But there is no simple, robust, and economical unit on the market with which it can be made: nor is there properly financed work on the design of such a unit. Until this work is done, LP will be thought of as an interesting possibility rather than as a recognized food.

1 N.W. Pirie, Expl Agric. *13*, 113 (1977).
2 J.B. Butler and N.W. Pirie, Expl Agric. *17*, 39 (1981).
3 N.W. Pirie, Indian J. Nutr. Dietetics *17*, 349 (1980).
4 D.B. Arkcoll, J. Sci. Fd Agric. *24*, 437 (1973).
5 N.W. Pirie, Leaf protein and other aspects of fodder fractionation. Cambridge University Press, London 1978.
6 N.W. Pirie, Appr. Technol. *5*, 22 (1978).
7 N.W. Pirie, Phil. Trans. R. Soc., Lond. B *281*, 139 (1977).
8 N.W. Pirie, in: UNESCO Symposium 'Wind and solar energy', p.216. UNESCO 1956.
9 N.W. Pirie, Chem. Ind. 442 (1953).

Algae and water plants as energy converters

In H.A. Wilcox' paper the concept of raising seaweeds on huge structures ('ocean farms') is presented and the use of this biomass for food, fibers, fertilizers, methane and other products is described.
The basic problems of collecting energy from the sun through microalgae are discussed by S. Aaronson and Z. Dubinsky. Applications of this technique may be suited for sewage purification plants or in saline ponds in the tropics or subtropics. Many products, such as pharmaceutics and chemical raw materials can be gained from phototrophic microorganisms. Furthermore, the cells can be used to eliminate toxic or polluting compounds.
A similar study is made by C. Santillan as he presents the particular case of *Spirulina* in Mexico, a protein source depended upon the Aztecs already centuries ago.
The paper by T.G. Tornabene reviews the potential that microorganisms have for producing lipids and hydrocarbons and the use of these as fuels.
The special case of the oil alga *Botryococcus braunii* is examined by R. Bachofen. Considerable basic research is still required before it will be possible to induce the state, as is occasionally found in nature, wherein 85% of the dry weight of the algal cells is hydrocarbons.
In the alga *Dunaliella* the main product of photosynthesis is glycerol. A. Ben-Amotz et al. discuss the biological glycerol production from CO_2 with sunlight in ponds. As a raw material for the chemical industry this seems to be a promising alternative for the future.

The ocean as a supplier of food and energy

by Howard A. Wilcox

Environmental Sciences Department, Code 5304(B), Naval Ocean Systems Center, San Diego (California 92152, USA)

Summary. This paper presents the concept of raising seaweeds and other valuable organisms with the aid of huge structures ('ocean farms') emplaced in the surface waters of the open oceans. Potential advantages from and difficulties to be expected in realizing the associated technologies are briefly set forth. Much of the published literature pertaining to the concept is referenced and summarized. Wave-powered upwelling of cool, nutrient-rich waters through vertical pipes extending to depths of 100–300 m is indicated as desirable. Technologies are outlined for using the harvested seaweeds to create foods and other valuable products such as animal feeds, fertilizers, fibers, plastics, synthetic natural gas (methane), and alcohol and gasoline fuels. Results from site selection studies and economic analyses are given. It appears that dynamically positioned farms orbiting with the surface current patterns typically found on the ocean will be most cost-effective. The general conclusion is stated that open ocean farming will become economically more feasible as the cheaper fossil fuels and food producing lands of the earth become increasingly consumed in the course of the next century.

The possibility of cultivating crops of vegetation on the vast areas of the open ocean, a possibility apparently first suggested (in 1968) by this author, has gripped the imaginations of a growing number of technicians and planners throughout the world. If successfully realized, this technology would enable the planet's oceans to become a huge new source of feeds, foods, fuels, and chemicals – fixed carbon and fixed nitrogen – for the benefit of humanity. Moreover, the resulting great oceanic aquacultural enterprise would supplement rather than be competitive with the already existing land based sources of these vital goods for all the world's peoples.

In late 1972, the US Navy initiated an experimental program to explore the ocean farm concept[1], and in 1977, the project was shifted to the management aegis of the General Electric Company, which continues to direct the program to this date.

The concept of open ocean aquaculture is grounded on the following combination of facts:

1. The rate of receipt of solar energy on the surface of the earth is huge, being more than 10,000 times greater than our present use rate of all other forms of energy.
2. The flow of solar energy is naturally maintained and highly reliable in both the technical and the political sense.
3. Most of the solar energy received by the surface of the earth is absorbed by the upper layers of the oceans because the oceans cover some 70% of the earth's surface area and possess relatively low average reflectivities.
4. Man's use of the earth's currently received solar energy need not upset the net balance of the planet's carbon dioxide, oxygen, water, and energy flow cycles, as calculated on a global average basis over time spans of a few months.
5. The sun is the only known energy source available for the large scale photosynthetic production of vegetation.
6. Vegetation can be converted by known technology into foods, fertilizers, fibers, plastics[2], synthetic natural gas (methane)[3], synthetic liquid fuels such as ethanol, methanol, and gasoline[4], etc. Indeed, vegetation is the only practically renewable source of such products that is available today.
7. Although nearly all areas of the surface waters of the major oceans are 'biological deserts' because they are almost devoid of the nitrate and phosphate compounds required for the growth of vegetation, ocean waters below depths of 100–300 m generally contain these nutrient materials at relatively high average concentrations[5], and these waters stand at a gravitational potential energy deficit relative to the surface of only about 3000 J per ton of water[6].
8. Plants immersed in the ocean can be highly efficient photosynthesizers[7] and are generally immune to the 2 main hazards of land farming, namely, drought and frost; yet, thus far, we have brought only a very small fraction of the ocean's area under systematic cultivation[8].
9. Most of the vegetation producing areas of the world's dry land surface have already been exploited[9], and the ocean appears to possess 5–10 times more 'potentially arable' area than the land.

Up until now, 3 primary problems have prevented farming of the open oceans: 1. the natural bottom is so far down in most places that the sunlight cannot reach it, thus preventing the reproduction and growth of attached seaweeds; 2. the natural surface waters are (as mentioned above) almost devoid of some of the nutrients required for the growth of plant life; 3. the hazards of storms at sea have seemed insurmountable.

Potential answers to these problems are: 1. emplace an open-work mesh of stout plastic lines some 15–30 m down from the ocean's surface, thus giving the seaweeds a substrate for attachment regardless of the depth of the natural ocean bottom; 2. extend the intake pipes of wave or wind powered pumps vertically down from the surface zone some 100–300 m in order to create artificial unwelling of the cool, nutrient rich deeper waters; 3. apply modern ocean engineering techniques plus judicious siting of the farm systems in order to withstand stresses from, and to reduce encounter frequencies with, the marine storms that cannot be avoided.

The major question for this concept at present concerns its economic feasibility, but the issue is more one of 'when' rather than 'whether' the concept will eventually pay off. As the world's population increases and as its vegetation and energy producing potentials are progressively diminished, there will necessarily be an improvement in the economic feasibility of ocean farming as compared to land farming or other methods of utilizing the energy- and food-producing powers of the sun.

Growing and harvesting the seaweeds

Figure 1 shows one concept which has been under

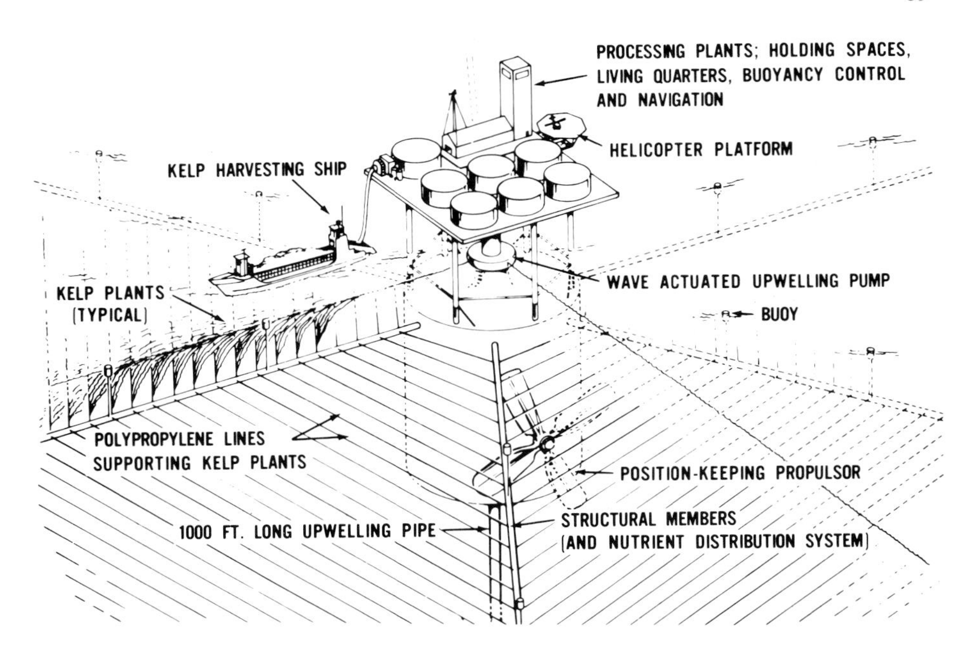

Figure 1. Conceptual design. 400-ha ocean food and energy farm unit.

study since 1973[10]. In this system the passing waves would cause floats to rise and fall, and these floats would be connected to lift pumps in upwelling pipes so as to force cool, nutrient laden water from the deeps to the surface zone. There the seaweeds – the giant California kelp *Macrocystis pyrifera* (fig. 2) – would use these nutrients plus high value photons from the sun to weld carbon dioxide and water drawn from the surrounding ocean into the kinds of energy-bearing molecules that form the basis of the entire world's food chain, namely, carbohydrates and proteins. The seaweeds would be attached by their 'holdfasts' to the mesh of plastic lines shown in figure 1, and they would be periodically harvested (coppiced), as they are today in California coastal waters, by ships bearing large clippers which cut the seaweed fronds at a level 1 or 2 m beneath the ocean's surface[11]. After each harvesting the seaweeds would continue to grow, thus replacing the previously cut fronds in readiness for a subsequent harvest. If a few plants were to be lost, natural recruitment of juveniles from neighboring plants would replace them, so planting of the farm is expected to be required only once.

Figure 3 shows the general flow of materials in this ocean farm system. All the carbon, oxygen, and other atoms involved would simply be recycled, going from the farm system's input point on the left into the products on the right and then returning around the system back to its input point again. High value photons would enter the system shown in figure 3 at the left, and low value photons (infrared radiation) would leave the system at the right.

Each hectare of cultivated ocean is expected to yield some 700–800 tonnes of whole seaweed (fresh weight basis) per year. This translates to a conversion efficiency relative to the incident solar energy of about 2%[12]. At first there was a fear in some quarters that the deep ocean waters might contain compounds toxic or otherwise inimical to the growth of kelp, but work by Prof. Wheeler J. North in 1976[13] demonstrated that there need be no anxiety in this regard. Indeed, his research work showed growth rate stimulation of juvenile kelp plants by flowing mixtures of waters drawn from the surface and from several hundred meters of depth, in both the Atlantic and the Pacific

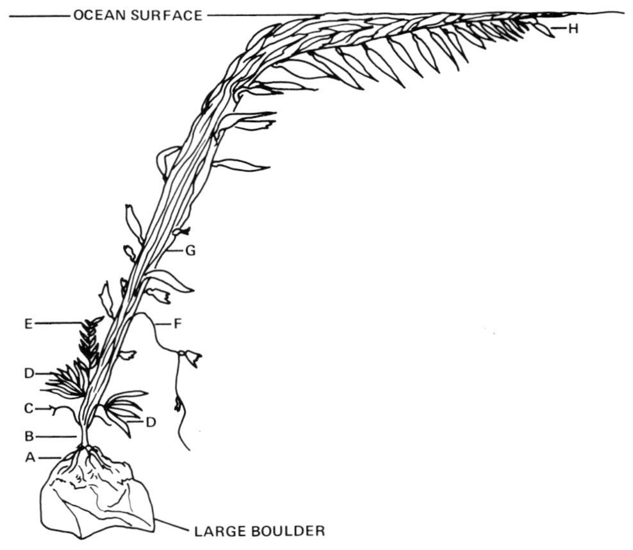

Figure 2. Diagram of young adult *Macrocystis* plant. A, Holdfast; B, primary stipe; C, stubs of frond; D, sporophyll cluster; E, juvenile frond; F, senile frond; G, stipe bundle; H, apical meristem. No root involved – plant takes all nutrient direct from surrounding water.

oceans, in approximate accordance with the following table:

Mix ratio (% deep/% surface)	Growth rate (% per day)
100/0	15.1
50/50	19.3
10/90	17.5
5/95	12.2
0/100	9.5
Control plant in flowing bay water at Corona del Mar, California	7.9

The upwelling of cool, nutrient-laden water (probably at about 10 °C) from the deeps is expected to make it possible for ocean farms to operate successfully even in the warmest waters of the tropics.

Kelp stands naturally shelter and support abundant faunal communities[14]. Hence the ocean farm system will be expected to encompass the harvesting of its fin fish, and it will probably include also the culturing of oysters and other organisms in order to utilize to the greatest possible extent the phytoplankton whose growth will inevitably be accelerated by the upwellings it produces.

Foods, feeds, and other products

Whole *Macrocystis pyrifera* has been used in experimental feeding trials with sheep, abalone, fish, chickens, and fly larvae[15]. Results from tests at the University of California at Davis showed sheep digestion efficiencies of some 58% of the organic matter in the dried kelp – about the same as for the basal ration composed of alfalfa hay, oat hay, barley and sodium phosphate. Juvenile abalone showed 10% conversion efficiencies, considered on a fresh weight of kelp to fresh weight of abalone basis. Experiments with the fish, chickens and fly larvae did not produce positive results.

Figure 4 shows a schematic processing diagram for producing methane gas and other desirable products

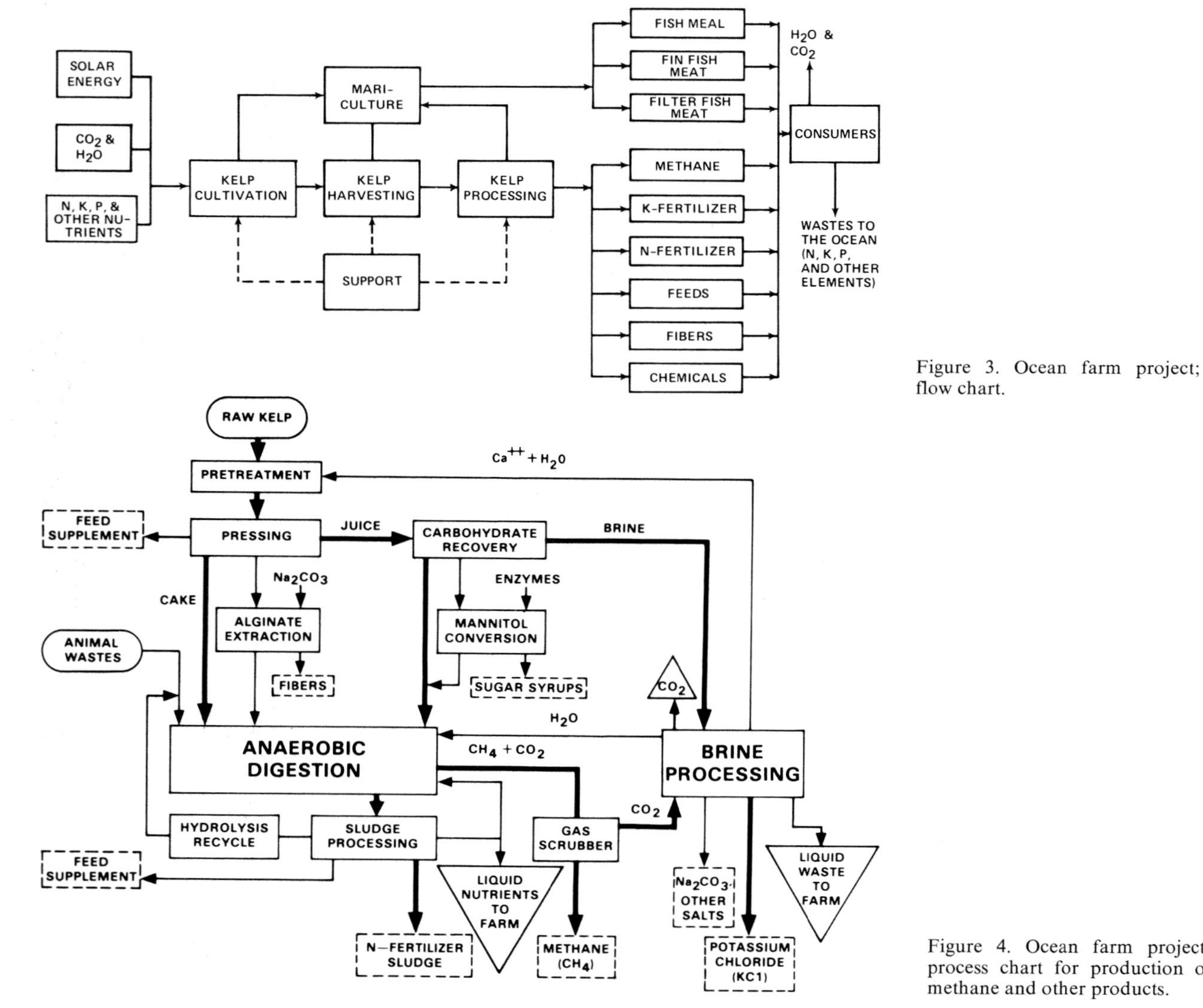

Figure 3. Ocean farm project; flow chart.

Figure 4. Ocean farm project; process chart for production of methane and other products.

from kelp[16]. Energy conversion efficiencies of about 50% have already been achieved with anaerobic digesters operating on a whole kelp feedstock to produce methane[17], and it is believed that efficiencies in the 60–80% range can probably be achieved for this process.

Site selection

It appears that optimal sites for future ocean farms will be in 3 relatively storm-free latitude bands: a) about 25 °N to about 40 °N of the equator, b) a similar zone south of the equator, and c) about 15 °S to about 15 °N of the equator[18].

Economics

Because of the high costs of deep ocean mooring lines, most of the commercially practical ocean farms of the future (not the small, experimental units of today) will probably need to be dynamically positioned by fuel powered propulsors (see figure 1). This holds true even for areas where the natural bottom is only 300 m or so beneath the surface. More than 90% of the ocean is more than 300 m deep[19]. However, if suitable arrangements can be worked out among the various ocean-owning nations – perhaps by the paying of appropriate rental fees for use of one another's waters – so that large ocean farms may move predominantly with the circulating currents which exist naturally on the ocean's surface, then the costs of ocean farming will be dramatically reduced (probably by as much as 20–40-fold) compared to the expenses that will be entailed if the farms are required to remain anchored over specified locations on the ocean floor below.

Assuming that such dynamically positioned farms can be emplaced and operated to produce kelp at the rate previously stated (corresponding to 2% conversion efficiency relative to the incident solar energy), economic studies[20] have showed that an ocean farm system using some 40,000 ha of ocean will be able to produce some 620 million m^3 of methane per year at a cost ranging from a low of about US$0.08 to a high of about US$0.25 per m^3. The range of costs given depends mainly on the assumptions used for food and byproduct credit values, distances to coasts, etc. (All dollar values are for the year 1975.) These studies also showed that 1. large systems of area 8000 ha or more, are required to be economically feasible, 2. oceanic structures will probably cost less than US$7000 per ha, 3. harvesting ship costs will probably amount to about US$3300 per ha of cultivated ocean, and 4. associated on-shore processing facility investment costs will amount to about US$4000 per ha of cultivated ocean area.

1 H.A. Wilcox, Project Concept for Studying the Utilization of Solar Energy via the Marine Bio-Conversion Technique, October 1972. This paper, as well as some of the following ones, is available from Code 5304(B), Naval Ocean Systems Center, San Diego, CA 92152, USA.
2 E.L. Saul, Wood Chemicals, in: Encyclopedia of Science and Technology, vol. 14, p. 526. McGraw-Hill, New York 1966.
3 T.C. Stadtman, Bacterial Methanogenesis, in: Encyclopedia of Science and Technology, vol. 8, p. 397. McGraw-Hill, New York 1971.
4 P.B. Weisz et al., Catalytic Production of High-Grade Fuel (Gasoline) from Biomass Compounds by Shape-Selective Catalysis. Science *206*, 57 (1979).
5 H.U. Sverdrup, M.W. Johnson and R.H. Fleming, The Oceans, p. 239. Pentice-Hall, New York 1942.
6 H.A. Wilcox, Artificial Oceanic Upwelling, July 1975. See 1.
7 J.H. Ryther, Sea-water Fertility, in: Encyclopedia of Science and Technology, vol. 12, p. 157. McGraw-Hill, New York 1971.
8 J.E. Bardach, The Harvest of the Sea, p. 157. Harper and Row, New York 1968.
9 President's Science Advisory Committee, The World Food Problem, vol. 2. The White House, Washington, DC, 1967.
10 H.A. Wilcox, The Ocean Food and Energy Farm Project, 26 January 1975. See 1.
11 H.A. Wilcox, Ocean Farm Project Kelp Harvesting Report, 1973. See 1.
12 H.A. Wilcox, Expected Yields and Optimal Harvesting Strategies for Future Oceanic Kelp Farms. Biosources Digest *1*, 103 (1979).
13 W.J. North, Marine Farm Studies Progress Reports for September and October, 1976. See 1.
14 W.J. North, ed., The Biology of Giant Kelp Beds (Macrocystis) in California, p. 52. Cramer, Lehre, Germany, 1971.
15 M.H. Beleau and D. van Dyke, Ocean Food and Energy Farm Project Food Subsystem Final Report, November 1976. See 1.
16 T.M. Leese, The Conversion of Ocean Farm Kelp to Methane and Other Products, 27 January 1976. See 1.
17 S. Ghosh et al., Research Study to Determine the Feasibility of Producing Methane Gas from Sea Kelp. Institute of Gas Technology, Chicago, IL, 1976. See 1.
18 P.F. Seligman, Survey of Oceanographic and Meteorological Parameters of Importance to the Site Selection of an Ocean Food and Energy Farm (OFEF) in the Eastern Pacific, NOSC TR 121, 1977. See 1.
19 H.U. Sverdrup, M.W. Johnson and R.H. Fleming, The Oceans, p. 19. Prentice-Hall, New York 1942.
20 V.S. Budhraja et al., Ocean Food and Energy Farm Project, Overall Economic Analysis. Integrated Sciences Corp., Santa Monica, 1976, 7 volumes. See 1. Also, an extensive computerized study by H.A. Wilcox, not yet published, some results of which, are available from Code 5304(B), Naval Ocean Systems Center, San Diego, CA 92152.

Mass production of microalgae*

by S. Aaronson and Z. Dubinsky

Biology Department, Queens College, City University of New York, Flushing (New York 11367, USA) and Department of Life Sciences, Bar Ilan University, Ramat-Gan (Israel)

The limits of the earth's arable lands[1], the continuing need for more agricultural products and/or raw materials for industry[2], and agricultural products for animal feed and human food[3], the growth of world population[4], and the increasing cost and depletion of fossil fuels[5] all point to the need for new sources of agricultural products that will not tax the earth's declining agricultural and energy resources. We suggest that microalgae may serve as a supplemental source of useful agricultural products[6-8] without making demands on land or mineral resources or requiring large amounts of scarce or depleted energy supplies[6] needed for conventional agriculture. Furthermore, the growth of microalgae in high rate domestic waste sewage oxidation ponds[9,10] can provide microalgal biomass for industrial materials[6] or biogas generation[11,12] from sunlight and CO_2 evolved during primary sewage oxidation or from industrial activity[13], and at the same time reduce the eutrophication potential of wastewater and provide reutilizable water for agriculture and/or industrial cooling[8]. Microalgae may also be grown on arid lands in the tropics or subtropics in saline or alkaline waters and at relatively high temperature to 45 °C[6]; conditions that are not useful for conventional agriculture. This type of microalgal biomass production may necessitate the addition of minerals (nitrates, ammonia and phosphate) and CO_2 unless this production is coupled to a source of these nutrients.

In terms of photosynthetic efficiency, microalgal yields are greater than those of macroalgae and similar to those of higher plants (table 1). Pirt[14] has recently estimated that up to 18% of the solar energy can be stored in algal cells in contrast to the 6% of higher plants in conventional agriculture[15]. Photosynthetic efficiencies of 36–46% (reflecting species differences) of the white light used were claimed for microalgae on continuous culture in the laboratory[16]. Unlike higher plants, the microalgal biomass has a uniform cell content and chemistry as there are no leaves, stems or roots with their different chemical composition like higher plants. Microalgal and metaphyte biomass usually have little ash content (less than 10% dry wt) in contrast to the larger amount of ash (up to 50%) of macroalgae. Microalgae may be selected for the richness of their protein, lipid or carbohydrate content (table 2) and the use to which their biomass may be put. The content of major cell biochemicals may be modified by a variety of environmental manipulations[17]. Microalgae can be grown on a large scale in a variety of outdoor ponds on different parts of the earth under varying light and temperature conditions (table 3). Goldman[10] recently reviewed outdoor mass culture of microalgae and suggested that yields of 15–25 g of dry wt m^{-2} day^{-1} could be attained for reasonably long periods of time.

Microalgae as human and animal food

Microalgae have served as human food in times of famine[18,19] and also in times of plenty[20,21]. The blue-green bacterium, *Spirulina*, is still eaten in the Lake Chad area of Africa when other foods are scarce[18,19] and the freshwater red alga *Lemanea mamillosa* is presently eaten, after frying, in India[20]. Microalgae were used as food by the Aztecs in Mexico[22,23] in the past and macroalgae, and probably microalgae as well, have served as food in Asia for millenia.

In recent years it has been proposed that microalgae, along with other microorganisms, supplement, as single-cell protein (SCP), human and animal foods[24,25]. They have proven unpleasant or even toxic for humans[26] but no attempt has been made to render algal SCP harmless or more palatable because it is

Table 1. Yields and photosynthetic efficiencies for several crops*

	Yield (+ ha^{-1} $year^{-1}$)	Total photosynthetic efficiency (%)
Theoretical maximum		
US average (annual)	224	6.6
Microalgae	17–92	0.8–2.3
Macroalgae	0.8–65	0.04–2.2
Higher plants	13–112	0.8–3.2
'Energy farm' (lumber)	25	

* See Dubinsky et al.[6].

Table 2. Range of major biomolecules in microorganisms and conventional foods*

Organism or food	Range (% cell dry wt)			Total nucleic acids
	Protein	Carbohydrates	Lipids	
Bacteria	47–86	2–36	1–39	1–36
Blue-green bacteria	36–65	8–20	2–13	3–8
Microalgae	46–60	2–7	1–76	3–6
Fungi	13–61	25–69	1–30	5–13
Egg	49	3	45	
Meat muscle	57	2	37	1
Fish	55		38	
Milk	27	38	30	
Corn	10	85	4	
Wheat	14	84	2	
Soy flour	47	41	7	

* See Aaronson et al.[7] for details.

presently not competitive with conventional human foods or food supplements. Microalgal SCP has, however, proven useful as an animal feed supplement and microalgae are used extensively as part of the food chain of invertebrates larvae and adults in aquaculture (table 4). As the price of fish and soybean meal currently used as a protein supplement in domestic animal feed continues to rise, microalgal SCP may become economically useful especially as it is a cost saving by-product of a necessary wastewater process[8,27]. Microalgal SCP is sufficient for proper nutrition for it contains adequate to rich amounts of the essential and non-essential amino acids as well as most fat- and water-soluble vitamins needed by animals[7]. Microalgae can be grown in large quantity in outdoor ponds or tanks in many climates and environments (table 3) and the annual protein yield is better than any other source (table 5). It may be argued that microalgae accumulate toxic materials such as pesticides and heavy metal ions which may render them toxic; these same toxic materials accumulate in widely accepted food crops and animal feed materials if they are exposed to air or water containing them[28a,b].

Lipids

Microalgae contain large quantities of fats and oils (table 6) for the manufacture of surfactants, fatty nitrogen compounds, rubber, surface coatings, grease, textiles, plasticizers, food additives, cosmetics, and pharmaceuticals. In the United States alone, industry used over 10^9 lbs in 1 year and the amount appears to be increasing. Algal lipids for industrial use could reduce the use of petroleum products for energy purposes and plant and animal fats for human consumption. (See Dubinsky et al.[6,8] and Aaronson et al.[8]

Table 3. Yields of microalgae in mass culture*

Alga	Place	Yield range (g dry wt · m^{-2} · day^{-1})
Chlorella	Cambridge, Mass. USA	2–11
Chlorella	Essen, Fed. Rep. of Germany	4
Chlorella	Tokyo, Japan	4
Chlorella	Tokyo, Japan	16–28
Chlorella	Tokyo, Japan	14
Chlorella	Jerusalem, Israel	12–16
Chlorella	Jerusalem, Israel	27–60
Chlorella	Japan	21
Chlorella	Taiwan	22
Chlorella	Taiwan	18–35
Chlorella	Rumania	22–36[41]
Diatoms	Woods Hole, Mass., USA	13
Diatoms	Fort Pierce, Fla., USA	25
Diatoms	Woods Hole, Mass., USA	10
Microactinum	Richmond, Cal., USA	13
Microactinum	Richmond, Cal., USA	32
Microactinum	Richmond, Cal., USA	12
Phaeodactylum	Plymouth, England	10
Scenedesmus	Tokyo, Japan	14
Scenedesmus	Dortmund, Fed. Rep. of Germany	28
Scenedesmus	Trebon, Czechoslovakia	12–25
Scenedesmus	Tylitz, Poland	12–16
Scenedesmus	Rupite, Rumania	23–30
Scenedesmus	Firebaugh, Cal., USA	10–35
Scenedesmus	Bangkok, Thailand	15–35
Spirulina	Bangkok, Thailand	15–18
Spirulina	Mexico City, Mexico	10–20
Selenastrum bibrajanum	Rumania	20–40[41]

* Adapted from data cited in Goldman[10] except as shown.

Table 4. Use of microalgae single-cell protein for animal feed*

Vertebrates	Invertebrates
Fish	Molluscs
Poultry	Silkworm larvae[42]
Swine	Bees[42]
Rabbits	

* See Aaronson et al.[7] for references except as shown.

Table 5. Protein productivity of microalgae compared with other protein sources*

Protein source	Protein yield (kg wt ha^{-1} $year^{-1}$)	Reference
Microalgae		
Chlorella (54% protein)	37,449	1
Diatoms (33% protein)	22,886	1
Scenedesmus (43% protein)	29,821	1
Spirulina (57% protein)	39,530	1
Clover leaf	1,680	43
Grass	670	43
Peanuts	470	43
Peas	395	43
Wheat	300	43
Milk from cattle on grassland	100	43
Meat from cattle on grassland	60	43

* Data compiled from mean yield 19 g dry wt m^{-2} day^{-1} of microalgae in table 2 of this paper, ha = 10,000 m^2, and mean protein for these algae in table 3 in Aaronson et al.[7].

Table 6. Total lipids of microalgae in vitro, in nature, and in sewage oxidation ponds

Algae	Range of total lipids (% dry wt)
Blue-green bacteria	2–13
Bacillariophyceae	1–39
Chlorophyceae	1–53
Chrysophyceae	12–39
Cryptophyceae	13
Dinophyceae	5–36
Euglenophyce	17
Haptophyceae	5–48
Phaeophyceae	1–9
Prasinophyceae	3–18
Rhodophyceae	tr–14
Xanthophyceae	6–16
Sewage oxidation (high rate oxidation ponds)	
Chlorella-Euglena (Israel)	23
Euglena (Israel)	11
Microactinum (Israel)	17
Oocystis (Israel)	20
Scenedesmus (Israel)	22
Scenedesmus, *Microactinum*, *Selenastrum* (Richmond, CA., USA)	24

for details on algal lipids, their value, and the economics of their production.)

Carbohydrates

Many microalgae accumulate large quantities of polysaccharides as reserve materials or to compensate for higher external osmotic pressures. A green microalga, *Dunaliella*, is currently being exploited for the production of glycerol[29]. Seaweed (macroalgae) currently supplies phycocolloids (polysaccharides such as agar, carrageenan, etc.) for food additives. If the world's supply of seaweed diminishes as the result of overexploitation and/or pollution, it may become necessary to look to the mass culture of microalgae such as the red alga, *Porphyridium*, which produces a sulfated galactan[30]. Polysaccharides are also used in the petroleum industry; microalgae may provide the long chain polymers with flocculating properties that are needed for oil drilling.

Pharmaceuticals

Microorganisms such as bacteria or fungi have been exploited for almost a century to provide useful drugs, antibiotics, and other pharmacologically active compounds[31,32]. Microalgae like macroalgae may produce a wide variety of pharmacologically-active compounds. Antibiotics, active against bacteria, fungi and even viruses, have been isolated from marine algae, especially macroalgae[33,34]. Antibacterial and antifungal agents have also been found in microalgae (table 7). Microalgae produce phycocolloids like macroalgae and these were reported to have hypocholesteremic properties[35]. Folk medicine contains several microalgal prescriptions to alleviate the symptoms of gall and other stones, gout, cancer, fistula, piles, and vaginitis[36]. Microalgae contain acetylcholine and similar molecules, amines and several alkaloids[36].

Miscellaneous substances

Microalgae may contain volatile compounds. Among these are organic acids, aldehydes, essential oils[7]. All microalgae contain significant amounts of carotenes and xanthophylls which could satisfy the needs for these pigments for coloring poultry, eggs, human food, animal feed, and carp and goldfish[7]. Microalgae contain plant growth factors and they have also been used in small quantity to prepare radioactive biochemicals for research from labeled CO_2, water, etc.

Microalgae as a source of useful molecules on a continuous or discontinuous basis

Microorganisms have proven useful for the production by secretion or excretion of a variety of large and small organic molecules for the food and pharmaceutical industries (Demain[31] and Woodruff[32], for reviews). No published work is available that might indicate the usefulness of microalgae in this area. We suggest, however, that microalgae may be harnessed to produce useful molecules. Microalgae may excrete large quantities of organic molecules (see Aaronson et al.[7], table XIV for details). Among these molecules are small molecules: sugars, nucleic acid derivatives, cAMP, amino acids, amines, fatty acids, volatiles and macromolecules: polysaccharides, nucleic acids, peptides, proteins (including enzymes) (see Hellebust[37] and Aaronson[38] for reviews). Microalgae may be induced to produce large quantities of extracellular molecules in the same way as other microorganisms but without the expenditure of expensive natural raw

Table 7. *Microalgae producing antibiotics**

Algae	Compound
Prokaryota	
Blue-green bacteria	
Hydrocoleus sp.	Terpene, carbohydrate
Lyngbya majuscula	Terpene, carbohydrate
Trichodesmium erythraeum	Terpene, carbohydrate
Eukaryota	
Bacillariophyceae	
Asterionella notata	Unidentified
Asterionella japonica	Nucleosides, fatty acids
Bacillaria paradoxa	Unidentified
Bacteriastrum elegans	Fatty acids
Chaetoceros lauderi	Polysaccharides
Chaetoceros lauderi	Fatty acids
Chaetoceros lauderi	Acid polysaccharide
Chaetoceros peruvianus	Fatty acids
Chaetoceros pseudocurvisetus	Unidentified
Chaetoceros socialis	Fatty acids
Cyclotella nana	Unidentified
Fragillaria prinata	Peptides
Gyrosigma spenceri	Unidentified
Liomophora abbreviata	Unidentified
Lithodesmium undulatum	Unidentified
Navicula incerta	Unidentified
Nitzschia longissima	Unidentified
Nitzschia ascicularis	Unidentified
Nitzschia seriata	Unidentified
Rhizosolenia alata	Unidentified
Skeletonema costatum	Fatty acids
Thalassiosira decipiens	Fatty acids
Thalassiosira nana	Fatty acids
Thalassiothrix frauenfeldi	Unidentified
Chlorophyceae	
Dunaliella sp.	Unidentified
Spirogyra sp.	Unidentified
Chrysophyceae	
Stichochrysis immobilis	Unidentified
Cryptophyceae	
Hemiselmis	Unidentified
Rhodomonas	Unidentified
Dinophyceae	
Gonyaulax tamarensis	Terpene, carbohydrates
Prorocentrum micans	Terpene, carbohydrates
Goniodoma sp.	Unidentified
Prymnesiophyceae	
Coccolithus sp.	Terpene, carbohydrates
Isochrysis sp.	Terpene, carbohydrates
Monochrysis (= *Pavlova*) sp.	Terpene, carbohydrates
Phaeocystis pouchetti	Acrylic acid
Prymnesium parvum	Terpene, carbohydrate

* See Aubert et al.[33] and Glombitza[34] for references to antibiotics in microalgae.

materials and energy. This production might become continuous with the continuous efflux of useful molecules and biomass at the expense of inorganic salts and solar energy. Furthermore, some of this production might be coupled to domestic wastewater treatment or smokestack efflux of CO_2 where the saleable end products might be useful biomass, products for industry, and reuseable water for agriculture in arid lands and/or cooling water for industry. Microalgae, as is true of other microorganisms, may be induced to excrete desired molecules under a variety of environmental or life cycle manipulations such as stage in life cycle, senescence, nutrient deprivation, chemical or physical stress. The production of useful molecules, as in other microorganisms, may be enhanced by the selection of deregulated mutants.

Microalgae as traps for toxic or polluting compounds

Microalgae, like other microorganisms, may prove useful in the uptake of heavy metals in industrial waste outfalls by accumulating the toxic metals in their cell bodies in a waste trap and then being harvested to remove the toxic compound(s) from the fresh or salt water. Among the metal ions that accumulate as much as several thousand-fold in microalgae are zinc, mercury, cadmium, copper, uranium, and lead. Microalgae also accumulate pesticides and other polluting hydrocarbons (Dubinsky and Aaronson[36]). This concentrating capability of microalgae may be useful in 'scrubbing' waste waters of industry or possibly smokestack effluent to remove and concentrate toxic materials. The algal product, however, may have no further economic use unless it concentrates useful amounts of toxic molecules or can be used as biomass for biogas production. Microalgae may also remove excess nitrate and/or phosphate or sulfite from domestic industrial or feed lot or paper mill waste water. This type of microalgal 'scrubbing' of organic pollutants is coupled with bacterial oxidation in the high rate sewage oxidation pond which has proven useful for the sewage treatment of domestic or feed lot wastes[40]. The resulting algal biomass may be used for any of the products mentioned in earlier sections of this review.

Economics of microalgal biomass and products

The products of microalgal biomass must compete in quality and price with conventional material. Because of our lack of experience with algal products, they must offer significant economic and/or quality benefits to induce the consumer to use them. At present, we think that there is not enough economic return from the production of microalgal biomass for a single product i.e., protein, lipid, etc. to warrant exploitation at current costs of the product unless that product commands unusually high prices, as for example *Chlorella* which is consumed a health food in Japan[39]. However, the production of microalgal biomass becomes economically, advantageous when all of its products such as lipids, defatted algal meal and reutilizable water, and its services (domestic or feedlot wastewater treatment) are viewed together. Based on 1978 prices, Dubinsky and co-workers[8] calculated that microalgal biomass would yield a profit only if algal oil and algal meal were sold separately and the value of sewage treatment and reutilizable water was factored into the calculation. In 1980 the price of soybean oil and meal (suggested as price references) increased 40% and 53%, respectively, while costs have probably increased about 30% in 2 years making the net yield from microalgal production more profitable. Thus it appears economically feasible at present to couple the use of microalgae in domestic and feedlot wastewater treatment with the production of useful compounds for industry and animal feeds. This should not, however, be construed to indicate that these are the only uses for microalgae. If the cost of production and the value of the product warrant it, other microalgal products may become competitive on the world market.

* Acknowledgments. This work was supported by grants to S.A. from the National Science Foundation No. PFR-7919669 and 5-SO5-RR-07064 to Queens College from the National Institutes of Health.

1 R.H. Whittaker and G.E. Likens, in: Primary Productivity of the Biosphere. Ed. H. Lieth and R.H. Whittaker. Springer, New York 1975.
2 I.S. Shapiro, Science *202*, 287 (1978).
3 Third World Food Survey, Freedom from Hunger Campaign Basic Study, FAO, Rome 1963, vol. 11.
4 National Academy of Sciences, Rapid Populations Growth, vols 1 and 2. John Hopkins Press, Baltimore.
5 M.K. Hubbert, in: The Environment and Ecological Forum, 1970–1971. US Atomic Energy Commission, Oak Ridge, TN 1972.
6 Z. Dubinsky, T. Berner and S. Aaronson, Biotechnol. Bioengng Symp. *8*, 51 (1978).
7 S. Aaronson, T. Berner and Z. Dubinsky, in: Algae Biomass. Ed. G. Shelef, C.J. Soeder and M. Balaban. Elsevier/North-Holland, Amsterdam 1980.
8 Z. Dubinsky, S. Aaronson and T. Berner, in: Algae Biomass. Ed. G. Shelef, C.J. Soeder and M. Balaban. Elsevier/North-Holland, Amsterdam 1980.
9 W.J. Oswald, Chem. Engng Prog. Symp. Ser. *65*, 87 (1969).
10 J.C. Goldman, Water Res. *13*, 1 (1980).
11 W.J. Oswald and G.G. Golueke, Adv. appl. Microbiol. *2*, 223 (1960).
12 P.H. Abelson, Science *208*, 1325 (1980).
13 E. Stengel, Ber. dt. bot. Ges. *83*, 589 (1970).
14 S.J. Pirt, Biochem. Soc. Trans. *8*, 479 (1980).
15 J.A. Bassham, Science *197*, 630 (1977).
16 S.J. Pirt, Y.K. Lee, A. Richmond and M.W. Pirt, J. chem. Technol. Biotechnol. *30*, 25 (1980).
17 T. Berner, Z. Dubinsky and S. Aaronson, Proc. IInd Int. Workshop on Biosaline Research. Plenum Press, New York, in press (1981).
18 M.Y. Brandily, Sciences Avenir *152*, 516 (1959).
19 J. Léonard and P. Compère, Bull. Jard. bot. nat. Belg. *37*, 1 (1967).
20 M. Khan, Hydrobiologia *43*, 171 (1973).
21 H.W. Johnston, Tuatara *22*, 1 (1976).
22 W.V. Farrer, Nature *211*, 341 (1966).
23 M.M. Ortega, Revta lat. Am. Microbiol. *14*, 85 (1972).

24 R.J. Matales and S.R. Tannenbaum, ed., Single-Cell Protein. M.I.T. Press, Cambridge, MA, 1968.
25 J.K. Bhattacharjee, Appl. Microbiol. *13*, 139 (1970).
26 V.R. Young and N.S. Scrimshaw, in: Single-Cell Protein II. Ed. S.R. Tannenbaum and D.J.C. Wang. M.I.T. Press, Cambridge, MA 1975.
27 R. Moraine, G. Shelef, A. Meadan and A. Levin, Biotechnol. Bioengng, in press (1981).
28 a) H. Egan, EQS Environm. Qual. Safety *2*, 78 (1973); b) Pesticides residues and radioactive substances in food: a comparative study of the problems, Environm. Qual. Safety *3*, 17 (1974).
29 A. Ben-Amotz and M. Avron, in: Algae Biomass. Ed. G. Shelef, C.J. Soeder and M. Balaban. Elsevier/North-Holland, Amsterdam 1980.
30 J. Ramus, in: Biogenesis of Plant Cell Wall Polysaccharides. Ed. F. Loewus. Academic Press, New York 1973.
31 A.L. Demain, Biotechnol. Lett. *2*, 113 (1980).
32 H.B. Woodruff, Science *208*, 1225 (1980).
33 M. Aubert, J. Auber and M. Gauthier, in: Marine Algae in Pharmaceutical Science, Ed. H.A. Hoppe, T. Levring and Y. Tanake. Walter de Gruyter, Berlin 1979.
34 K.W. Glombitza, in: Marine Algae in Pharmaceutical Science. Ed. H.A. Hoppe, T. Levring and Y. Tanake. Walter de Gruyter, Berlin 1979.
35 G. Michanek, in: Marine Algae in Pharmaceutical Science. Ed. H.A. Hoppe, T. Levring and Y. Tanake. Walter de Gruyter, Berlin 1979.
36 Z. Dubinsky and S. Aaronson, Proc. IInd int. Workshop on Biosaline Research. Plenum Press, New York, in press.
37 J.A. Hellebust, in: Algal Physiology and Biochemistry. Ed. W.D.P. Stewart. Univ. of California Press, Berkeley 1974.
38 S. Aaronson, Chemical Communication in Microorganisms. CRC Press, Boca Raton, FL, in press.
39 O. Tsukada, T. Kawahara and S. Miyachi, in: Biological Solar Energy Conversion. Ed. A. Mitsui, S. Miyachi, A. San-Pietro and S. Tamura. Academic Press, New York 1977.
40 A.S. Watson, ed., Aquaculture and Algae Culture. Noyes Data Corp., Park Ridge, NJ 1979.
41 L. Polesco-Ianasesco, Acta bot. Horti bucur. 183 (1974).
42 A.M. Muzafarov, M.I. Mavlani and T.T. Taubaev, Mikrobiologiya *47*, 179 (1978).
43 W.A. Vincent, Symp. Soc. gen. Microbiol. *21*, 47 (1971).

Mass production of *Spirulina*

by Claudio Santillan

Biochemical Research Department, Sosa Texcoco SA, Reforma 213, Mexico 5 (Mexico)

Introduction

Among the lower plants, the blue-green alga *Spirulina* (fig. 1) has been the subject of a number of basic and applied investigations[1,2]. This alga can be harvested, processed and used for food. Attention has been directed to *Spirulina platensis*, which some tribes in the Lake Chad area have been eating since ancient times[3-5], as well as *Spirulina Geitleri* J. de Toni, which was consumed by the Aztecs that lived around Lake Texcoco, near Mexico City[6-9].

Production of Spirulina

Spirulina belongs to the family of Oscillatoriaceae and grows in alkaline waters in Africa, Asia, North and South America[10], in latitudes between 35°S and 35°N, areas of incident solar irradiation from 600 to 850 $KJ/cm^2 \cdot$ year and total insolation from 3000 to 4000 h/year[11]. Like other microorganisms, *Spirulina* has a higher specific growth rate than higher plants. It has been cultivated in a semicontinuous system and harvested continuously all year round. *Spirulina* as other cyanobacteria possesses the following properties:

a) a short life cycle, approximately 1 day under optimal laboratory conditions and 3-5 days under natural conditions, depending on season and meteorological conditions[12];

b) a high specific growth rate ($0.3\ d^{-1}$) under optimal laboratory conditions[13-15], $0.2\ d^{-1}$ in natural conditions during the summer[16,17] and $0.1\ d^{-1}$ in winter[17,18];

c) growth in an aquatic medium which allows growth to a dense culture of algae biomass, consequently a good efficiency of solar energy conversion is obtained (3-4.5%)[19];

d) a high yield in good quality protein (28 ton/ha · year)[12];

e) the tendency to float and stick together thus facilitating the harvesting;

f) besides the high content of protein, substantial amounts of vitamins, carotenoids, minerals and moderate quantities of lipids and carbohydrates[19,20] can be isolated.

With current technology, 2 methods for cultivating *Spirulina* are known: the artificial culture and the seminatural culture. The 1st method, named syphogas, has been developed by the French Institute of Petroleum, which permits agitation, homogenization and supplementation with CO_2 as the carbon source through the injection of air enriched with carbon dioxide with diffusors[21]. This method has been tested on small plants with an area up to 1000 m^2 located in the Caribbean Martinique Island. It has been demonstrated that technical and economic problems limit industrial production[22].

A 2nd method, called seminatural, has been developed by the Mexican company Sosa Texcoco, SA, and consists of using the natural alkaline brines in raceway ponds supplemented with fertilizer to increase biomass production. This method has been very successful during the last 9 years, resulting in a production of approximately 3000 tons during THIS period[23,24].

Table 1. Chemical composition of *Spirulina* spray dried

Raw protein	70.0%
Essential aminoacids	
Isoleucine	4.13%
Leucine	5.80%
Lysine	4.00%
Methionine	2.17%
Phenylalanine	3.95%
Threonine	4.17%
Triptophan	1.13%
Valine	6.00%
Non essential aminoacids	
Alanine	5.82%
Arginine	5.98%
Aspartic Acid	6.43%
Cystine	0.67%
Glutamic Acid	8.94%
Histidine	1.08%
Proline	2.97%
Serine	3.18%
Available lysine	85%
Lipids	7.00%
Fatty acids	5.70%
Lauric	229 mg/kg
Myristic	644 mg/kg
Palmitic	21,141 mg/kg
Palmitoleic	2035 mg/kg
Heptadecanoic	142 mg/kg
Stearic	353 mg/kg
Oleic	3009 mg/kg
Linoleic	13,784 mg/kg
γ-Linolenic	11,970 mg/kg
α-Linolenic	427 mg/kg
Others	699 mg/kg
Ash	9.00%
Calcium	1,315 mg/kg
Phosphorus	8,942 mg/kg
Iron	580 mg/kg
Sodium	412 mg/kg
Chloride	4,400 mg/kg
Magnesium	1,915 mg/kg
Manganese	25 mg/kg
Zinc	39 mg/kg
Potassium	15,400 mg/kg
Others	57,000 mg/kg
Carotenoids	4,000 mg/kg
β-Carotene	1,700 mg/kg
Xanthophylls	1,600 mg/kg
Carbohydrates	16.50%
Ramnose	9.00%
Glucane	1.50%
Cyclitols	2.50%
Glucosamine and muramic acid	2.00%
Glycogen	0.50%
Sialic acid and others	0.50%
Nucleic acids	4.50%
Ribonucleic acid	3.50%
Deoxyribonucleic acid	1.00%
Vitamins	
Biotin	0.40 mg/kg
Cyanocobalamin	2.00 mg/kg
d-Ca-pantothenate	11.0 mg/kg
Folic acid	0.50 mg/kg
Inositol	350.00 mg/kg
Nicotinic acid	118.00 mg/kg
Pyridoxine	3.00 mg/kg
Riboflavine	40.00 mg/kg
Thiamine	55.00 mg/kg
Tocopherol	190.00 mg/kg

Large scale *Spirulina* culture requires all nutrients essential for life. The mineral content of *Spirulina* biomass and culture conditions of the alga are given in table 2.
The optimization of an algae production system draws problems such as species selection, nutrient selection, growth unit design and harvesting methods into a practical and economical overall design[23,24]. *Spirulina* has its filaments arranged in an elongated helix; this shape facilitates harvesting and permits filtration, e.g. by screens. It also allows the use of different depths of ponds and lower cell concentrations compatible with economical harvesting (see fig. 2). Zarrouk[13] and others[14,25,26] developed special nutrient solutions for the cultivation of *Spirulina*.
In the growth ponds, depth is usually less than 0.5 m for the algae production, with a baffle system and recirculation and mixing equipment. The mixing by flow is usually more practical and economical than the paddle agitator. The former also induces the water mass to circulate through the raceways at a speed of 0.03–0.06 m/sec, thus preventing thermic stratification. The maximal length of the channels depends upon the depth and the surface properties of these channels.

Processing and use of the harvested biomass

Spirulina is ruptured by homogenization or sonication of the cells[20,27,36]. The fluid is pasteurized to eliminate microbiological contamination and spray-dried to the final product.
So far, industrial plants have produced such a dry product from which its chemical composition, its nutritional value and its toxicity have been investigated. The material has been shown to have an exceptionally high content of vitamins, especially of the B group, E and H (Biotin) (see table 1). A total of 4.2–4.4% of nucleic acids[28], as well as of 6.2–7.0% of unsaturated lipids have been found. Fatty acids represent 83% of this fraction and the rest is made up of the insaponifiable fraction[29]. Among the pigments, chlorophyll is present in 0.8%, β-carotene in 0.23% and xanthophylls in 0.12–0.15%. Phycobillins amount up to 12–15%[30,31].

Table 2. Mineral content of *Spirulina* and culture conditions

Minerals (to be supplied in growth medium)	Dry weight (g/kg)	Growth conditions	
Carbon (from CO_2)	550	pH	9–11
Nitrogen (from NO_3)	100	Salinity	3–8%
Sulfur	30	Average temperature in ponds	7–30 °C
Potassium	30		
Phosphorus	16	Average sunlight	8–60 klx
Magnesium	2		
Calcium	1		
Micronutrients	–		

The product is nontoxic and has no side effects in animals or humans[32]. Nutritional tests have demonstrated that the *Spirulina* algae has a protein efficiency ratio (PER) of 2.2–2.6 (74–87% that of casein), a net protein utilization (NPU) of 53–61% (85–92% that of casein) and a digestibility of 83–84%.

In animal feeding experiments *Spirulina* protein is perfectly digested in weanling pigs[37] and in broilers[39]. In hens it has a good pigmentation effect on the skin and on egg yolks[38,40]. For fish it has an excellent effect in cripinids[41].

In human feeding experiments, *Spirulina* protein is adequately taken up when fed to adults and results in a low level of uric acid of the serum and a moderate increase in fecal nitrogen[42]. With children suffering from third degree malnutrition, *Spirulina* proved to be better than soya, but not as good as whole cow's milk and human milk[43], with regard to nitrogen which is retained.

More than 50 food products made with *Spirulina* or complemented by it have been explored in commercial systems. The most successful ones are health foods (capsules, tablets and powder); especially promising are the ones mixed with cereals and peanuts, e.g. cookies and nutritious candies where it has been possible to increase the protein content up to 12 and 18%, respectively[20]. Actually, in the USA it is widely used as a health food to help people lose weight without being hungry or suffering malnutrition.

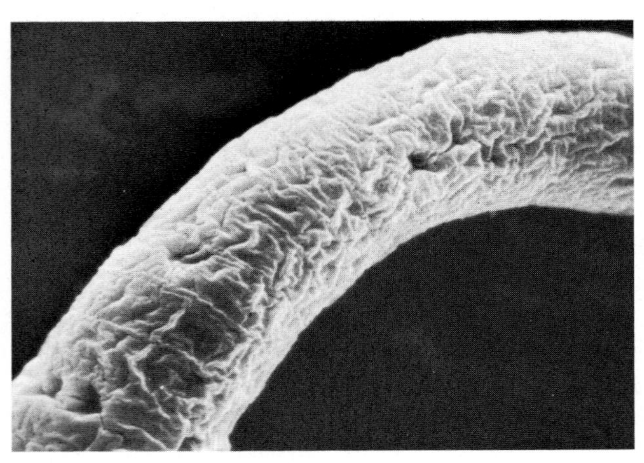

Figure 1. Alga *Spirulina*.

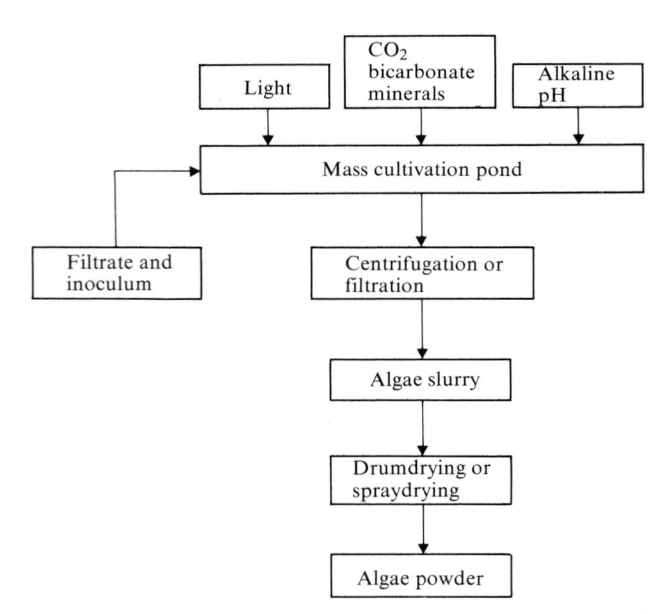

Figure 2. Flow sheet of the industrial cultivation and processing of *Spirulina*.

Conclusions

A large industrial production of *Spirulina* and its products is possible already now, thanks to technological development. This production capacity and its excellent properties as a food have justified the effort to overcome the barriers of sanitary regulations in countries such as Mexico, Japan, Canada, the United States of America, France, Great Britain, the Federal Republic of Germany, New Zealand, Australia, Korea etc., where its consumption is expanding.

In the forecoming years, the production of *Spirulina* is expected to reach several thousands of tons yearly, which will initiate a massive production and consumption program. *Spirulina*, which Farrar[9] calls 'glimpse of the Aztec food technology', will undergo a broad distribution and its alimentary benefits will be within everyone's reach.

At present, Sosa Texcoco is ready to market 500 tons more in 1982, and is working on another expansion of 2000 tons per year which should be realized by 1983.

1 E. Guérin-Dumartrait and A. Moyse, Ann. Nutr. Alim. *30*, 489 (1976).
2 P. Chanthavong, Ultrastructure de Spirulina platensis. Diplôme d'Etudes aprofondies. Université de Dijon, France, 1974.
3 P. Dangeard, Act. Soc. linn. Bordeaux *91*, 39 (1940).
4 P. Compere, Bull Jard. bot. Nat. Belg. *37*, 109 (1967).
5 M.Y. Brandily, Sci. Aven. *152*, 516 (1967).
6 F. López de Gomara, The Conquest of the West Indies, New York 1940, Scholars' Facsimiles & Repr., XXI+231 408; Historia de la Conquista de México, con una introducción y notas por D. Joaquín Ramírez Cabañas. Pedro Robredo, Mexico 1943; 2 vols.
7 F. Toribio Motolinía, Memoriales (Historic Documents of Mexico 1541), vol. 1, p. 327, Mexico, edn 1903.
8 F.B. Sahagun, Historia General de las Cosas de la Nueva España 1580, Ed. Robredo. Mexico, edn 1938.
9 W. Fárrar, Nature *211*, p. 341 (1966).
10 F. Busson, Spirulina platensis (Gom.) Geitler et Spirulina Geitleri J. de Toni, Cyanophycees Alimentaires. Service de Santé. Parc du Pharo, Marseille 1971.
11 H.E. Landsberg, Global Distribution of Solar and Sky Radiation in World Maps of Climatology, 3rd edn. Ed. E. Rondenwalt and H.J. Jusatz. Springer, Berlin 1966.
12 H. Durand-Chastel, El Tecuitlatl (Spirulina). II Coloquio Franco-Mexicano de Alga Spirulina, México, D.F., 1975.
13 C. Zarrouk (1966), Contribution à l'Etude d'une Cyanophycée sur la Croissance de la Photosynthèse de Spirulina maxima. (Stech et Gardner) Geitler. Thèse, Paris 1966.

14 T. Ogawa and G. Terui, J. Ferment. Technol. *50*, p.? (1971).
15 R.E. Martínez and C. Santillán, Estudio del Efecto de Hormonas Vegetales en el Alargamiento y División Celular del Alga Cianofícea Spirulina, II Coloquio Franco-Mexicano de Alga Spirulina, México 1975.
16 V.J. Luna, Tesis Prof., E.N.C.B.-I.P.N., México 1979.
17 M. Craules, Tesis Prof., E.N.C.B.-I.P.N., México 1978.
18 A. Richmond and A. Vonshak, Spirulina Culture in Israel. Applied Algology Institute for Desert Research Ben-Gurion, University of the Negev Sede Boqer, Israel 1976.
19 H. Durand-Chastel and M. David, The Spirulina Algae. European Seminar on Biological Solar Energy Conversion Systems. Grenoble-Autrans, France 1977.
20 S.C. Santillán, Progresos con el Alga Spirulina en la Alimentación de Animales y Humanos. VIII Congreso Interamericano de Ingeniería Química. Bogotá, Colombia 1979.
21 A. Buisson, P. Trambouze, H. Van Landeghem and M. Rebeller, Process for the Culture of Algal and Apparatus Therefore, U.S. Patent No. 3468057, 1969.
22 Anonymous, Culture Expérimentale d'Algues Spirulines. France-Antilles Quotidien d'Information, 13 October 1976.
23 H. Durand-Chastel, Procedimiento para Acelerar el Desarrollo de Algas en Medios Naturales. Mexican Patent No. 22619, 1970.
24 S.C. Santillán, El Desarrollo y Proyección del Alga Spirulina, Un Alimento del Mañana. Semin. de Alim. Proteinicos no Tradic. en México, Centr. Investig. y Estud. Avanz., México, 1974.
25 H. Nakamura, Spirulina, Ed. Keimo Sha, Tokyo, Japan, 1978.
26 N.H. Kosaric, H.T. Nguyen and M.A. Bergougnou, Biotechn. Bioengng *16*, 881 (1974).
27 L. Enebo, Conference on Preparing Nutritional Protein from Spirulina. Swedish Council for appl. Research, Stockholm 1968.
28 Y. Jassey, C.r. Acad. Sci., Paris 1356 (1971).
29 B.J.F. Hudson and I.G. Karis, J. Sci. Fd Agric. *23*, 759 (1974).
30 C. Paoletti, G. Florenzano and W. Balloni, Ann. Microbiol. 71 (1971).
31 S. Boussiba and A.E. Richmond, Arch. Microbiol. *120*, 155 (1979).
32 C.G. Chamorro, Estudios Toxicológicos de Alga Spirulina, Proyecto VC/MEX/76/090, DP/MEX/77/009, UF/MEX/78/048, UNIDO, 1980.
33 H. Burges, Nutr. Report Int. *4*, 32 (1971).
34 S.K. Kim, Protein Value of Spirulina Maxima Growing Rats, XI Int. Congr. Nutr., Rio de Janeiro, Brazil, 1978.
35 TNO, Nr. R3193 Report, Zeist, Holland, 1970.
36 E. Bujard, V. Bracco, J. Maurow, R. Mottu, A. Nabholz and J. Wuhrmann, Nestle Res. News 59 (1971).
37 C. Février and B. Seve, Ann. Nutr. Alim. *29*, 625 (1976).
38 A. Bezares, C. Arteaga and E. Avila, Valor Pigmentante y Nutritivo de Alga Spirulina en Dietas para Gallinas en Postura, 1976.
39 J.C. Blum and C. Calet, Valeur Alimentaire des Algues Spirulines pour la Croissance du Poulet de Chair. Ann. Nutr. Alim. *29*, 651 (1975).
40 J.C. Blum, S. Guillaumin and C. Calet, Valeur Alimentaire des Algues Spirulines pour la Poule Pondeuse. Ann. Nutr. Alim. *30*, 675 (1976).
41 J. Stanley, Utilization of Algae by Fish. Final Report US Department of the Interior, US Fish and Wildlife Service. Stuttgart, Arkansas, USA 1975.
42 C. Sautier and A. Trémolieres, Valeur Alimentaire des Algues Spirulines chez l'Homme. Ann. Nutr. Alim. *30*, 517 (1976).
43 R.R. Galvan, Experimentación Clínica con Spirulina Colloque sur la Valeur Nutritionnelle des Algues Spirulines. Rueil Malmaison, France, 1973.

Microorganisms as hydrocarbon producers

by T.G. Tornabene

School of Biology, Georgia Institute of Technology, Atlanta (Georgia 30332, USA)

Multifaceted and diverse energy sources will replace our once massive accumulations of energy reserves. One of these energy sources will be biomass and its natural products; in fact, it will most certainly be one of the essential elements in the complex of the future energy structure.

Solar and chemical energy conversion, through biology as a practical energy conversion mechanism, has been extensively documented and reviewed; therefore, this discussion will be restricted to microbial fermentations with specific evaluations of the potentials for microorganisms to synthesize oily hydrocarbons as fermentation products. In biosynthesis, the acyclic hydrocarbons are referred to as fermentation products on the basis of the strict definition of fermentation as being those chemical energy yielding reactions that require organic components as electron acceptors. A generalized fermentation scheme is given in the figure. The scheme is purposely restrictive to emphasize products that are potential fuels. Each of the fermentation products represents a valuable energy form. The most efficient of these fermentation products, in terms of cost of production, cannot be fairly evaluated at this time because of the differences in cell cultivation requirements, product recovery, and most importantly, since many of these products via microbial fermentations are not yet sufficiently developed for commercial consideration. With increasing awareness of microorganisms which grow well or adapt to marginal, extreme or waste environments (taking into account the benefit value of these environments and the rising expenses of waste treatment) the distinct probability exists that the production costs in developing fermentation systems for fuel will become increasingly feasible and attractive. Although the compounds listed in the figure are acceptable fuels and are accessible through microbial processes, the obvious selection of a biochemical fuel for development cannot be determined at this time because not all systems have been adequately investigated. The competitive readiness of the different fermentation systems and the economics of producing each product as they become developed will automatically map out our course of action in years to come.

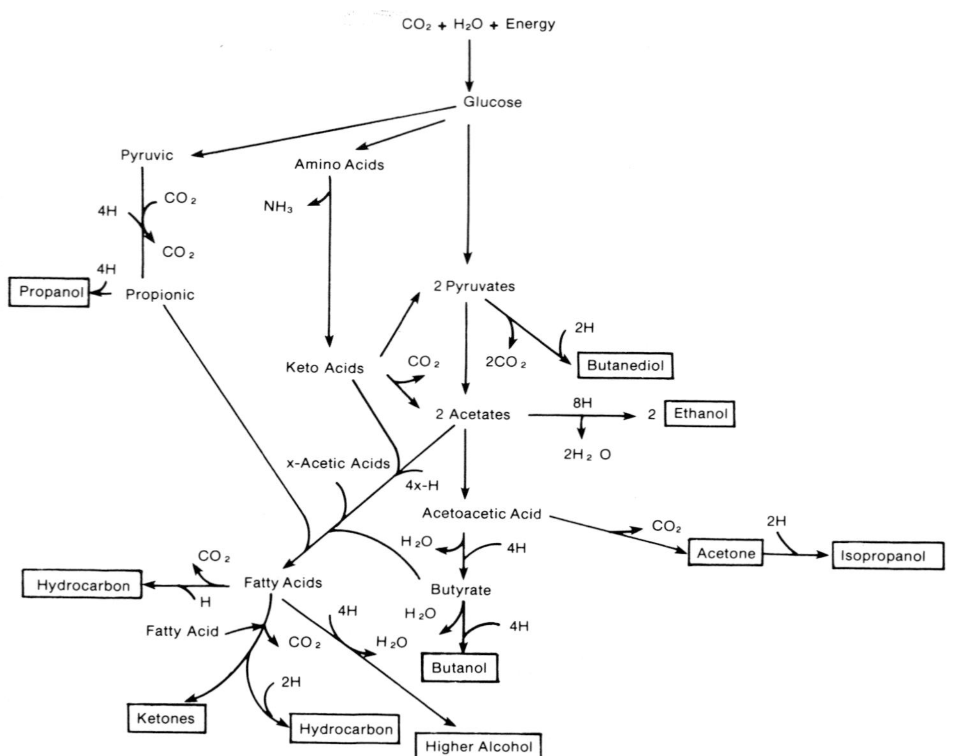

Fermentation end product of metabolism.
Hydrocarbons are natural metabolic products of many microorganisms. Intensive cultivation systems and regulatory metabolic mechanisms will have to be developed, however, before microorganisms can be used as hydrocarbon producers.

Historically the global energy source has been, for the most part, the hydrocarbon compounds. The reasons for this are clear; however, the massive availability of hydrocarbons is rapidly becoming reduced. Should the substitutes for these naturally occurring hydrocarbons be replaced with biologically produced hydrocarbons? Both hydrocarbons and oxygenated compounds (fig.), either of which could serve as an excellent source of liquid fuels, are synthesized by microorganisms. Some liquid fuels are given in the table so that the properties of hydrocarbons and oxygenated compounds may be compared. While the standard free energy of combustion and boiling points of the hydrocarbons and alcohols become more asymptotic with increasing chain length, the boiling point values of the corresponding hydrocarbons are always lower and therefore, are more easily volatilized. The properties of a liquid fuel that can be converted to a gaseous one provide greater handling, storage and combustion capabilities and thus explaining a wider utilization in today's market. The routes of formation of hydrocarbons and oxygenated compounds are mechanistically related, both being synthesized through fatty acid precursors (fig.); therefore, whatever one's preference, research on the development of either one is intertied with the other. The most important question that must be dealt with first, however, is whether microbial systems are capable of providing us with such compounds as hydrocarbons in sufficient quantities to be used as a commercial fuel directly, or to be used as a substrate for fuel production. As is typical of fermentation systems, only specific organisms have the potential to produce hydrocarbons; however, unlike the relatively established industrial microbial fermentation systems, applications of acyclic hydrocarbon syntheses are still in their infancy. The exact biosynthetic mechanism of hydrocarbon formation remains unknown, except that the routes of formation are obviously different in different microorganisms. Hydrocarbons are the end product of the reduction of organic compounds derived from decarboxylation, elongation-decarboxylation, or decarboxylation-condensation reactions of

Combustion properties of some liquid fuels

Name, formula (mol.wt)	Density (g/cm^3)	Boiling point (°C)	Heat of combustion kJ/mole	kJ/g
Oxygenated compounds				
Methanol CH_4 (32.04)	0.793	64.6	−727.3	−22.7
Ethanol C_2H_6O (46.1)	0.798	78.5	−1368.7	−29.7
Propanol (iso) C_3H_8O (60.1)	0.785	82.3	−2007.2	−33.4
Butanol (t) $C_4H_{10}O$ (74.1)	0.789	82.8	−2645.8	−33.7
Butanol (n) $C_4H_{10}O$ (74.1)	0.810	117.7	−2679.3	−36.2
Acetone C_3H_6O (58.1)	0.792	56.5	−1791.2	−30.8
Hydrocarbons				
Pentane C_5H_{12} (72.1)	0.626	36.2	−3511.9	−48.7
Hexane C_6H_{14} (86.2)	0.660	69.0	−4166.2	−48.3
Octane C_8H_{18} (114.2)	0.704	125.8	−5474.8	−47.9
Dodecane $C_{12}H_{26}$ (170.3)	0.766	214.5	−8092.4	−47.5

fatty acids[1–5]. 4 previously published reviews[6–9] outline what is known about these hydrocarbon producing microorganisms; therefore, only a brief overview of microbial hydrocarbons will be presented here. In a variety of marine and freshwater algae including red, green and brown, diatoms and phytoplankton, n-heneicosahexaene (C-21:6) hydrocarbons exist in amounts inversely correlated with the abundance of the long-chain highly unsaturated fatty acids (C-22:6)[3,10,11]. Structural assignments of the C-21:6 hydrocarbons from different sources are developing into a concise taxonomic picture[12,13]. This hydrocarbon is produced in quantities that make up more than 1% of the total dry weight of some species. In contrast, nonphotosynthetic diatoms and dinoflagellates contain traces of aliphatic hydrocarbons but no C-21:6. Other studies[14,15] found that n-pentadecane (C-15) predominates in brown algae, n-heptadecane (C-17) in red algae; that olefins predominate in red algae, more so than in brown or green algae; and that C-19 polyunsaturated olefins exist in red, brown, and green algae. The concentrations of these hydrocarbons are generally in minute to trace quantities. The occurrences of mixtures of monoenes, dienes and trienes with a C-15 to C-18 range in algae are relatively common, and, for the most part, exist in relatively small quantities[6]. Most hydrocarbon distribution patterns in blue-green bacteria represent a small percentage of the organic material and consist of predominantly C-17 components and the unique internal methyl branched 7,9-dimethyl hexadecane and a mixture of 7- and 8-methyl-heptadecane[6]. The lipid composition of the green algae *Dunaliella salina* comprised some 50% of the cellular organic material; more than 30% of the total lipids consisted of acyclic and cyclic hydrocarbons[16]. Carotenes accounted for 21% of the cell mass with another 3.5% being saturated and unsaturated C-17 straight chain hydrocarbons and internally branched ones identified as 6-methyl hexadecane and 4-methyl octadecane[16]. Further studies[21] demonstrated that temperature, light intensities and specific light spectra greatly influence the amount and type of lipids synthesized in this alga. In the chlorophyte, *Botryococcus braunii*, there exist 3 carbon distribution ranges of hydrocarbons, C-17, C-27 to C-31 and C-34 to C-36, depending on the algal growth phase. Approximately 0%, 17% and 86% of dry cellular weight is hydrocarbons of the green resting, green active and brown resting growth phases, respectively[6]. Initial reports on the hydrocarbon composition of this alga obviously contained contradictory data since different studies were performed on cells in different growth phases. The relationships of growth phase vs hydrocarbon synthesis suggest the possibility that inaccuracies may exist in all previous reports on the evaluation of hydrocarbon biosynthesis in microorganisms when culture age and environmental parameters were not considered.

Our understanding of the acyclic hydrocarbon composition of bacteria is no further advanced than that of algae. The quantities of hydrocarbons range from 35% of the 2% total lipids in some micrococci[6,17], 17.4% of the 7.4% of the cellular lipids of *Pseudomonas maltophilia*[18], 25% of the 5.9% cellular lipids of *Desulfovibrio*[19] to generally insignificant quantities in all bacteria[6] studied for hydrocarbon formation. The most thoroughly analyzed of the non-isoprenoid hydrocarbons are those of the taxonomic family Micrococcaceae. Their hydrocarbons are in the range from C-16 to C-32[6,7,17]. These hydrocarbons are a unique mixture of symmetrical and asymmetrical terminal branched monoene isomers that depict a specific chemotaxonomy profile[17]. The hydrocarbons are clearly a product of carboxyl end to carboxyl end condensations of 2 fatty acids with one of the fatty acids being decarboxylated prior to, or simultaneous with, the condensation[4,5]. The blocking of hydrocarbon synthesis in micrococci with Pb^{2+} resulted in long chain ketones that corresponded exactly to the chemical structure of hydrocarbons in carbon number as well as branching configuration[20,21]. All lines of evidence obtained support the proposal that the ketones are a metabolic end-product and not a precursor in the hydrocarbon synthesis[21]. These data demonstrated that the formation of hydrocarbons and ketones are important compounds in the regulation of the cellular fatty acid pool. This may well be the underlying basis for the formation of hydrocarbons in specific microorganisms and a fundamental key with which to unlock the mystery of the formation and regulation of hydrocarbons in microorganisms in general. In other bacteria, such as *Pseudomonas maltophilia*[18], there also exists a complex family of aliphatic hydrocarbon isomers in the range from C-22 to C-32 that are terminally methyl branched. These complex hydrocarbons have not been fully characterized, but it appears that the biosynthetic pathway follows the same general pathway as that described for micrococci. Numerous other studies have analyzed the hydrocarbon compositions of many more diverse bacteria with different nutritional properties and from different habitats[6,7]. Most of the quantities of hydrocarbon compositions found in bacteria were small (< 1.0%) with the components being straight chains in the range from C-15 to C-31 with generally no predominance of odd over even numbered carbons. For example, the distribution of components of *Desulfovibrio* was in the range from C-15 to C-28 with the components intensities peaking at C-19, C-20 and C-21[19]. Many of the bacteria contain, in addition to the straight chain components, relatively trace amounts of isoprenoid hydrocarbons. For example, phototrophic bacteria contain pristane, (C-19), phytane (C-20), squalene (C-30) and/or carotenoid (C-40)[6,7]. There are diverse types of bacteria that have the isoprenoid hydrocarbons as the predominant ones and the non-isoprenoid hydrocarbons as the minor ones[6,22–26]. The principal isoprenoid found most commonly is squalene (C-30); however, isoprenoids, hydroisoprenoids and isopranoids of different chain lengths exist in the range from C-15 to C-30 in a variety of bacteria, including phototrophs, halophiles, methanogens, thermophiles, and acidophiles[6,22–26]. These isoprenoids generally account for a maximum of around 1% of the cell mass. A most interesting relationship was observed between the isoprenoid content of some of these bacteria and the similar distribution of these components in ancient sediments and petroleum[23–25]. The contemporary bacteria which synthesize these isoprenoids exist in unusual environmental conditions similar to those thought to exist in various evolutionary stages of the archaen ecology. These findings were proposed to have major implications for biological and biogeochemical evolution[23–26]. These types of hydrocarbons produced by bacteria from unusual or extreme environments exist in the hydrophobic regions of the cells, namely the cytoplasmic membrane, and represent the type of compounds one might expect to enhance membrane stability and function of cells that must exist in such habitats. It has also been demonstrated that the isoprenoids function in some capacity as hydrogen storage compounds[26].

Fungal acyclic hydrocarbon biosynthesis is readily detectable in virtually all samples but it generally provides only small quantities of the total cellular constituents and metabolic products[6,8]. Hydrocarbons have been reported in yeast, fungal spores, and fungal mycelia. In general, the fungal hydrocarbons in the range from C-14 to C-37 are n-alkanes with no special distribution features that relate to a chemotaxonomic picture[6,8]. Hydrocarbon biosynthesis by several yeasts has been reported. Hydrocarbons accounting for more than 1% of the cell dry weight were reported for *Debaryomyces*[27]; but the hydrocarbon fractions in the range of C-16 to C-39 were not fully characterized. Yeast hydrocarbons are more typically alkanes and alkenes in the range from C-14 to C-34 with squalene often representing a significant portion of the less than 0.1% lipid of the dry cell weight composition.

This overview attests that hydrocarbons are a metabolic product of specific organisms, Hydrocarbon formation by cells is mechanistically no different from any other specific fermentation product produced by select organisms. Many years of intense research were spent to discover and develop strains which would maximize the production of existing commercial fermentation products. Similar efforts also will be required to develop a microbial hydrocarbon producing system. However, in spite of the potential for developing intensive cultivation programs and cell manipulation of such good neutral lipid producers as *Botyrococcus* and *Dunaliella*, we have not yet found the organism that is suited to develop into a hydrocarbon producer on a commercial scale. The reasons for the widespread excitement in this field lie in the existing but as yet undetermined genetic regulation and metabolic parameters that control hydrocarbon producing capacities.

The exploitation of photosynthetic microorganisms affords the only feasible approach to bioproduction of hydrocarbon oils. Photosynthetic microalgae: a) use sunlight as energy for biochemical synthesis from inorganic compounds; b) can be intensively cultivated in bodies of water generally considered unusable for domestic purposes; and c) accumulate trace metals and synthesize proteins, carbohydrates and vitamins thereby making excellent livestock feed or fertilizers. Particularly important is the fact that the cultivation of microalgae for the purpose of producing fermentation products is unlike other biomass programs in that all endproducts increase our biomass rather than decrease it. Such rewards make the search for the hydrocarbon producing algae a most exciting one.

1 J. Han, H.W.S. Chan and M. Calvin, J. Am. chem. Soc. *91*, 5156 (1969).
2 P. Blanchardue and C. Cassagne, C.r. Acad. Sci. Paris Ser. D. *282*, 227 (1976).
3 R.F. Lee and A.R. Loeblich, Phytochemistry *10*, 593 (1971).
4 P.W. Albro and J.C. Dittmer, Biochemistry *8*, 3317 (1969).
5 P.W. Albro, T.D. Mechan and J.C. Dittmer, Biochemistry *9*, 1893 (1970).
6 T.G. Tornabene, in: UNITAR-BMFT Gottingen Seminar on Microbial Energy Conversion, p.281. Ed. H.G. Schlegel and J. Barnea. Pergamon Press, Oxford 1976.
7 P.W. Albro, in: Chemistry of Natural Waxes, p.419. Ed. P.E. Kolattukudy. Elsevier, Amsterdam 1976.
8 J. Weete, in: Chemistry of Natural Waxes, p.350. Ed. P.E. Kolattukudy. Elsevier Amsterdam 1976.
9 J.R. Sargent, R.F. Lee and J.C. Nevenzel, in: Chemistry and Biochemistry of Natural Waxes, p.49. Ed. P.E. Kolattukudy. Elsevier, Amsterdam 1976.
10 S. Caccamese and K.L. Rinehart, Jr, Experientia *34*, 1129 (1978).
11 W.W. Youngblood and M. Blumer, Mar. Biol. *21*, 163 (1973).
12 R.P. Gregson, R. Kazlauskas, P.T. Murphy and R.J. Welis, Aust. J. Chem. *30*, 2527 (1977).
13 J.L.C. Wright, Phytochemistry *19*, 143 (1980).
14 R.C. Clark and M. Blumer, Limnol. Oceanogr. *12*, 79 (1967).
15 M. Blumer, R.R.L. Guillard and T. Chase, Mar. Biol. *8*, 183 (1971).
16 T.G. Tornabene, G. Holzer and S.L. Peterson, Biochem. biophys. Res. Commun. *96*, 1349 (1980).
17 W.E. Kloos, T.G. Tornabene and K.H. Schleiffer, Int. J. syst. Bact. *24*, 78 (1974).
18 T.G. Tornabene and S.L. Peterson, Can. J. Microbiol. *24*, 525 (1978).
19 J.B. Davis, Chem. Geol. *3*, 155 (1968).
20 S.L. Peterson, L.G. Bennett and T.G. Tornabene, Appl. Microbiol. *29*, 669 (1975).
21 T.G. Tornabene, unpublished results.
22 T.G. Tornabene, M. Kates, E. Gelpi and J. Oro, J. Lipid Res. *10*, 294 (1969).
23 T.G. Tornabene, T.A. Langworthy, G. Holzer and J. Oro, J. molec. Evol. *13*, 73 (1979).
24 G. Holzer, J. Oro and T.G. Tornabene, J. Chromat. *186*, 795 (1979).
25 T.G. Tornabene, R.E. Lloyd, G. Holzer and J. Oro, in: COSPAR, Life Sciences and Space Research, vol. VIII, p.104. Ed. R. Holmquest. Pergamon Press, New York 1980.
26 T.G. Tornabene, J. molec. Evol. *11*, 253 (1978).
27 E. Merdinger and R.H. Frye, J. Bact. *91*, 1831 (1966).

The production of hydrocarbons by *Botryococcus braunii*

by Reinhard Bachofen

Institute of Plant Biology, University of Zürich, CH–8008 Zürich (Switzerland)

As Tornabene[1,2] has already discussed in this and in an earlier volume, many microorganisms – photosynthetic or not – are capable of producing hydrocarbons which can account for up to 1% of the dry mass (see tables I–V in ref. 1). At the moment, it seems that there is only 1 exception described in the literature which produces a larger proportion: the unicellular green alga *Botryococcus braunii.*

Botryococcus braunii was first described some 130 years ago by Kützing[3]. The alga is found widely, in fresh water lakes as well as in brackish water on all continents. In some rare cases water blooms of *Botryococcus* are noted[4], the algae float on the surface on the lake and form combustible sediments on the shore. Such sediments are regarded as the origin of the boghead coals and tar-like deposits found in different locations, known as torbanite, coorongite or balkaschite[5–9].

Botryococcus is unicellular, but the cells form aggregates of up to 0.5 mm in diameter. The single cells are embedded in a gelatinous mass containing oils and carotenoids[10]. This is seen in the different fluorescence properties of the chloroplast within the cells and the carotenoids in the gel between them (figure). 3 different growth states are known[11]. During exponential growth in cultures the alga is green having the chlorophylls a and b, and a hydrocarbon content of around 20%[11–15]. In algal blooms however, the cells change to a resting state of yellow-orange color due to massive accumulation of carotenoids. At the same time the lipid composition is drastically altered, the unsaponifiable lipids increase to up to 80% of the dry weight[11,16]. There is also a shift in chain length of the hydrocarbons produced from C_{27} to C_{31} during growth to mainly C_{34} (botryococcene and isobotryococcene) in the yellow state (tables 1 and 2)[17]. Furthermore, a green resting state exists in which the hydrocarbon production is very low, yielding less than 1% of the dry weight of the cells.

Growth in laboratory cultures as well as in natural environments is rather slow, with generation times of roughly 1 week[4,12]. Recently generation times of 2 days have been obtained in laboratory cultures[18].

With electron microscopy it can be seen that oil droplets accumulate predominantely outside the cell walls[14], containing around 95% of the hydrocarbons of the cells. Yet oil droplets are also found within the cytoplasm. Feeding experiments with radioactive precursors demonstrate that the 2 pools are not in equilibrium. The bulk of the hydrocarbons seems to be produced at the outer wall[19,20]. Interestingly enough the cells are not able to metabolize the stored hydrocarbons[19]. Both pools have a similar composition of hydrocarbons; however, in the internal pool

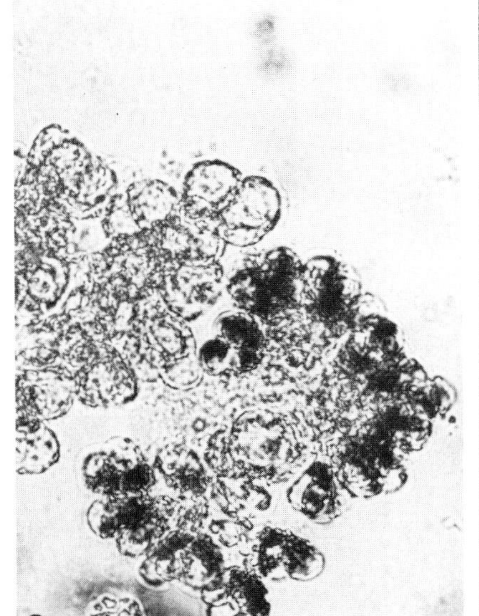

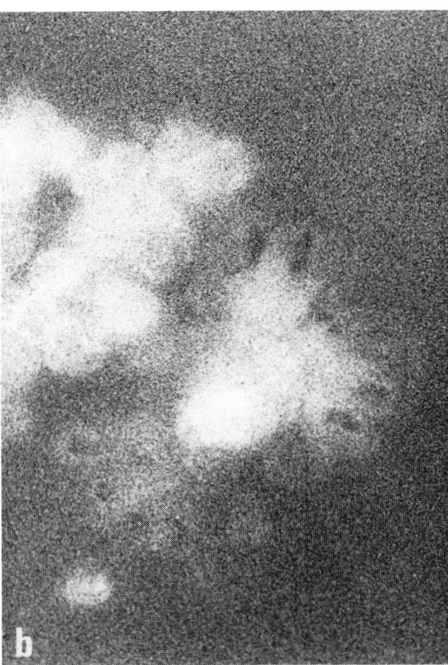

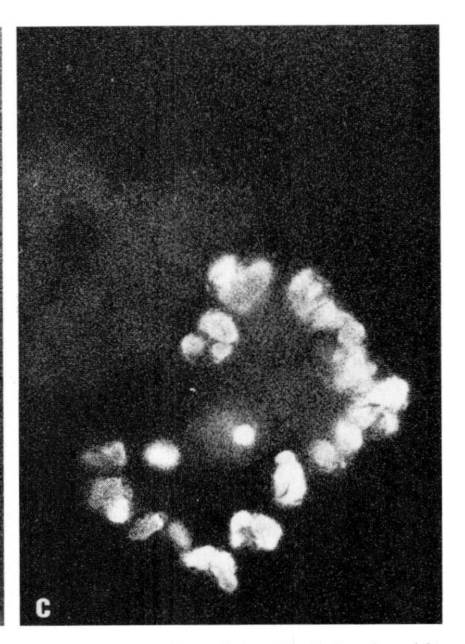

Micrographs of a colony of *Botryococcus braunii;* bar indicates 10 µm (photos made by Dr W. Egger). *a* in blue light: the light absorbing chloroplasts appear dark; *b* fluorescence visible through an optical filter transmitting green light after irradiation with light of 366 nm. The light areas indicate the fluorescence of the carotenoids which are solubilized in the hydrocarbons outside the cells; *c* fluorescence visible through an optical filter transmitting red light after irradiation with light of 366 nm. The light areas indicate the fluorescence of chlorophyll which gives an identical image as in *a*.

C_{27} and C_{29} chains are accumulated predominantely, in the external pool C_{29} and C_{31}. The C_{34} forms of the yellow resting state are not found during growth[14]. Unfortunately nothing is known about the factors responsible for the shift from the green growth state to the yellow resting state and so far this change has never been observed under laboratory conditions[12].

Table 1. Hydrocarbons from *Botryococcus braunii*[9]

Hydrocarbon	Gelpi et al.[13]	Maxwell et al.[15]	Brown et al.[11]	Belcher[12]
$C_{17}H_{34}$	1.52	–	–	–
$C_{23}H_{46}$	0.14	–	–	–
$C_{25}H_{46}$	0.10	–	–	–
$C_{25}H_{48}$	0.65	–	–	–
$C_{27}H_{52}$	11.10	–	7.2	–
$C_{28}H_{54}$	0.65	–	–	–
$C_{29}H_{54}$	5.54	–	23.0	–
$C_{29}H_{56}$	50.40	–	32.6	–
$C_{31}H_{60}$	27.90	–	25.1	–
$C_{34}H_{58}$	–	83.5	–	–
$C_{34}H_{58}$	–	83.5	–	–
$C_{34}H_{58}$ (iso)	–	8.2	–	–
other oils	–	8.3	12.1	–
Total hydrocarbons of dry weight	0.3%	75%	to 17%	15–22%
Source of alga	Laboratory culture 1–2 weeks	Wild from Oakmere Cheshire	Laboratory culture	Laboratory culture 16 weeks
State of algal development	Exponential green	Algal bloom resting state	Exponential green	Resting state green

Table 2. Dominant hydrocarbons from *Botryococcus*[9,15]

1 Heptacosa-1,18-dien ($C_{27}H_{52}$);
$CH_2=CH-(CH_2)_{15}-CG=CH-(CH_2)_7-CH_3$

2 Nonacosa-1,20-dien ($C_{29}H_{56}$);
$CH_2=CH-(CH_2)_{17}-CH=CH\cdot(CH_2)_7-CH_3$

3 Hentaconta-1,22-dien ($C_{31}H_{60}$);
$CH_2=CH-(CH_2)_{19}-CH=CH-(CH_2)_7-CH_3$

4 Botryococcen

Table 3. Change of lipid content of algal cells by environmental manipulation (from Dubinsky et al.[21])

Environmental variable	Organisms	Variation	Change in lipid composition (in % of dry wt)
Light intensity	*Spirulina*	10 → 40 klux	4.2 → 6.2%
Temperature	*Ochromonas*	15 → 30 °C	39 → 53
Nitrogen depletion	*Chlorella*	With → without	10 → 70
Salinity	*Botryococcus*	0 → 6% NaCl	36 → 51
Senescence	*Cyanobacterium* strain 92	Young → old	9 → 26
Combination of factors	*Chlorella*	Normal → stress	4.5 → 86

Several environmental changes are known, however, to affect the lipid composition in the cell and to induce lipid synthesis in general. As seen in table 3, a variation of factors such as light intensity, temperature and salinity increase the lipid content in various algae tested so far[21].

The yields in hydrocarbons calculated from the water bloom in Oakmere[4] would amount to about 2.6 t/ha · year[9], a value similar to yields for terrestrial plants synthesizing hydrocarbons. If the growth rate of *Botryococcus* could be increased to the values obtained from *Chlorella* and *Scenedesmus*, hydrocarbon yield could be 10–15 times higher[9]. Studies on the growth and physiology of *Botryococcus* may lead to great improvements.

It seems rather unlikely that *Botryococcus* is the only organism with such a high hydrocarbon content. In screening other photosynthetic microorganisms producing hydrocarbons, species with similar capabilities as *Botryococcus* may be found. The first positive results in this direction have been reported by Lien[22] and Mitsui (personal communication).

1 T.G. Tornabene, Microbial formation of hydrocarbons, in: Microbial Energy Conversion, p.281–299. Ed. H.G. Schlegel and J. Barnea. Pergamon Press, Oxford 1976.

2 T.G. Tornabene, Microorganisms as hydrocarbon producers. Experientia *38*, 43–46 (1982).

3 F.T. Kützing, Species Algarum. Lipsiae 1849. Cited from: K.B. Blackburn, Botryococcus and the algal coals. Trans. R. Soc. Edinburgh *58*, 841–853 (1935).

4 E.M. Swale, The phytoplankton of Oakmere, Cheshire, 1963–1966. Br. phycol. Bull. *3*, 441–449 (1968).

5 R.F. Cane, Coorongite, Balkaschite and related substances – an annotated bibliography. Trans. R. Soc. Aust. *101*, 153–164 (1977).

6 A.G. Douglas, K. Douraghi-Zadeh and G. Eglinton, The fatty acids of the alga Botryococcus braunii. Phytochemistry *8*, 285–293 (1969).

7 B. Fott, Algenkunde. Fischer, Stuttgart 1971.

8 E. Gelpi, H. Schneider, J. Mann and J. Oro, Hydrocarbons of geochemical significance in microscopic algae. Phytochemistry *9*, 603–612 (1970).

9 L.W. Hillen and D.R. Warren, Hydrocarbon fuels from solar energy via the alga Botryococcus braunii. Report 148, Dept of Defence, Aeronautical Research Laboratories, Australia.

10 E. Schnepf and W. Koch, Über den Feinbau der «Ölalge» Botryococcus braunii Kützing (Chlorococcales). Bot. Jahrb. Syst. *99*, 370–379 (1978).

11 A.C. Brown, B.A. Knights and E. Conway, Hydrocarbon content and its relationship to physiological state in the green alga Botryococcus braunii. Phytochemistry *8*, 543–547 (1969).

12 J.H. Belcher, Notes on the physiology of Botryococcus braunii Kützing. Arch. Mikrobiol. *61*, 335–346 (1968).

13 E. Gelpi, J. Oro, H.J. Schneider and E.O. Bennett, Olefins of high molecular weight in two microscopic algae. Science *161*, 700–701 (1968).

14 J. Murray and A. Thomson, Hydrocarbon production in Anacystis montana and Botryococcus braunii. Phytochemistry *16*, 465–468 (1977).

15 B.A. Knights, A.C. Brown, E. Conway and B.S. Middleditch, Hydrocarbons from the green form of the freshwater alga Botryococcus braunii. Phytochemistry *9*, 1317–1324 (1970).

16 A.C. Brown, Some aspects of hydrocarbon formation in the green alga Botryococcus braunii, Kützing Br. phycol. J. *4*, 211 (1969).

17 J.R. Maxwell, A.G. Douglas, G. Eglinton and A. McCormick, The botryococcenes-hydrocarbons of novel structure from the

alga Botryococcus braunii, Kützing. Phytochemistry *7*, 2157–2171 (1968).
18 C. Largeau, E. Casadeval and D. Dif, Renewable hydrocarbon production from the alga Botryococcus braunii. Int. Conf. Energy from Biomass, Brighton 1980.
19 C. Largeau, E. Casadeval and C. Berkaloff, The biosynthesis of longchain hydrocarbons in the green alga Botryococcus braunii. Phytochemistry *19*, 1081–1085 (1980).
20 C. Largeau, E. Casadeval, C. Berkaloff and P. Dhamelincourt, Sites of accumulation and composition of hydrocarbons in Botryococcus braunii. Phytochemistry *19*, 1043–1051 (1980).
21 Z. Dubinsky, T. Berner and S. Aaronson, Potential of large scale algal culture for biomass and lipid production in arid lands. Biotechnol. bioeng. Symp. *8*, 51–68 (1978).
22 S. Lien, Photobiological production of fuels by microalagae. Int. Conf. Energy from Biomass, Brighton 1980.

Glycerol production by *Dunaliella*

by Ami Ben-Amotz, Ilene Sussman and Mordhay Avron

Israel Oceanographic and Limnological Research, Tel Shikmona, Haifa (Israel), and Biochemistry Department, The Weizmann Institute of Science, Rehovot (Israel)

Summary. Species of the unicellular alga *Dunaliella* possess outstanding tolerance of a wide range of salinities. They can adapt to grow in salt media which range from less than 0.5 M to saturated salt solutions and withstand enormous osmotic shocks through a unique osmotic adaptation. The osmoregulating mechanism depends on photosynthetic production of glycerol, whose intracellular concentration varies in direct proportion to the extracellular salt concentration and reaches values in excess of 50% of the total dry weight of the cells. *Dunaliella,* and another halotolerant glycerol producing alga, *Asteromonas gracilis,* osmoregulate biochemically by controlling glycerol biosynthesis and degradation. 3 new enzymes, NADPH-dihydroxyacetone-reductase, dihydroxyacetone kinase and glycerol-1-phosphatase seem to be involved in the osmoregulatory response via glycerol in *Dunaliella* and *Asteromonas.* A hypothetical scheme of glycerol metabolism in these algae utilizing these enzymes is presented. Growth studies of *Dunaliella* indoors and outdoors showed that salt concentrations favoring maximal glycerol productivity are not identical with those required for maximal algal productivity. Maximal yield of glycerol occurred around 2 M NaCl while maximal algal productivity occurred below 0.5 M NaCl. Observed yields of glycerol in *Dunaliella* culture outdoors are compared with theoretically calculated maximal yield.

1. *Introduction*

Utilization of the photosynthetic machinery for the production of energy, chemicals and food has a particular appeal because it is the most abundant energy storing and life-supporting process on earth. Starting with the photosynthetic reaction converting carbon dioxide and water into organic carbon and oxygen with solar irradiation as the energy source, photosynthetic plants and algae utilize intricate biochemical pathways to produce a variety of organic metabolites. Serveral potential crops have been suggested in recent years as possible candidates for converting solar energy via photosynthesis into biofuels and/or valuable organic compounds[1]. These include fast-growing tree species grown at high densities, conventional crops such as corn, sugar beet, sugarcane, and plants native to arid environments. Though each has its own specific advantages and disadvantages, there are several drawbacks common to almost all terrestrial plants, such as low solar conversion efficiency due to reflection and partial absorption, storage capacity limitations, investment of much of the photosynthate in nonrecoverable parts of the plant, competition with high economic value agricultural land and a high consumptive use of fresh water. The halotolerant unialga *Dunaliella*[2] has only a few of the drawbacks and in addition offers advantages not found in the other systems. It can be grown in a population density resulting in the presentation of an optimum absorbing surface area-to-unit land area ratio throughout the year. It can grow in salt water on arid land where there is maximum availability of solar energy and where the land is not utilizable for any other kind of potential crop. Lacking a typical polysaccharide cell wall, *Dunaliella* invests a much smaller fraction of the photosynthetic products in difficult to utilize structural constituents than do other algae and plants. Most significantly, the major photosynthetic end product in *Dunaliella* is glycerol, the concentration of which varies in direct proportion to the extracellular salt concentration reaching a maximum of around 80% of the algal dry weight[3,4]. The use of *Dunaliella* for direct conversion of solar energy into a useful chemical product is therefore of particular interest. Also of interest are the rather unique metabolic pathways which exist in this alga, permitting the synthesis and regulation of the massive intracellular concentrations of glycerol. The purpose of this manuscript is to examine the basic biochemistry of glycerol production and regulation in *Dunaliella* and to describe indoor and outdoor experiments in which glycerol and algal material yield have been measured and compared.

2. *Growth of Dunaliella*

The genus *Dunaliella* contains species whose normal habitats range from seawater of around 0.4 M NaCl to salt lakes containing NaCl at concentrations up to saturation (>5 M). Moreover, algae originated from seawater can be adapted to high salt concentration and vice versa[3,5]. Figure 1 illustrates the effect of the NaCl concentration of the medium on the growth of *D. salina* as measured by chlorophyll content. The remarkable adaptation to a wide range of salt concentrations is evident. Algae grown at 4 M NaCl multiply at a rate which approximates only about one third that of algae grown under optimal conditions. A lack of requirement for high salt concentration for growth coupled with the halotolerance adaptability to a wide range of salinities differentiate these eukaryotic algae from the obligate halophilic bacteria[6].

3. *Intracellular composition*

The unique ability of *Dunaliella* to survive in highly saline water bodies was found to depend on the photosynthetic production and accumulation of high intracellular concentrations of glycerol[3,7,8]. A linear relation between the concentrations of intracellular glycerol and extracellular salt is maintained over a broad range of salt concentrations from 0.5 M to 4.5 M (fig. 2). This has been observed in various species of *Dunaliella* and in 1 species of *Asteromonas*[9]. All produce and accumulate glycerol, the intracellular content of which depends on the salt concentrations of the medium and the algal volume. Thus all available data suggest that glycerol is the major intracellular solute which serves to osmotically balance the medium salt concentration.

4. *Osmoregulatory mechanism*

Several lines of evidence provide information regarding the underlying mechanism which enables *Dunaliella* to display its unique halotolerance and adaptability.

Microscopic observations show that *Dunaliella* cells behave like perfect osmometers rapidly shrinking or swelling under hypertonic or hypotonic conditions, respectively (fig. 3). The absence of a rigid polysac-

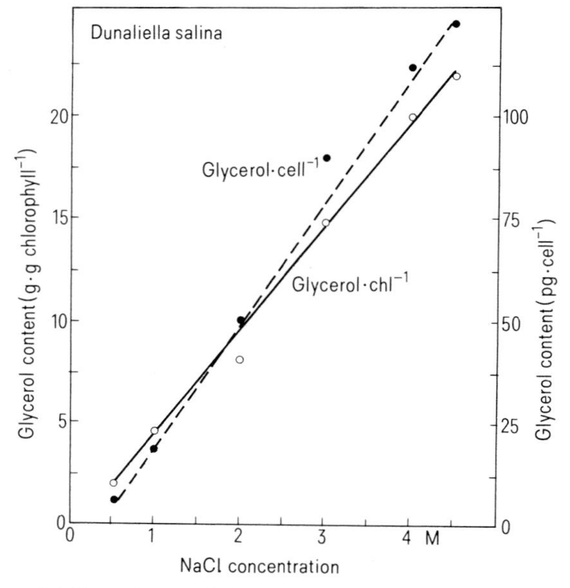

Figure 2. Effect of extracellular salt concentration on the intracellular glycerol content in *Dunaliella salina*. Assay conditions were as described previously[3].

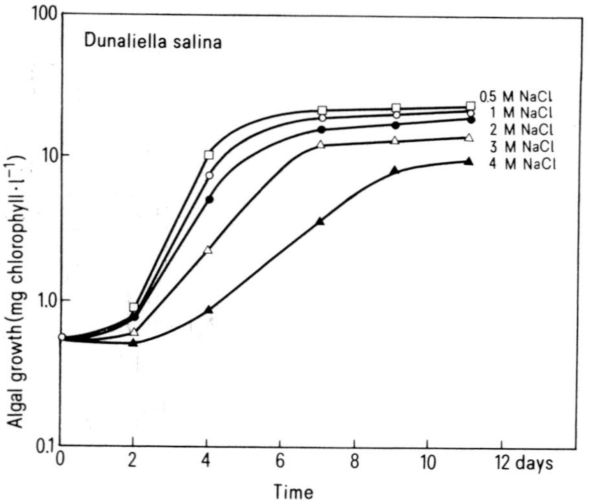

Figure 1. Growth of *Dunaliella salina* at several concentrations of salt. Growth conditions and medium were as described previously[11].

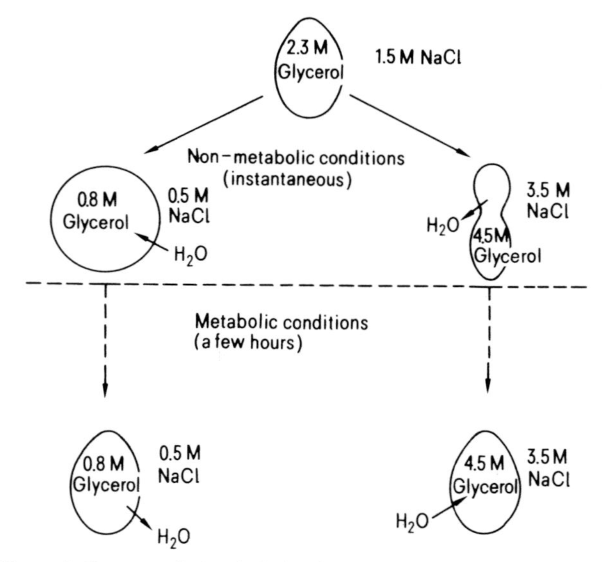

Figure 3. Osmoregulation in halotolerant wall-less algae. Schematic representation of the adjustment of *Dunaliella* to hypertonic and hypotonic conditions.

charide cell wall permits a rapid adjustment of the intracellular osmotic pressure by fluxes of water through the cytoplasmic membrane. Thereafter the cells slowly return to their original ellipsoid-like shape through a phase of metabolic adjustment. During this metabolic adjustment period under hypertonic conditions the algae produce and accumulate glycerol above the original level, while under hypotonic conditions the algae reduce the glycerol content below the original level. In either case water flows through the cytoplasmic membrane in response to the new level of intracellular glycerol so that at the steady state the original cell volume is regained[7,10]. Cellular osmoregulation in *Dunaliella* can be defined, therefore, as the ability of the cell to maintain approximately constant volume in the face of changing water potential.

The kinetics of synthesis and elimination of glycerol upon transition from low to high salt concentration or vice versa have been studied in detail[3,5,7,8] and indicate that: a) the process is very rapid; glycerol synthesis or elimination can be detected within minutes after the transition, and b) such synthesis is independent of protein synthesis or of illumination; thus, the mechanism of response is ever present and rapid in responding.

Three unique enzymes have been described for *Dunaliella* which are likely to be involved in its osmoregulatory response (table 1). The 1st is an $NADP^+$-dependent dihydroxyacetone reductase which catalyzes the interconversion of dihydroxyacetone and glycerol[8,11]. The classical NAD^+-dependent glycerol-3-phosphate dehydrogenase has also been described in *Dunaliella*[12]. The 2nd unique enzyme is dihydroxyacetone kinase which is highly specific toward dihydroxyacetone[13], and the 3rd is glycerol-1-phosphatase which specifically dephosphorylates α-glycerol phosphate.

Taking into consideration the presence of these enzymes in *Dunaliella*, a reasonable hypothetical scheme of the osmoregulatory metabolism of *Dunaliella* may be as shown in figure 4. Glycerol may accumulate by production of triose phosphate via photosynthesis or from polysaccharide degradation followed by reduction to α-glycerol phosphate and dephosphorylation. Conversion of glycerol back to polysaccharides may proceed via oxidation to dihydroxyacetone and phosphorylation to dihydroxyacetone phosphate.

Finally, it is of interest to note that these same unique enzymes were found in another halotolerant alga which osmoregulates with glycerol, *Asteromonas gracilis* (table 1)[9].

Table 1. The activity of several enzymes involved in glycerol metabolism in halotolerant algae

Algal species	Dihydroxy-acetone reductase	Dihydroxy-acetone kinase	Glycerol-1-phosphatase
	(μmoles substrate consumed · min^{-1} · mg chl^{-1})		
Dunaliella salina	0.89	0.54	4.27
Dunaliella bardawil	0.25	0.35	1.75
Asteromonas gracilis	2.22	1.01	5.18

Algae were grown in growth medium containing 3 M NaCl in an illuminated room as previously described[10,11]. Enzymes in the crude algal extract were assayed: dihydroxyacetone reductase[11]; dihydroxyacetone kinase[13], as previously described, and glycerol-1-phosphatase by following the release of inorganic phosphate in the presence of 5mM $MgCl_2$.

5. *Glycerol yield optimization*

Much has been learned about photosynthesis in terrestrial plants, aquatic plants and algae. All share the same basic biochemical machinery for converting carbon dioxide and water into organic carbon and oxygen, and so their energy conversion efficiency is governed by the same basic principles.

Solar energy strikes the earth at a low flux of 2000 kcal m^{-2} day^{-1}, hence requiring very large collection systems for capturing the light. Moreover, photosynthetic conversion efficiencies are rather low. Therefore, under the most ideal conditions the most efficient plant can convert at best about 8% of solar irradiation into stored energy in the form organic matter[14,15]. In reality, photosynthetic conversion efficiencies of natural terrestrial systems are considerably lower and seldom exceed 1–2%, primarily because other factors such as light availability, nutrients, water, etc., are limiting. Aquatic plants including microalgae are among the most efficient converters of

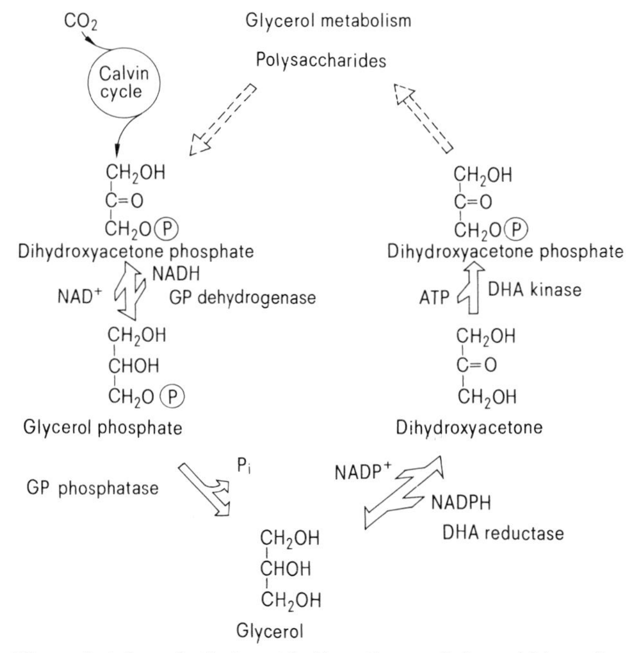

Figure 4. A hypothetical metabolic pathway of glycerol biosynthesis and degradation in *Dunaliella*.

radiant energy; photosynthetic efficiencies under laboratory conditions with low light intensity have been reported to approach the theoretical limit. Goldman[14,15] has recently summarized the theoretical and practically observed light conversion efficiency of large-scale algal culture grown under natural and light limiting conditions. Maximal yield of around 30 g m^{-2} day^{-1} have been calculated and in practice, a similar high yield data has been reported in various locations in the world for short periods[15].

Assuming solar conversion efficiency of 8% for calculating the potential for production of glycerol by *Dunaliella,* algae containing 40% glycerol on a dry weight basis can yield 16 g glycerol m^{-2} day^{-1} [4]. Table 2 illustrates the effect of salt concentration on the actual production of glycerol in open culture of *Dunaliella.* Maximal long term productivity of about 4.5 g glycerol m^{-2} day^{-1} has been observed at a salt concentration of 3.5 M. However, since the conditions for optimal growth are not necessarily those which maximize glycerol production, laboratory experiments have been undertaken to check the effect of salt on glycerol and algal yield production (fig. 5). Indeed, optimization for high yield of glycerol occurred around 2 M NaCl while the conditions favoring maximal algal productivity were in the low range of salt concentration.

Table 2. The effect of salt concentration on the productivity of glycerol in outdoor cultures of *Dunaliella bardawil*

NaCl concentration (M)	Average level of glycerol at harvest ($g \cdot l^{-1}$)	Productivity of glycerol ($g \cdot m^{-2} \cdot day^{-1}$)
3.5	0.18	4.4
4.0	0.14	3.4
4.5	0.16	3.0
5.0	0.16	2.4

Dunaliella bardawil were grown in 10-cm-depth miniponds outdoors for about 60 days between May and June. When the algae content of the pond reached the indicated level of glycerol half of the culture volume was harvested by centrifugation and the remaining algae diluted with fresh medium to the original volume.

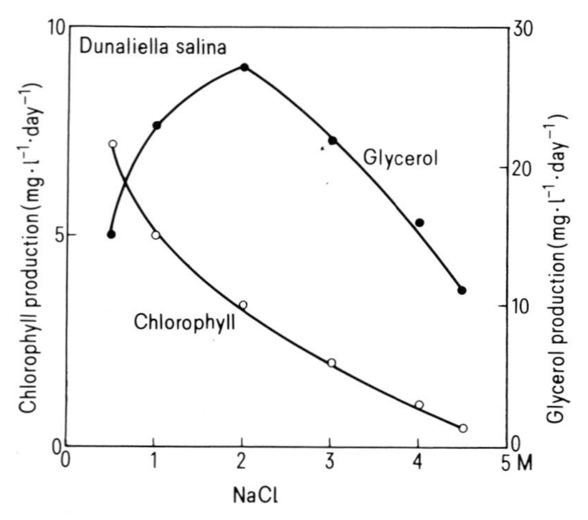

Figure 5. Effect of extracellular salt concentrations on the chlorophyll and glycerol production by *Dunaliella.* Algae were grown in a constant temperature growth room as described previously[10,11].

A variety of considerations come into play when we wish to optimize growth conditions for the production of a particular product, such as glycerol. For example, the higher the salt concentration, the lower is the interference by other organisms and predators. Since these conditions are not necessarily optimal for glycerol production, maximum yield potential may need to be sacrificed. In evaluating optimal production capability for a chemical derivative of *Dunaliella* or other microorganisms, a quantitative determination of the yield potential as a function of a variety of conditions in addition to maximal growth rate needs to be carried out. Thereafter we can assess the economic value of the algal product.

1 M. Calvin, Chem. Engng News *56,* 30 (1978).
2 A.D. Brown and L.J. Borowitzka, in: Biochemistry and Physiology of Protozoa, vol. 1, p. 139. Ed. M. Levandowsky and S.H. Hunter. Academic Press, New York 1979.
3 A. Ben-Amotz and M. Avron, in: Energetics and Structure of Halophilic Microorganisms, p. 529. Ed. S.R. Caplan and M. Ginzburg. Elsevier, Amsterdam 1978.
4 A. Ben-Amotz and M. Avron, in: Algal Biomass: Production and Use, p. 603. Ed. G. Shelef and C.G. Soeder. Elsevier, Amsterdam 1980.
5 L.J. Borowitzka, D.S. Kessly and A.D. Brown, Archs Microbiol. *113,* 131 (1977).
6 H. Larsen, Adv. microbial Physiol. *59,* 15 (1967).
7 A. Ben-Amotz and M. Avron, Pl. Physiol. *51,* 875 (1973).
8 L.J. Borowitzka and A.D. Brown, Archs Microbiol. *96,* 37 (1974).
9 A. Ben-Amotz and M. Avron, in: Genetic Engineering of Osmoregulation, p. 91. Ed. D.W. Rains and C. Valentine. Plenum, New York 1980.
10 A. Ben-Amotz, J. Phycol. *11,* 50 (1975).
11 A. Ben-Amotz and M. Avron, Pl. Physiol. *53,* 628 (1974).
12 K. Wegmann, Biochim. biophys. Acta *234,* 317 (1971).
13 H.R. Lerner, I. Sussman and M. Avron, Biochim. biophys. Acta *615,* 1 (1980).
14 J.C. Goldman, Water Res. *13,* 1 (1979).
15 J.C. Goldman, Water Res. *13,* 119 (1979).

Biological photoproduction of hydrogen and ammonia

Ammonia and hydrogen are energy-rich products which can be used as fuels. They are produced by photosynthetic microorganisms under certain limited conditions, especially stress situations. The organisms and the environmental conditions leading to a high ammonia production from nitrate or nitrogen gas are discussed by M.G. Guerrero et al., while the systems leading to gaseous hydrogen from different electron donors are presented separately for algae (H. Bothe) and photosynthetic bacteria (H. Zürrer).

Photosynthetic production of ammonia*

by Miguel G. Guerrero, Juan Luis Ramos and Manuel Losada

Departamento de Bioquímica, Facultad de Biología y C. S. I. C., Universidad de Sevilla, Sevilla (Spain)

Introduction

The conversion of solar energy into suitable redox energy through photosynthesis of the water-splitting type carried out by intact or reconstituted systems is of great interest and significance, and a major effort is presently under way to use this process to provide 'biofuels' on a renewable basis. The fact that some photosynthetically generated metabolites are highly reduced compounds of immediate practical interest has, however, been overlooked until now. Such is the case of ammonia, a compound generated by green cells from oxidized inorganic nitrogenous substrates in light-driven reactions, and that, in itself, is a very valuable derivative, both as a fuel and as a fertilizer[1,2].

Ammonia represents the primary source of the nitrogen fertilizers presently used in agriculture to improve crop productivity, which is usually limited by the availability of nitrogen in the soil. The manufacture of fertilizers requires vast inputs of energy, and it should be appreciated that nitrogen fertilizers often contribute up to 50% of the energy input in modern agriculture[3]. Ammonia for fertilizers is made through a catalytic reduction of atmospheric nitrogen with molecular hydrogen in a reaction which requires high pressures and temperatures (Haber-Bosch process). The world's industrial production of nitrogen fertilizers in 1974–75 reached about 42 megatons N with an energy cost of about 2 million barrels of oil per day[4,5]. These values represent more than a 10-fold increase since 1950, and it has been estimated that the world's annual requirement for nitrogen fertilizers will rise to 150–200 megatons N by the end of this century[4].

In addition to its value as a fertilizer ammonia is an excellent and powerful fuel and has been prominently mentioned as a constituent of various types of mixtures both for internal combustion engines and for jet propulsion[1,6]. It can react with molecular oxygen and be oxidized either to dinitrogen or to nitrate, according to the following highly exergonic equations:

$$NH_3 + {}^3/_4\, O_2 \xrightarrow{3e} {}^1/_2\, N_2 + {}^3/_2\, H_2O$$

($\Delta E'_o$, pH 7 = +1.10 V; $\Delta G'_o$ = −318 kJ · mole^{-1})

$$NH_3 + 2\, O_2 \xrightarrow{8e} HNO_3 + H_2O$$

($\Delta E'_o$, pH 7 = +0.47 V; $\Delta G'_o$ = −360 kJ · mole^{-1})

These reactions reveal that 1 atom of nitrogen reduced to the state of ammonia can supply either 3 electrons at the potential level of −0.28 V and be oxidized to molecular nitrogen, or 8 electrons at the potential level of +0.35 V and be oxidized to nitrate. In addition, ammonia can undergo thermal dissociation in the presence of certain catalysts to yield molecular nitrogen and molecular hydrogen (E'_o, pH 7, −0.42 V). Decomposition into its elements can also be effected by photochemical means and by passing a silent electrical discharge through the gas. The reaction is weakly endergonic, as shown by the following equation:

$$2\, NH_3 \xrightarrow{6e} N_2 + 3\, H_2$$

($\Delta E'_o$, pH 7 = −0.14 V; $\Delta G'_o$ = +80 kJ · mole^{-1})

Here, 2 moles of ammonia yield 3 moles of hydrogen. The hydrogen resulting from the lysis of ammonia can be utilized, for example, in the oxy-hydrogen blowpipe for producing intensely hot flame for welding metals and special steels, taking advantage of the strong exergonicity of its reaction with oxygen:

$$H_2 + {}^1/_2\, O_2 \xrightarrow{2e} H_2O$$

($\Delta E'_o$, pH 7 = +1.24 V; $\Delta G'_o$ = −238 kJ · mole^{-1})

For the above reasons, liquid ammonia can be used for storing and transporting hydrogen in a convenient and compact way.

Aside from its industrial synthesis, a much more significant production of ammonia is being carried out by the plant kingdom. The extent of this biological synthesis of ammonia is as much as 2×10^4 megatons N per year, about 1% of this figure corresponding to the process of dinitrogen fixation, and the rest being accounted for by the assimilatory reduction of nitrate[7-9]. The energy demand for both processes is ultimately supplied by the sun via photosynthesis.

Photosynthesis is usually defined as the light-driven formation of carbohydrates and oxygen from CO_2 and water. This formulation ignores, however, the basic fact that in the photosynthetic process not only CO_2, but also the oxidized forms of other primordial bioelements are reduced and incorporated into cell material. Actually, photosynthesis drives several biosynthetic pathways involved in the assimilation of inorganic carbon, nitrogen, and sulfur. At the expense of sunlight energy, unstable products – cell material and oxygen – are synthesized from fully oxidized substrates with no useful chemical potential, namely water, carbon dioxide, nitrate or dinitrogen, sulfate and phosphate[1,2,9].

Ammonia is an obligate intermediate in the assimilation of oxidized inorganic nitrogen, either nitrate or dinitrogen, since these compounds have to be reduced and converted to ammonia prior to the incorporation of N into cell material. Whereas practically all of the phototrophs, either prokaryotes or eukaryotes, are able to use nitrate as a nitrogen source, the ability to photoassimilate dinitrogen is restricted to some groups of the photosynthetic prokaryotes, including representatives of both photosynthetic bacteria and cyanobacteria.

The photosynthetic reduction of nitrate to ammonia does not require ATP and involves 8 reducing equivalents of photosynthetic origin. The process takes place in 2 steps: nitrate is first reduced to nitrite in a 2-electron reaction catalyzed by the molybdoprotein nitrate reductase, and then nitrite is reduced to ammonia in a 6-electron reaction catalyzed by the iron-protein nitrite reductase[7-9].

$$NO_3^- \xrightarrow[\text{nitrate reductase}]{2e} NO_2^- \xrightarrow[\text{nitrite reductase}]{6e} NH_3$$

The process of nitrate reduction as it takes place in the blue-green algae, with reduced ferredoxin acting as the immediate electron donor for both partial reactions, is one of the simplest examples of photosynthesis[7-10].

Some photosynthetic bacteria and cyanobacteria (blue-green algae) are also able to carry out the fixation of dinitrogen, in a process which converts N_2 to ammonia. This reduction of molecular nitrogen involves a 6-electron reaction that requires in addition the input of a large quantity of energy in the form of ATP, at least 12 molecules of ATP being hydrolyzed per molecule of dinitrogen reduced:

$$N_2 \xrightarrow[\text{nitrogenase}]{6e,\ \geqslant 12 \sim P} 2\ NH_3$$

The reaction is catalyzed by the oxygen-labile enzyme complex nitrogenase, which is composed of 2 different iron-sulfur proteins, one of them containing molybdenum in addition to non-heme iron[11]. The assimilatory power for this reaction in phototrophic prokaryotes is ultimately provided by photosynthesis.

Assimilatory nitrate reduction and dinitrogen fixation by photosynthetic organisms are thus processes which result in the conversion of solar energy into redox energy and its storage in the molecule of ammonia, as shown by the following exergonic reactions corresponding to the photosynthetic reduction of nitrate and N_2 with water as the electron donor:

$$HNO_3 + 4\ H_2O \xrightarrow{\text{light}} NH_3 + 3\ H_2O + 2\ O_2$$

$(\Delta E'_o,\ pH\ 7 = -0.47\ V;\ \Delta G'_o = +360\ kJ \cdot mole^{-1})$

$$2\ N_2 + 3\ H_2O \longrightarrow 2\ NH_3 + {}^3/_2\ O_2$$

$(\Delta E'_o,\ pH\ 7 = -1.10\ V;\ \Delta G'_o = +318\ kJ \cdot mole^{-1})$

Two potentially valuable systems wherein ammonia is photoproduced are worthy of attention:

1. The type where preparations of active water-photolysing photosynthetic membranes are able to generate the assimilatory power required for the reduction of nitrate or dinitrogen. These preparations should contain, or be supplemented with, the required enzymes and cofactors, and they might in turn be replaced by other biological or synthetic systems able to transduce light energy into suitable redox and chemical energy. This type of system presents several problems, especially in the case of dinitrogen as substrate.

2. The type where whole photosynthetic organisms, ideally microorganisms, carry out photosynthesis of the water-splitting type and which, after chemical or genetic manipulation, become suited for the continuous production and release of a significant fraction of the ammonia resulting from the reduction of either nitrate or dinitrogen. The achievement of this goal rests on effectively impeding both the incorporation of ammonia into carbon skeletons and the expression of any of the several antagonistic effects of ammonia on inorganic nitrogen metabolism.

The present status and future prospects of this novel field of biological ammonia photoproduction are described below.

Photoproduction of ammonia from nitrate

The photosynthetic nature of the process of nitrate reduction in blue-green algae was conclusively demonstrated in 1976 by Candau et al.[10]. These authors reported the ability of unsupplemented membrane preparations from the unicellular blue-green alga *Anacystis nidulans* to carry out the light-dependent steady reduction of nitrate to ammonia coupled to the evolution of molecular oxygen. The process took place under aerobic or anaerobic conditions alike, and the activity of the system was linear for at least 90 min, at a rate of ammonia production of about 2 μmoles per mg chlorophyll and h. A reconstituted aerobic system including ferredoxin and the nitrate-reducing enzymes of *Anacystis* together with spinach grana was also able to photoreduce nitrate to ammonia at a rate of about 40 μmoles per mg chlorophyll and h[2,12].

The stability problems inherent in the in vitro photosynthetic systems impose at present a serious limitation to their use on a continuous basis. For this reason, whole photosynthetic microorganisms, when conveniently manipulated, are in the short run better candidates as solar energy converters for achieving a prolonged photoproduction of ammonia.

The steady production of ammonia from nitrate with whole living cells requires the prevention of both ammonia assimilation and the negative effects which ammonia exerts on nitrate utilization, namely inhibition of nitrate uptake, inactivation of nitrate reductase and repression of the 2 enzymes of the nitrate-reducing pathway[7-9]. This seemed at first a difficult task to achieve, but recent evidence obtained in our own and in other laboratories[7-9] has shown that many (if not all) of these antagonistic effects of ammonia are in fact not caused by ammonia itself, but rather result from its metabolism. This means that prevention of ammonia assimilation should allow, at least to a certain extent, the production of ammonia from nitrate by whole cells without further interference by the accumulated ammonia on the process.

A series of previous observations found in the literature indicated the viability of such an approach for achieving the production and excretion of ammonia. They corresponded in general to experiments carried out with illuminated suspensions of green algal cells that, under conditions in which the generation of the carbon skeletons required to incorporate ammonia is impeded, i.e. in the absence of CO_2 or another utilizable carbon source, can reduce nitrate to ammonia and export it to the medium[13-17]. These experiments did not try to achieve ammonia photoproduction but were designed to study physiological aspects of nitrate reduction by green algae. The ability of such systems to continuously produce ammonia for long periods of time under the established conditions of prolonged carbon starvation could, however, be questioned, particularly since an essential for nitrate utilization by many species of green and blue-green algae is a carbon source[8].

For our purpose we have selected the approach of interrupting the early stages of ammonia assimilation by using specific inhibitors of the incorporation of ammonia into carbon skeletons, allowing carbon assimilation to proceed unimpaired. Recent experiments carried out with *Anacystis* have shown that the prevention of ammonia assimilation through inactivation of glutamine synthetase with L-methionine-D,L-sulfoximine (MSX)[18] abolish the inhibitory effects of ammonia on both nitrate uptake[19] and nitrate reductase synthesis[20]. Glutamine synthetase is the first enzyme involved in the main route of ammonium assimilation in *Anacystis*[21] as is also the case for many other photosynthetic organisms where ammonia incorporation into carbon skeletons rests primarily on the operation of the glutamine synthetase/glutamate synthase pathway[22]. MSX-treated *Anacystis* cells are able to take up and reduce nitrate at rates much higher than those in normal untreated cells, a significant fraction of the ammonia resulting from nitrate reduction being exported to the medium.

We have characterized the effects of MSX on *Anacystis* cells using nitrate in order to maximize the extent and the duration of the process of ammonia production, and with the aim of using these MSX-treated cell suspensions as effective ammonia photoproducing systems on a continuous basis. For these experiments, unless otherwise indicated, cell suspensions containing about 10 µg chlorophyll per ml culture medium were illuminated with white light (25 $W \cdot m^{-2}$). Some of the main events following the addition of 10 µM MSX to such a suspension of *A. nidulans* cells are shown in the figure. After addition of the inhibitor, cellular glutamine synthetase is rapidly and completely inactivated, and ammonia appears on the medium concomitantly with this activity loss. Cell growth is prevented, and the rates of nitrate uptake and reduction increase 2- to 3-fold. Ammonia accumulates in the medium at a constant rate of about 25 µmoles per mg chlorophyll and h, which corresponds to about 90% of the rate of nitrate consumption by the cells. The process keeps going on for about 30 h without further additions provided that nitrate is not limiting, and nitrate reductase activity remains at its original high level during this period[23]. The rate of ammonia photoproduction by MSX-treated *Anacystis* cells can be improved by increasing either the pH of the medium, the cell density or the intensity of illumination. Thus, at pH 9.5 the rate of the process is about 1.5-fold higher than at pH 7.0. By using a cell density equivalent to 15 µg chlorophyll per ml medium and a light intensity of 200 $W \cdot m^{-2}$, under conditions otherwise similar to those described in the figure, a rate of ammonia production of 60 µmoles per mg chlorophyll and h has been achieved[24].

With 10 µM MSX (the optimal initial concentration of this compound), the effective period of ammonia production lasts for about 30 h, but its effectiveness is lost thereafter. Readdition of 10 µM MSX to the cell suspension every 24 h allows the initial rate of ammonia production to remain constant for at least 3 days[24]. At this stage the addition of MSX no longer results in the recovery of the process. Reinitiation of ammonia production can, however, be achieved if a pulse of

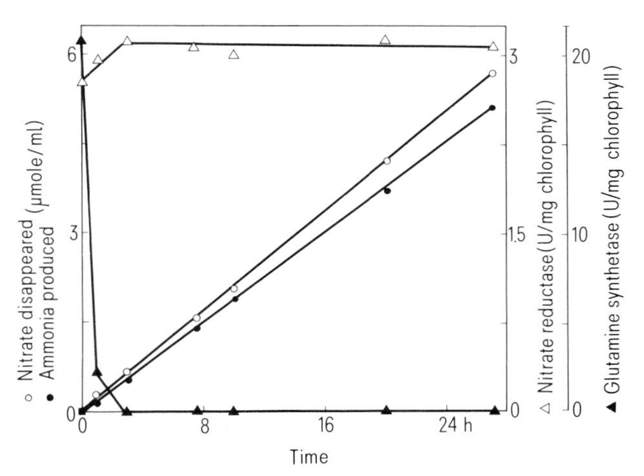

Effect of MSX on nitrate uptake and ammonia production and on glutamine synthetase and nitrate reductase cellular activity levels in *Anacystis nidulans*[25]. Nitrate-grown *A. nidulans* cells were harvested, washed and suspended in normal growth medium[19] containing 20 mM KNO_3 up to a cell density of 8 µg chlorophyll per ml. MSX to reach a final concentration of 10 µM was added at t = 0, and the cell suspension was illuminated (25 $W \cdot m^{-2}$, white light) at 40 °C and sparged with a mixture air:CO_2 (97:3, v/v). Aliquots were withdrawn at the times indicated, and nitrate and ammonia, in the cell-free supernatants, and nitrate reductase and glutamine synthetase, in the cells, were estimated after centrifugation[19,34].

glutamine is added together with MSX. In fact, after 3 days of active production the MSX-treated cells exhibit negligible levels of glutamine and low levels of glutamate. The metabolic situation which leads to the cessation of ammonia production seems therefore to correspond to glutamine starvation rather than to a generalized nitrogen depletion; this is corroborated by the otherwise normal content of phycobiliproteins, good indicators of the general nitrogen status in cyanobacteria. An alternative and more practical way of prolonging the active phase of ammonia production by MSX-treated cells is to allow them to recover from the glutamine depletion by keeping them recurrently for 8 h in the absence of MSX after every 48-h period of production in the presence of the inhibitor. With this procedure the effective period of ammonia photoproduction by *Anacystis* can be extended for at least 10 days, the rate of the process remaining constant throughout this period of time[24,25].

A production of ammonia lasting for several hours in response to the addition of 1 mM MSX has also been reported to occur in nitrate-utilizing glutamate-grown cells of *Cyanidium caldarium*, a thermophilic red alga[26]. The presence of MSX prevented the inactivation of nitrate reductase which otherwise took place in the presence of ammonia, indicating that a product of ammonia metabolism, but not ammonia itself, is responsible for this effect in *Cyanidium*. On the other hand, Florencio et al.[27] have reported that the addition of MSX to cell suspensions of the green alga *Chlamydomonas reinhardii* results in a transient accumulation of ammonia in the medium. This accumulated ammonia causes the inactivation of nitrate reductase even in the presence of MSX, supporting the idea that ammonia plays a direct role in the reversible inactivation of nitrate reductase in *Chlamydomonas*[7–9].

Photoproduction of ammonia from dinitrogen

Effective in vitro photobiological systems for the generation of ammonia from dinitrogen have, to the best of our knowledge, not yet been reported. In addition to being inherently unstable, such systems would require strict anaerobic conditions to be operative since the nitrogenase complex is extremely sensitive to molecular oxygen. The limitations on the effective operation of these types of systems are comparable to those currently encountered in the photoproduction of hydrogen through 'water biophotolysis' with in vitro systems[28].

The in vivo photoproduction of ammonia from N_2 by cells of photosynthetic diazotrophs offers, however, much better perspectives. Anaerobic or microaerophilic conditions are still required for in vivo N_2-fixation by photosynthetic bacteria and a range of blue-green algae. Contrary to the cyanobacteria, the N_2-fixing photosynthetic bacteria cannot use water as the electron donor for photosynthesis, but they require substrates with a more negative potential, usually reduced inorganic compounds. Certain blue-green algae, the filamentous cyanobacteria able to differentiate heterocysts, are unique in that they can fix N_2 to ammonia under aerobic conditions with light as the unique source of energy, and water as the source of reductant. Moreover, they (as other cyanobacteria) are also able to carry out anoxygenic photosynthesis similar to that found in the photosynthetic bacteria, being thus able to develop in various habitats.

As is the case with nitrate photoreduction, ammonia behaves as an antagonist of photosynthetic N_2-fixation, leading to repression of nitrogenase in photosynthetic bacteria and cyanobacteria and to a reversible inactivation of nitrogenase activity in the former group of organisms. Both of these negative effects of ammonia on N_2-fixation have been shown to require ammonia metabolism, indicating again an indirect involvement of this compound. In fact, the accumulation of ammonia is harmless for N_2-fixation provided that its further metabolism is prevented[29–31].

Following the demonstration by Gordon and Brill[32] with the chemotrophic bacterium *Azotobacter* that the inhibition of ammonia assimilation by certain glutamate analogs caused derepression of N_2-fixation and excretion of ammonia to the medium, a similar behavior was reported for the photosynthetic diazotrophs, both cyanobacteria[29] and photosynthetic bacteria[30], in response to treatments with the glutamine synthetase inactivator MSX. With the filamentous blue-green algae *Anabaena cylindrica*, Stewart and Rowell[29] presented evidence that MSX alleviated the inhibitory effect of exogenous ammonia on heterocyst production and nitrogenase synthesis. In the presence of 1 μM MSX, illuminated *Anabaena* cell suspensions excreted ammonia at a rate of about 3 μmoles per mg chlorophyll and h. Subsequently, Weare and Shanmugam[30] presented similar results for the photosynthetic bacterium *Rhodospirillum rubrum*. The production of ammonia by the bacterial suspensions was significatively increased when a fixed nitrogen source was added to the cells together with MSX. In the presence of 55 mM MSX and 5 mM glutamate, a net production of 11 mM ammonia after 6 days with a maximal rate of 0.7 μmoles per mg cell protein per h was attained.

Using a marine strain of the filamentous blue-green alga *Anabaena* (*Anabaena* sp. strain ATCC 33047), we have recently achieved an efficient and continued photoproduction of ammonia from N_2[33]. The addition of 35 μM MSX to suspensions (about 10 μg chlorophyll per ml medium) of this cyanobacterium causes a rapid inhibition of cellular glutamine synthetase and an increase in nitrogenase activity of about 2-fold over the normal derepressed activity level found in the absence of MSX. About 90% of the N_2 fixed by the cells, which do not grow any longer, is exported to the medium as ammonia at a rate of about 25 μmoles per mg chlorophyll and h (equivalent to 1.4 μmoles per mg cell protein and h) when illumination is 25 $W \cdot m^{-2}$ (white light). With successive additions of 35 μM MSX every 20 h, and intercalating 8-h periods in the absence of the inhibitor every 40 h, a steady generation of ammonia lasting for 5 days with a net production of 20 mM ammonia has been achieved. Increasing conveniently the cell density and the light intensity allows the achievement of higher production rates: 60 μmoles per mg chlorophyll and h for a cell density of 11 μg chlorophyll per ml and a light density of 200 $W \cdot m^{-2}$ [25,33,34].

A further step in optimizing conditions for an effective photoproduction of ammonia from N_2 is reached by the use of mutant strains of photosynthetic diazotrophs blocked in the early stages of ammonia assimilation and derepressed for nitrogenase. In this context, Weare[35] has reported the isolation of a mutant of *Rhodospirillum* with low glutamate synthase activity and significant derepression of nitrogenase which exports ammonia to the medium at a rate of about 0.02 μmoles per mg cell protein per h. More recently, Wall and Gest[36] have isolated glutamine auxotrophs of *Rhodopseudomonas capsulata,* another photosynthetic bacterium. These mutant strains have negligible levels of glutamine synthetase activity and exhibit derepressed nitrogenase activity, thus allowing a continued production of ammonia from N_2 leading to a net ammonia production of about 15 mM after 4 days. Mutants of filamentous blue-green algae similar to those described above are currently being sought in our laboratory. The interest in obtaining such mutant strains is manifest since they represent solar energy converters which, by themselves, would generate ammonia under aerobic conditions with ubiquitous substrates (air and water) by drawing energy from an unlimited energy source (sunlight).

Concluding remarks

Research on the biochemical, physiological and genetic aspects of inorganic nitrogen metabolism in microorganisms has provided relevant information on the mechanism and the regulation of the uptake and the reduction of oxidized inorganic nitrogenous substrates and the incorporation of the resulting ammonia into carbon skeletons. The information discussed here indicates the feasibility of using photosynthetic organisms, preferably microalgae, with chemically or genetically deregulated inorganic nitrogen metabolism, for the steady photoproduction of ammonia in significant amounts. In its presently most relevant version, using whole cells of N_2-fixing filamentous blue-green algae, the valuable biofuel ammonia is generated at the expense of only light, air and water in a remarkable example of biological solar energy conversion. The state of the art in this field is still in its infancy, but very interesting perspectives can be envisaged after further improvement of the in vivo biological systems considered here or the development of other in vitro or artificial systems. Certainly much more investigation is needed for the installment of practical ammonia-generating systems, and emphasis should also be laid on both environmental and bioengineering aspects.

With regard to the in vivo systems, the microalgae seem the most interesting organisms, with obvious advantages over the photosynthetic bacteria, which carry the extra requirement for reduced compounds and also for anaerobic or microaerobic conditions. Till adequate mutant strains of these algae are available, chemical manipulation with specific inhibitors, especially MSX, seems to be the preferred procedure for inducing excretion of a significant fraction of the biologically generated ammonia. It has to be pointed out that not every alga would be suitable for the production of ammonia after treatment with MSX. An unavoidable requisite to this end is that ammonia assimilation takes place solely or predominantly through glutamine synthetase, the enzyme target of this inhibitor. Obviously, the cell wall should be permeable to MSX, a condition that is not shared by all algal species. A careful screening of different algal strains is needed to select those exhibiting sustained high rates of ammonia production. In this selection, high activity levels of the corresponding enzymes involved in the generation of ammonia and high rates of photosynthetic electron flow are also important aspects to be taken into account, since these factors ultimately determine the rate of ammonia production. The maximum expected efficiency of light conversion into chemical energy stored in ammonia by whole plant or algal photosynthetic nitrate reduction under 'ideal' conditions is about 13% of the radiant energy involved in the process[1,2]. This corresponds to a maximum expected ammonia production rate of about 2 moles per m^2 and day (120 tons ammonia per ha and year) for a solar energy value of 200 $W \cdot m^{-2}$, when allowance is made for factors such as the fraction of photosynthetic active radiation with respect to total solar radiation, and losses by reflection, absorption and transmission.

In experiments carried out in 50-l containers with 0.1 m^2 surface exposed to white light (500 $W \cdot m^{-2}$) using MSX-treated *Anacystis* cell suspensions (3 μg chlorophyll per ml of nitrate medium), values of net ammonia production of 1.5 mM ammonia in 12 h, corresponding to 0.75 moles ammonia per m^2 and day, have been obtained[25]. From a dual extrapolation of both area and time, a production of 46 tons ammonia per ha and year can be calculated from these values. This represents as much as 30% of the maximum expected value under our experimental conditions. In fact, we have estimated that the rest of the available electron flow is used for CO_2 fixation, which results in an accumulation of carbohydrates by the MSX-treated cells[24,25]. Should this electron flow to carbon be avoided and channeled to nitrate reduction, the values of ammonia production would be increased by a factor of 3.

Ammonia photoproduction from N_2 might result in even higher yields since the quantum requirement for the production of ammonia by photosynthetic N_2 fixation can be estimated at 12 quanta per molecule of ammonia instead of the 16 quanta which are required with nitrate as the substrate.

In addition to possible applications at the industrial level, the photoproduction of ammonia by appropriate N_2-fixing microorganisms might prove useful for the in situ generation of nitrogen fertilizers in rural areas. If, for example, the process were carried out in small ponds, it might provide, with the help of a simple technology, an ammonia-enriched fluid to be used for direct watering of the field.

The possibility of achieving effective photoproduction of ammonia by in vitro photosynthetic systems depends on the development of stable components. A range of model systems which might in principle also be adequate for ammonia photoproduction can be found in the recent review paper by Bolton and Hall[28]. These authors have also considered the possibility of using artificial systems in solar energy conversion to mimic the photosynthetic systems by the use of synthetic catalysts, and Losada[1] has also stressed the potential of flavin photosystems for practical uses in this field. In this respect it is worth mentioning the use of a deazaflavin photosystem for the reduction of nitrate to nitrite with ferredoxin and nitrate reductase of the blue-green alga *Anacystis*[37]. Since nitrite reductase of this and other photosynthetic organisms uses also reduced ferredoxin as physiological reductant, an enzyme system composed of nitrate reductase, nitrite reductase and ferredoxin may be envisaged that, upon illumination in the presence of a suitable electron donor and catalytic amounts of a flavin, might also generate ammonia from nitrate. The photoproduction of ammonia from N_2 might also be achieved with the help of flavin photosystems. In fact, a deazaflavin mediated photoreduction of acetylene with a reconstituted nitrogenase system from *Azotobacter* has already been reported[38]. The non-enzymatic photochemical reduction of N_2 to ammonia has also been achieved with titanium dioxide catalysts, these systems being particularly effective when illuminated with UV-light[39].

* Acknowledgments. Research cited from the authors' laboratory was supported in part by grants from the Centro de la Energía del Ministerio de Industria, Comisión Asesora de Investigación, and the Philips Research Laboratories.

1 M. Losada, Bioelectrochem. Bioenerg. *6*, 205–225 (1979).
2 M. Losada and M.G. Guerrero, in: Photosynthesis in Relation to Model Systems, p.365. Ed. J. Barber. Elsevier, Amsterdam 1979.
3 D.O. Hall, Solar Energy *22*, 307–328 (1979).
4 L. Fowden, in: Nitrogen Assimilation of Plants, p.1. Ed. E.J. Hewitt and C.V. Cutting. Academic Press, London 1979.
5 G.C. Sweeney, Symposium on Dinitrogen Fixation. Ed. W.E. Newton and C.J. Nyman. Washington State University Press, Pullman 1976.
6 K. Jones, The Chemistry of Nitrogen. Pergamon Press, Oxford 1975.
7 M.G. Guerrero, J.M. Vega and M. Losada, A. Rev. Pl. Physiol. *32*, 169–204 (1981).
8 M. Losada, M.G. Guerrero and J.M. Vega, in: Biology of Inorganic Nitrogen and Sulfur, p.30. Ed. H. Bothe and A. Trebst. Springer, Berlin 1981.
9 M. Losada, J.M. Vega and M.G. Guerrero, in: Proc. 5th Int. Photosynthesis Congress. Int. Science Services, Jerusalem, in press.
10 P. Candau, C. Manzano and M. Losada, Nature *262*, 715–717 (1976).
11 L.E. Mortenson and R.N.F. Thorneley, A. Rev. Biochem. *48*, 387–418 (1979).
12 C. Manzano, P. Candau and M.G. Guerrero, Eur. Seminar Biological Energy Conversion Systems, Grenoble-Autrans, 1977; abstr. 14.
13 O. Warburg and E. Negelein, Biochem. Z. *110*, 66–115 (1920).
14 L.H.J. Bongers, Neth. J. agric. Sci. *6*, 79–88 (1958).
15 P.J. Syrett and I. Morris, Biochim. biophys. Acta *67*, 566–575 (1963).
16 A. Thacker and P.J. Syrett, New Phytol. *71*, 423–433 (1972).
17 R. Eisele and W.R. Ullrich, Pl. Physiol. *59*, 18–21 (1977).
18 R.A. Ronzio, W.B. Rowe and A. Meister, Biochemistry *8*, 1066–1075 (1969).
19 E. Flores, M.G. Guerrero and M. Losada, Archs Microbiol. *128*, 137–144 (1980).
20 A. Herrero, E. Flores and M.G. Guerrero, J. Bact. *145*, 175–180 (1981).
21 J.L. Ramos, E. Flores and M.G. Guerrero, 2nd Congress Federation European Societies Plant Physiology, Santiago de Compostela, 1980; abstr. 226.
22 P.J. Lea and B.J. Miflin, in: Encyclopedia of Plant Physiology, New Series, vol. 6, p.445. Ed. M. Gibbs and E. Latzko. Springer Verlag, Berlin 1979.
23 J.L. Ramos, M.G. Guerrero and M. Losada, 2nd Congress Federation European Societies Plant Physiology, Santiago de Compostela, 1980; abstr. 227.
24 J.L. Ramos, F. Madueño, M.G. Guerrero and M. Losada, I Congresso Luso-Espanhol de Bioquímica, Coimbra, 1980a; abstr. P241.
25 J.L. Ramos, M.G. Guerrero and M. Losada, unpublished results.
26 C. Rigano, V.D.M. Rigano, V. Vona and A. Fuggi, Archs Microbiol. *121*, 117–120 (1979).
27 F.J. Florencio, J.M. Vega and M. Losada, I Congresso Luso-Espanhol de Bioquímica, Coimbra, 1980; abstr. P163.
28 J.R. Bolton and D.O. Hall, A. Rev. Energy *4*, 353–401 (1979).
29 W.D.P. Stewart and P. Rowell, Biochem. biophys. Res. Commun. *65*, 846–856 (1975).
30 N.M. Weare and K.T. Shanmugam, Arch. Microbiol. *110*, 207–213 (1976).
31 B.L. Jones and K.J. Monty, J. Bact. *139*, 1007–1013 (1979).
32 J.K. Gordon and W.J. Brill, Biochem. biophys. Res. Commun. *59*, 967–971 (1974).
33 M. Losada, J.L. Ramos and M.G. Guerrero, 5th Int. Photosynthesis Congress, Halkidiki, 1980; abstr. 351.
34 J.L. Ramos, M.G. Guerrero and M. Losada, Proc. 5th Int. Photosynthesis Congress, Int. Science Services, Jerusalem, in press.
35 N.M. Weare, Biochim. biophys. Acta *502*, 486–494 (1978).
36 J.D. Wall and H. Gest, J. Bact. *137*, 1459–1463 (1979).
37 P. Candau, C. Manzano, M.G. Guerrero and M. Losada, Photobiochem. Photobiophys. *1*, 167–174 (1980).
38 G.H. Scherings, H. Haaker and C. Veeger, Eur. J. Biochem. *77*, 621–630 (1977).
39 G.N. Schrauzer and T.D. Guth, J. Am. chem. Soc. *99*, 7189–7193 (1977).

Hydrogen production by algae

by Hermann Bothe*

Botanical Institute, The University of Cologne, D-5000 Köln 41 (Federal Republic of Germany)

I. The enzymes catalyzing formation or uptake of molecular hydrogen

A variety of microorganisms can evolve H_2 according to the following equation: $2H^+ + 2e \rightleftarrows H_2$. These include strict or facultative anaerobic bacteria, aerobic bacteria, blue-green and green algae. In aerobic bacteria and in blue-green algae H_2 formations are restricted to N_2-fixing species. Strict and facultative anaerobic bacteria as well as green algae *(Chlamydomonas, Scenedesmus, Chlorella)* form the gas only under O_2 exclusion in the cultures. There is no clear-cut demonstration for H_2-formation by mosses, ferns and higher plants. Lists of the H_2-forming organisms are compiled in Mortenson and Chen[1] and Schlegel and Schneider[2].

Since the redox potential of the couple $2H^+/H_2$ is -413 mV at pH 7.0, a low potential reductant is required for H_2-formation to proceed in the cells. The reaction is also enzyme mediated. Cells may contain 3 clearly distinguishable enzymes catalyzing either uptake or evolution of H_2 under physiological conditions (for a more detailed account and the references see Bothe and Eisbrenner[3]).

a) *Reversible, 'classical' hydrogenase*

This soluble enzyme has been characterized best from the anaerobic bacterium *Clostridium pasteurianum*. In the isolated state it catalyzes both uptake and evolution of the gas which are independent of ATP and severely affected by CO and O_2. The enzyme has a mol. wt of about 60,000 and a prosthetic group consisting of three 4 Fe:4 acid labile sulphur centres among which only one is believed to undergo oxidation/reduction during catalysis. Carriers that supply the electrons for H_2-formation by hydrogenase are the iron-sulphur protein ferredoxin or, under iron-deficiency in the culture medium, the flavoprotein flavodoxin. In vitro, ferredoxin and flavodoxin can artificially be substituted by viologen dyes (methyl or benzyl viologen). Continuous H_2-formation requires a continuous supply of reducing equivalents for the reduction of H^+ and ferredoxin. In *Clostridium*, the reducing equivalents are supplied by the electron donors pyruvate or NADH. In the presence of coenzyme A, pyruvate is split to acetylcoenzyme A and CO_2, and the remaining 2 electrons reduce ferredoxin or flavodoxin. This so-called pyruvate phosphoroclastic reaction is catalyzed by the enzyme pyruvate:ferredoxin oxidoreductase. Alternatively, reduced ferredoxin can be generated from NADH in a reaction catalyzed by NADH: ferredoxin oxidoreductase which is allosterically regulated by acetylcoenzyme A in *C. pasteurianum*. Under physiological conditions, the formation of H_2 is the favored reaction in *C. pasteurianum*. Hydrogenase functions by removing excess reducing equivalents generated during fermentation. Since this bacterium cannot degrade carbohydrates completely to CO_2 and H_2O due to the absence of the respiratory chain, it must produce large amounts of H_2 in order to avoid overreduction.

The occurrence of this soluble hydrogenase has been established for saccharolytic *Clostridia* and for facultative anaerobic bacteria (e.g. *Bacillus polymyxa*). A similar enzyme is present in the photosynthetic *Chromatium*. Hydrogenase of the Enterobacteriaceae (e.g. *Escherichia coli*) is part of the membrane-bound formate:hydrogen lyase complex which has not yet been fully characterized. The same complex has recently been demonstrated in the photosynthetic Rhodospirillaceae (e.g. *Rhodospirillum rubrum*). The hydrogenase from green algae is apparently soluble and possibly couples with ferredoxin.

In regard to blue-green algae, a hydrogenase catalyzing H_2-formation under anaerobic conditions in the cells has been demonstrated unequivocally only for the halophytic *Oscillatoria limnetica*. Its occurrence has also been suggested in aerobic, N_2-fixing blue-green algae *(Nostoc muscorum, Anabaena cylindrica)*; but this is likely an artifact of cell-free preparations[4]. A soluble, reversible hydrogenase has not been found in aerobic bacteria, including the N_2-fixing species.

b) *Nitrogenase*

The enzyme catalyzes the reduction of N_2 to ammonia, of C_2H_2 to C_2H_4 and of other substrates which have a triple bond in common[5]. In addition, it converts protons and electrons to molecular hydrogen. H_2-formation by nitrogenase is irreversible, insensitive to CO, dependent on a supply of electrons from reduced ferredoxin and requires large amounts of ATP. In vitro, 3–4 molecules of ATP are hydrolyzed for the formation of one molecule of H_2, and in vivo H_2-productions are probably even more energy consuming. The mechanism of H_2-formation by nitrogenase is not understood at present. In the absence of any other substrate, e.g. under argon, all the electrons flowing to isolated nitrogenase reduce H^+ to H_2 despite the low H^+-concentration in such assays normally performed at pH 7–8. Even in the presence of N_2 in the vessels, H_2-evolution is still substantial. The measured stoichiometry between N_2-reduction and H_2-formation is often 1:1, indicating that both reactions are coupled according to the following equation: $8H^+ + 8e^- + N_2 \rightarrow 2NH_3 + H_2$. H_2-production is always observed with isolated nitrogenases but is often marginal in intact organisms.

c) *Uptake hydrogenase*

Although described already in the early 1940's, the biochemical properties of this enzyme are largely unknown at present. It is an integral protein of membranes and therefore difficult to characterize. It is virtually insensitive to oxygen and has a high affinity for H_2. In nitrogen-fixing cells, it recycles all or most of the H_2 lost by the ATP-dependent formation of H_2 catalyzed by nitrogenase. This explains the low net H_2-formation rates of most aerobic N_2-fixing organisms. The recycling of H_2 has at least 3 beneficial functions for the cells: 1. it provides the organisms with extra ATP. H_2-consumptions proceed by an oxyhydrogen (Knallgas) reaction which is coupled to the respiratory electron transport and to ATP-formation. 2. there is experimental evidence that the oxyhydrogen reaction removes oxygen from the nitrogenase site and thereby protects the enzyme from damage by this gas. 3. H_2 and the uptake hydrogenase can supply electrons for the reduction of N_2 to ammonia by nitrogenase or for the conversion of CO_2 to carbohydrates by the Calvin cycle. The latter reaction has recently been demonstrated in *Rhizobium* and *Derxia gummosa.* H_2-supported nitrogen fixation (C_2H_2-reduction) is particularly pronounced in the heterocysts of blue-green algae where H_2 is an effective electron donor in a strictly light-dependent reaction. Experimentally unverified is a 4th possible function of the uptake hydrogenase. N_2-reduction catalyzed by nitrogenase is affected by high concentrations of H_2. The uptake hydrogenase may, therefore, remove the deleterious H_2 unevitably formed with N_2-reduction by nitrogenase. It is, however, questionable whether the high inhibitory concentrations of H_2 are reached at the nitrogenase site.
Growth under N_2-fixing conditions drastically enhances the activity levels of the uptake hydrogenase in the organisms. The enzyme is, however, not restricted to N_2-fixing cells, since it can be demonstrated in non N_2-fixing blue-green algae and in aerobic, H_2-oxidizing bacteria *(Alcaligenes, Paracoccus, Xanthobacter).* The electron acceptor for H_2-utilization by the membrane-bound uptake hydrogenase has not been clearly identified in any of the organisms. Inhibitor studies indicate that the electron entry is at or close to the quinone site in respiration and photosynthesis and at a redox level of about 0 volt. This reflects the unidirectional nature of the enzyme; the potential gap between the quinone/hydroquinone and the H^+/H_2 couples prevents the formation of H_2 by this enzyme under physiological conditions. In the isolated state, the enzyme is, of course, able to catalyze H_2-formation provided high concentrations of strong reductants (methyl viologen reduced by excess of $Na_2S_2O_4$) are supplied to the assays. Strains of *Alcaligenes eutrophus* are unique in containing 2 different uptake hydrogenases[2]. In addition to the membrane-bound hydrogenase, these bacteria form a soluble, flavin containing enzyme catalyzing the reduction of NAD^+ by H_2.

II. *Comparison of the capabilities of organisms to produce H_2*

As has already been mentioned, many obligate or facultative anaerobic bacteria ferment organic substrates to H_2. However, none of them is able to degrade organic matter completely to CO_2 and H_2[6]. The highest yield ever measured was 4 mole of H_2 formed from 1 mole of hexose (e.g. glucose). Such findings are in accord with theoretical considerations arguing that 4 mole is the maximal achievable amount[6]. This value decreases to about 2.6 mole of H_2 per mole of glucose when cultures are growing under a H_2-pressure of ≥ 1 at. 4 mole of H_2 contain only 33% of the combustible energy of glucose and 2.6 mole approximately 20%. These figures have to be compared with those for CH_4-formation. 85% of the energy is conserved when CH_4 is the end product of glucose degradation. Energetically it is, therefore, much more efficient to produce CH_4 instead of H_2 when organic matter is to be converted to energy by microorganisms (see Thauer[6]).
The situation may, however, be different with photosynthetic organisms which use solar radiation to build up chemical energy. Photosynthetic bacteria use either inorganic sulphur compounds (Chromatiaceae) or organic substrates (Rhodospirillaceae) as the source of electrons for the photoreduction of CO_2 or, alternatively, photoproduction of H_2. They evolve H_2 in relatively high amounts in a nitrogenase-dependent reaction. Since they do not produce O_2 photosynthetically, H_2 and O_2 must not be separated from each other. Rhodospirillaceae are easily manipulated genetically. However, since their nitrogenase is rather sensitive to exposure to O_2, all photohydrogen production must be performed under strict anaerobic conditions. Photosynthetic bacteria are ubiquitous in nature but appear seldom in blooms and usually show sluggish growth in natural environments. They may be considered when waste material is to be converted to produce H_2. For such a purpose, they have to be grown on a large scale basis under anaerobic conditions and at a defined supply with substrates and light. They do not appear, therefore, to be likely candidates for solar energy conversion programs. Somewhat contrary views have recently been expressed in review[31,32] where photosynthetic bacteria were claimed to currently show the most promise for short-term applied systems.
Green and blue-green algae appear to be more rewarding, at least at first glance. Both groups of organisms are very different in cytological respect;

blue-green algae are of prokaryotic nature and green algae are true eukaryotes like the higher plants. The only property which they share is the capability to perform plant-type photosynthesis. They are able to utilize sunlight and CO_2 for carbohydrate formation at the expense of water as the electron donor. They use two photosystems to generate a strong reductant with a redox potential of about -500 mV (see fig. 1). The chemical nature of this electron carrier X has not been entirely resolved, but it is likely a membrane - bound iron-sulphur protein. This compound X reduces ferredoxin, which is the reductant in several reactions (fig. 1). Ferredoxin donates electrons either to NADPH:ferredoxin oxidoreductase and $NADP^+$ for CO_2-fixation or to nitrite and thiosulphonate reductases in assimilatory nitrate and sulphate reductions. In blue-green algae, a plant-type ferredoxin reduces nitrogenase for the conversion of N_2 to NH_4^+ or H^+ to H_2 [7]. It is conceivable that photoreduced ferredoxin may transfer electrons also to hydrogenase, since many of the reversible hydrogenases couple with ferredoxin (see section Ia). Indeed, H_2-formation by green algae may be dependent on ferredoxin and a classical hydrogenase[8], although the biochemistry of this reaction awaits further elucidation.

The energetic efficiency of solar energy conversion by the photosynthetic electron transport (in the generation of reduced ferredoxin from water) ranges between 8 and 10%, referred to the radiation energy reaching the earth surface[9,10]. The efficiency of the conversion to plant biomass is only 1.0–1.3%. Such low figure is mainly due to energy losses which inevitably occur when CO_2 is reduced to carbohydrates in the Calvin cycle. A great portion of the energy is also used up to supply the plants with water and nourishment. It is the hope to couple the photosynthetic electron transport and reduced ferredoxin to hydrogenase to liberate the energy as H_2 with maximal output. Obviously only artificial systems can proceed with an efficiency of 8–10%, however, they suffer from their inherent instability. Living organisms are stable but have to be manipulated in order to release the captured energy as H_2 with maximal possible yield. For this final goal, blue-green algae offer a better starting position than green algae.

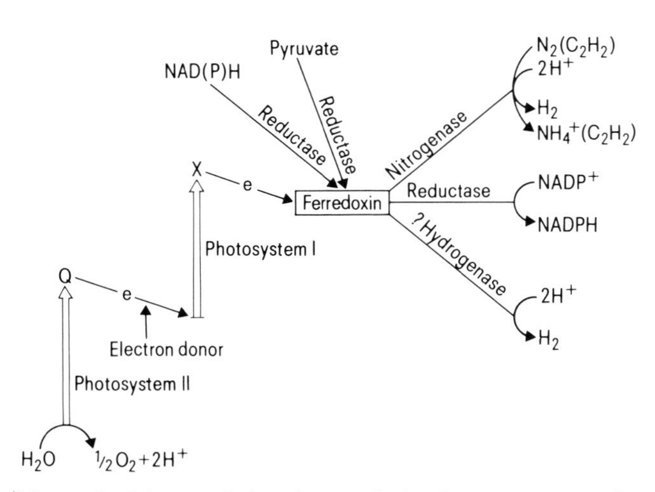

Figure 1. Scheme of the photosynthetic electron transport in thylakoids of green plants and blue green algae.

Green algae produce H_2 only when the cells have been adapted to anaerobic conditions[11-13]. The length of this adaptation process for maximal H_2-formation varies from organism to organism. Adaptation may cause the activation of a constitutive hydrogenase or the synthesis of new enzyme molecules, depending on the algal strain used. In the dark, green algae evolve only minuscule amounts of H_2. The production of the gas is stimulated by organic substrates and is accompanied by a release of CO_2 with a stoichiometry of 2.2:1 between H_2 and CO_2. H_2-formation in the dark is inhibited to uncouplers indicating the involvement of an energy-dependent reverse electron flow in the degradation of carbohydrates to H_2. Light stimulates H_2-evolution of anaerobically adapted green algae. The photoproduction of H_2 is enhanced by uncouplers and is therefore energy-independent. The source of electrons in the cells for the light-dependent H_2-formation is not entirely clear at present. H_2-evolution is not accompanied by a stoichiometric release of O_2. Thus reducing equivalents may be generated partly from endogenous carbon reserves and partly from the photosynthetic water splitting reaction.

Any program for solar energy conversion by green algae is immediately faced with the extreme sensitivity of H_2-formation and hydrogenase towards O_2. H_2-formation capability is entirely lost when the level or O_2 in the assays exceeds 1% of the atmospheric concentration. Since the algae evolve O_2 photosynthetically, H_2 is only formed at low light intensities where hydrogenase synthesis can compete with its destruction by O_2 from photosynthesis. This means that the efficiency of solar energy conversion is very low.

Many blue-green algae (cyanobacteria) need only light, water, CO_2 and mineral salts for growth. The N_2-fixing species thrive without combined nitrogen and have, therefore, the simplest nutrient requirements among all organisms. They are often abundant in nature, although the reasons for their seasonal fluctuations and their sudden blooms are not fully understood. Photoproduction of H_2 catalyzed by hydrogenase has unambiguously been shown for the halophytic *Oscillatoria limnetica*[14]. The following findings show that H_2-formation by all other species is catalyzed solely by nitrogenase[7]: The addition of NH_4^+ to the cultures obliterates N_2-fixation (C_2H_2-reduction) and H_2-formation activities in a parallel fashion. Under aerobic conditions, H_2-formation is strictly light-dependent and sensitive to uncouplers indicating the requirement for energy. Low activities are also observed in the dark when optimal amounts of O_2 are provided for the generation of ATP by respiration. H_2-formations are reduced by N_2 or C_2H_2 which compete with H_3O^+ for electrons in nitrogenase. Adaptation to anaerobic conditions is not necessary for H_2-evolution to begin.

N_2-fixation and thus H_2-formation is found in unicellular, filamentous non-heterocystous and in filamentous, heterocystous species[5]. Unicellular forms are the coenobial *Gloeocapsa* (now designated as *Gloeothece*) and *Aphanothece* which are slow-growing algae. A number of filamentous, non-heterocystous forms (e.g. *Plectonema boryanum*) perform N_2-fixation under very low O_2-tensions. The best-known examples for N_2-fixing blue-green algae are *Anabaena* species and *Nostoc muscorum*. These filamentous forms contain 2 cell-types, the vegetative cells and the heterocysts (fig. 2). The vegetative cells perform photosynthetic CO_2-fixation and O_2-evolution and provide the heterocysts with fixed carbon compounds. Heterocysts lack the photosynthetic water splitting reaction and are therefore not exposed to O_2 produced photosynthetically. Under aerobic conditions, nitrogenase was shown to be located exclusively in these specialized cells. This means that nitrogenase must be protected against damage by O_2 diffusing into the heterocysts. The protection mechanisms have not yet been fully elucidated, although respiration and the oxyhydrogen reaction seemingly are of major importance in removing O_2 from the nitrogenase site.

III. *The extent of H_2-production by blue-green algae*

In our own experiments, always very small amounts of H_2 were produced by aerobically grown blue-green algae (*Anabaena* species, *Nostoc muscorum*) assayed either aerobically or under strict O_2-exclusion and under limiting or saturating light conditions[7]. Maximal rates were approximately 1% of those obtained for photosynthetic CO_2-fixation and 10% of the C_2H_2-reduction capability. H_2-formation was increased to some extent by incubating the cells with optimal concentrations of CO and C_2H_2 which block the reutilization of H_2 by the oxyhydrogen reaction and hydrogenase (see section Ic and Bothe et al.[15]). Higher rates of H_2-evolution can also be obtained by artificially increasing the number of heterocysts. This can be achieved by treating the cultures with 7-azatryptophan[16]. However, maximal H_2-production never exceeded $\frac{1}{5}$ of the rate of C_2H_2-reduction. Other researchers[17-19], too, were unable to obtain high H_2-formation rates by blue-green algae. In a more systematic survey, Berchtold and Bachofen[20] found only very little H_2-production by a whole series of new isolates from the Zürich area as well as by known laboratory strains under stationary conditions. In contrast, a number of investigators reported high and long lasting H_2-productions when the gasses formed (H_2 and O_2) are constantly removed[21-24]. The absolute maximum is probably the 7500 µl H_2 produced/h × mg chlorophyll for *Anabaena cylindrica*[20]. Unfortunately, any experimental detail for such a high value is missing in this publication. Using severely nitrogen starved *Anabaena*, Weissman and Benemann[21] and Jeffries et al.[22] reported long lasting H_2-formations where the ratio between H_2-formation and photosynthetic O_2-evolution approached one. The efficiency of converting solar energy to H_2 was maximally 0.4%[21] or ranged between 0.35 and 0.85%[22] and the cells produced H_2 up to 19 or 30 days, respectively. A marine, non-bacteria free strain was able to evolve 250 µmoles H_2/h × mg chlorophyll[25].

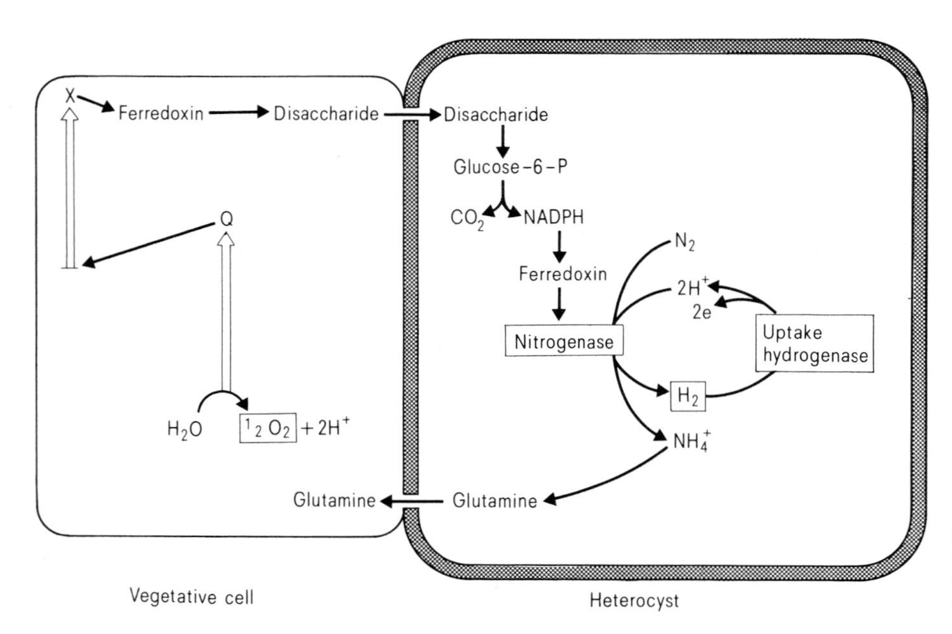

Figure 2. Metabolism of photosynthesis in vegetative cells and of nitrogen fixation, hydrogen production and hydrogen uptake in heterocysts of blue green algae.

It is difficult to judge the reliability of all these divergent data. It has become clear, however, that H_2-formation rates depend not so much on the algal strain used but on the culture and assay conditions employed. A major factor influencing H_2-formation capability is the nitrogen content of the cells. Prolonged nitrogen starvation, achieved by incubating the algae under argon in the absence of combined nitrogen for days, leads to the synthesis of additional nitrogenase (probably at the expense of phycocyanin) and increases the C_2H_2-reduction[26] activities and H_2-formation[21] capabilities. Also inhibition of the oxyhydrogen reaction by the uptake hydrogenase is prerequisital for a high H_2-production[15,27]. Incubation of *Nostoc* under an atmosphere of argon plus C_2H_2 has recently been shown to enhance H_2-formation considerably[23]. H_2-production is also influenced by other factors. These include during growth: the temperature, the supply of the culture with iron and CO_2, the O_2 tensions and the light intensities. During the assays rates largely depend on the duration of the experiments, the concentration of O_2, N_2 and CO_2 and of algal cells in the vessels and on the nutritional status of the cells, particularly on their reserves of organic carbon and nitrogen.

Obviously the optimal physiological conditions for maximal H_2-production have not yet been established. It must be pointed out that H_2-formation capability is limited by the nitrogenase content of the cells. With the strains commonly used in the laboratories, C_2H_2-reductions (as a measure of nitrogenase activity) vary between 5 and 20 µmoles C_2H_4 formed/$h \times mg$ chlorophyll. This is $\frac{1}{5}$ to $\frac{1}{10}$ of the rate of photosynthetic CO_2-fixation, in accord with the requirements of the cells for fixed carbon and nitrogen compounds. A rate of 20 µmoles H_2 formed/$h \times mg$ chlorophyll (~ 400 ml $H_2/h \times g$ chlorophyll $= \sim 10$ ml/$h \times g$ dry weight) means that all the electrons flowing to nitrogenase must reduce H^+. Only cells devoid of any regulatory mechanism to switch off the energy consuming reaction would sustain such high production over a longer period. Then they would have to be supplied with combined nitrogen. The addition of combined nitrogen (ammonia or nitrate), however, represses the synthesis of new nitrogenase molecules which should cause a gradual decline in the H_2-formations. To conclude, I am not convinced about H_2-productions which account to 20 or more, µmoles/$h \times mg$ chlorophyll and which last over a longer period.

On the other hand, autotrophic blue-green algae clearly have the potential to produce H_2 in a light-dependent reaction. We are only beginning to understand the physiology of the process. Basic research is necessary to find out the maximal capability for H_2-production and solar energy conversion. The search for new strains may be rewarding. The blue-green algae commonly used in the laboratory are probably not the fastest growing strains. The newly isolated *Anabaena CA*[28] and *Anabaena TA 1*[29] show considerably shorter generation times and higher N_2-fixation activities and are promising candidates for further investigation of H_2-formation. Genetic manipulations may be rewarding. Mutants that lack the uptake hydrogenase could be selected. N_2-fixation and consequently H_2-formation rates exceeding 20 µmoles H_2/$h \times mg$ chlorophyll would require additional amounts of reductant and ATP which would have to be provided by photosynthesis. Mutants could possibly be constructed which have higher photosynthetic capabilities or which furnish carbohydrates to heterocysts with higher rates but still survive. It may take a long time until a fair judgement can be made on whether blue-green algae are of value in projects of solar energy conversion programs.

In the near future, additional investigation of the hydrogenase-nitrogenase relationship is conceivable[30]. *Rhizobium* strains that possess the uptake hydrogenase were shown to fix N_2 more efficiently and to grow faster than strains without the enzyme. In nature, *Rhizobia* of the nodules of leguminous plants often do not possess an active hydrogenase. The productivity of plants could be improved by the applications of newly constructed *Rhizobium*-legume symbioses. These should have an active uptake hydrogenase and would have to compete with those currently existing in nature. Such projects could help save energy in a more indirect way.

Notes added in proof. Kayano et al.[33] have now immobilized an *Anabaena* strain in 2% agar gel where it produces H_2 with higher rates (≤ 0.5 µmoles/$h \times g$ gel) than in the free state. A photo-current of 15–20 mA was continuously produced for 7 days by a photochemical system consisting of the immobilized *Anabaena*, an oxygen removing 'reactor' containing aerobic bacteria and a hydrogen-oxygen fuel cell[33].

Hallenbeck et al.[34] have reported H_2-formation catalyzed by *Anabaena*. Their maximal rate (~ 2 µmoles/$h \times g$ dry weight, see table 1 in their paper) is self-evident. Any uptake hydrogenase should catalyze the reverse reaction at such an extremely low rate. Alternatively, bacterial contaminations producing these quantities of H_2 are difficult to dismiss in such experiments with large scale batch cultures. Houchins and Burris[35,36] have separated and biochemically characterized a reversible, soluble hydrogenase and the uptake hydrogenase from blue-green algae. Much of their data can not be simply reconciled with our own findings. Definitive proof for the existence of two different hydrogenases can only come from immunological studies. An excellent detailed account on hydrogenase by microorganisms has now been published by Mortenson's group[37].

* The author is indebted to the Deutsche Forschungsgemeinschaft for financial support of his own research.

1 L.E. Mortenson and J.-S. Chen, in: Microbial Iron Metabolism, p.231. Ed. J.B. Neilands. Academic Press, New York 1974.
2 H.G. Schlegel and K. Schneider, in: Hydrogenases: Their Catalytic Activity, Structure and Function, p.15. Ed. H.G. Schlegel and K. Schneider. Goltze Verlag, Göttingen 1978.
3 H. Bothe and G. Eisbrenner, in: Metabolism of inorganic nitrogen and sulphur compound, p.141. Ed. H. Bothe and A. Trebst. Springer, Berlin 1981.
4 G. Eisbrenner and H. Bothe, J. gen. Microbiol. *125*, in press (1981).
5 H. Bothe, M.G. Yates and F.C. Cannon, in: Encyclopedia of Plant Physiology, New Series, vol.12. Ed. A. Läuchli and R. Bieleski. Springer, Berlin, in press, 1981.
6 R. Thauer, in: Microbial energy conversion, p.201. Ed. H.G. Schlegel and J. Barnea. Goltze Verlag, Göttingen 1978.
7 H. Bothe, E. Distler and G. Eisbrenner, Biochemie *60*, 277 (1978).
8 D. King, D.L. Erbes, A. Ben-Amotz and M. Gibbs, in: Biological Solar Energy Conversion, p.69. Ed. A. Mitsui, S. Miyachi, A. San Pietro and S. Tamura. Academic Press, New York 1977.
9 P. Böger, Naturwissenschaften *65*, 407 (1978).
10 C. Fedtke, Ber. dt. Bunsenges. *84*, 973 (1980).
11 E. Kessler, in: Algal Physiology and Biochemistry, p.456. Ed. W.D.P. Stewart. Botanical Monograph, vol.10. Blackwell, Oxford 1974.
12 N.I. Bishop, M. Frick and L.W. Jones, in: Biological Solar Energy Conversion, p.3. Ed. A. Mitsui, S. Miyachi, A. San Pietro and S. Tamura. Academic Press, New York 1977.
13 A. Ben-Amotz, in: Encyclopedia of Plant Physiology, New Series, vol.6, II, p.497. Springer, Berlin 1979.
14 S. Belkin and E. Padan, FEBS Lett. *94*, 291 (1978).
15 H. Bothe, J. Tennigkeit, G. Eisbrenner and M.G. Yates, Planta *133*, 237 (1977).
16 H. Bothe and G. Eisbrenner, Biochem. Physiol. Pfl. *171*, 323 (1977).
17 E. Tel-Or, L.W. Luijk and L. Packer, Archs Biochem. Biophys. *185*, 185 (1978).
18 R.B. Peterson and R.H. Burris, Archs Microbiol. *116*, 125 (1978).
19 R.M. Tetley and N.I. Bishop, Biochim. biophys. Acta *546*, 43 (1979).
20 M. Berchtold and R. Bachofen, Archs Microbiol. *123*, 227 (1979).
21 J.C. Weissman and J.R. Benemann, Appl. environm. Microbiol. *33*, 123 (1977).
22 T.W. Jeffries and H. Timourian, R.L. Ward, App. environm. Microbiol. *35*, 704 (1978).
23 S. Scherer, W. Kerfin and P. Böger, J. Bact. *141*, 1037 (1980).
24 A. Daday, G.R. Lambert and G.D. Smith, Biochem. J. *177*, 139 (1979).
25 A. Mitsui and S. Kumazawa, in: Biological Solar Energy Conversion, p.23. Ed. A. Mitsui, S. Miyachi, A. San Pietro and S. Tamura. Academic Press, New York 1977.
26 R.V. Smith and M.C.W. Evans, Nature, Lond. *225*, 1253 (1970).
27 H. Spiller, A. Ernst, W. Kerfin and P. Böger, Z. Naturf. *33c*, 541 (1978).
28 J.W. Gotto, F.R. Tabita and C.H. van Baalen, J. Bact. *140*, 327 (1979).
29 P. Antarikanonda, H. Berndt, F. Mayer and H. Lorenzen, Archs Microbiol. *126*, 1 (1980).
30 S.L. Albrecht, R.J. Maier, F.J. Hanus, S.A. Russell, D.W. Emerich and H.J. Evans, Science *203*, 1255 (1979).
31 P.F. Weaver, S. Lien and M. Seibert, Solar Energy *24*, 3 (1980).
32 H. Zürrer, Experientia *38*, 64–66 (1982).
33 H. Kayano, I. Karube, T. Matsunaga, S. Suzuki and O. Nakayama, Eur. J. appl. Microbiol. Biotechnol. *12*, 1 (1981).
34 P.C. Hallenbeck, L.V. Kochian and J.R. Benemann, Z. Naturf. *36c*, 87 (1981).
35 J.P. Houchins and R.H. Burris, J. Bact. *146*, 209 (1981).
36 J.P. Houchins and R.H. Burris, J. Bact. *146*, 215 (1981).
37 M.W.W. Adams, L.E. Mortenson and J.S. Chen, Biochim. biophys. Acta *594*, 105 (1980).

Hydrogen production by photosynthetic bacteria

by H. Zürrer

Institute of Plant Biology, University of Zürich, CH–8008 Zürich (Switzerland)

Photosynthetic bacteria utilize hydrogen as electron donor for autotrophic CO_2 assimilation. Many of these organisms also evolve hydrogen under dark anaerobic conditions and, in large quantities, anaerobically in the light in the absence of ammonia and molecular nitrogen. Hydrogen photoproduction in photosynthetic bacteria is largely or completely associated with the action of nitrogenase. It is not inhibited by CO, an inhibitor of hydrogenase and is dependent on ATP. The conventional hydrogenase catalyzes the reversible reaction $H_2 \rightleftarrows 2\,H^+ + 2\,e^-$. It seems however that in photosynthetic bacteria this enzyme catalyzes mainly hydrogen uptake in vivo. It has been suggested that a function of hydrogenase is to reutilize the hydrogen which is evolved as a byproduct of the nitrogenase reaction, retaining reducing equivalents for N_2 or CO_2 reduction[1]. In contrast to aerobic bacteria, energy conservation in a Knallgas reaction is not possible for photosynthetic bacteria growing anaerobically in the light[2]. Besides molecular hydrogen, a variety of organic and inorganic electron donors are known in bacterial photosynthesis. Most of them are effective also for hydrogen production in the light.

Hydrogen production and utilization in vivo are catalyzed by different enzymes. A genetic or regulatory linkage between nitrogenase and hydrogenase has been proposed in a study with nif$^-$ mutants of *Rhodopseudomonas acidophila*[2]. It has recently been reported that in *Rhodopseudomonas capsulata* although nitrogenase may influence hydrogenase synthesis by supplying inducers (e.g., H_2), there is no strict correlation between hydrogenase synthesis and nitrogenase synthesis[3].

The exact mechanism of electron transfer in hydrogen metabolism and nitrogen fixation is not resolved so far. The figure shows a possible scheme of electron transport and hydrogen metabolism in the photosynthetic bacterium *Rhodospirillum rubrum*. A light driven electron flow generates ATP. It is assumed that NAD and other substances of negative redoxpotential are reduced in a reversed electron flow utilizing ATP.

When reducing power and energy for N_2 fixation in the cell is produced in excess, nitrogenase evolves hydrogen, which can be recycled probably via the ferredoxin-hydrogenase.

Thermodynamical aspects of hydrogen production in photosynthetic bacteria

Table 1 shows the theoretically possible ways in which glucose may be decomposed to H_2 or CH_4 and CO_2. Hydrogen formation in the dark is not regarded as an efficient process to transfer energy. Only 33% of the combustible energy of organic compounds is conserved in the fermentation products. In contrast to hydrogen, methane formation is more efficient. Approximately 85% of the energy is conserved assuming that 3 moles of methane are formed from 1 mole of glucose. In photosynthetic bacteria complete dissimilation of carbohydrates through a light dependent anaerobic Krebs cycle is thought to operate[4]. The decomposition of 1 mole of glucose yields 12 moles of hydrogen. Almost 100% of the combustible energy of glucose would be conserved. In practice about 70% energy conservation efficiency from lactate is reported[5,6].

Electron donors for hydrogen production

Hydrogen photoproduction has been found in species of all families of photosynthetic bacteria (Rhodospirillaceae, Chromatiaceae and Chlorobiaceae). In general, photosynthetic bacteria utilize a wide spectrum of organic substances such as carbohydrates, lipids, fatty acids and some inorganic sulfur compounds. The substrate specificity for hydrogen production varies from species to species. Some of the common electron donors used for hydrogen production by photosynthetic bacteria are summarized in table 2 (for complete data and references see Kumazawa and Mitsui[7]).

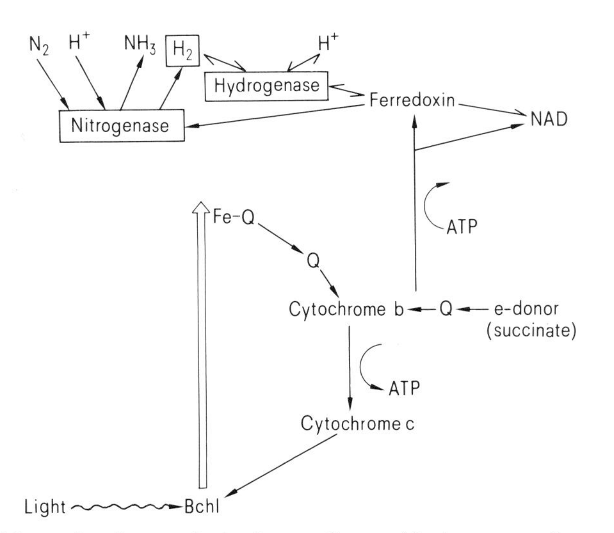

Scheme for photosynthetic electron flow and hydrogenase-nitrogenase catalyzed hydrogen metabolism in *Rhodospirillum rubrum*.

Agricultural byproducts and various wastes have been successfully treated with photosynthetic bacteria to produce biomass and to eliminate waste[8-10]. Prolonged hydrogen evolution from whey or other lactic acid containing wastes have been demonstrated[6]. Hydrogen production is limited by the availabilities of appropriate waste materials.

Hydrogen production rates and biomass potential

Hydrogen production rates reported by different researchers vary considerably. They do not always represent maximum rates that could be achieved under optimal conditions, and they are difficult to compare due to different experimental conditions. Hillmer and Gest[11] reported a rate of 130 ml of hydrogen per h and l culture with *Rhodopseudomonas capsulata* using lactate or pyruvate as electron donor. With *Rhodospirillum rubrum* a rate of 160 ml of hydrogen per h and l continuous culture was achieved (lactate as electron donor). The dilution rate was 0.05 h^{-1} and about 0.12 g of bacterial dry substance was produced per h and l culture[12].

Cells of photosynthetic bacteria are composed of about 65% protein, containing large quantities of essential amino acids and vitamins[13]. Mass culture of *Rhodopseudomonas capsulata* and their single cell protein values are reported and discussed referring to their effect on laying-hens upon mixing the bacterial

Table 1. Combustible energy of glucose compared to the combustible energy of the formed methane or hydrogen

Reaction	
Glucose + 2 $H_2O \rightarrow$ 2 acetate$^-$ + 2 H^+ + 2 CO_2 + 4 H_2 4 H_2 + 2 $O_2 \rightarrow$ 4 H_2O	$\Delta G^{\circ\prime} = -227$ kcal/mole
Glucose + 6 $O_2 \rightarrow$ 6 CO_2 + 6 H_2O	$\Delta G^{\circ\prime} = -686$ kcal/mole
Glucose $\rightarrow$ 3 CO_2 + 3 CH_4 3 CH_4 + 6 $O_2 \rightarrow$ 3 CO_2 + 6 H_2O	$\Delta G^{\circ\prime} = -586$ kcal/mole
Glucose + 6 $H_2O \rightarrow$ 6 CO_2 + 12 H_2 12 H_2 + 6 $O_2 \rightarrow$ 12 H_2O	$\Delta G^{\circ\prime} = -680$ kcal/mole

Table 2. Electron donors used for hydrogen production by photosynthetic bacteria

Rhodospirillaceae	Chromatiaceae	Chlorobiaceae
Acetate	Acetate	Citrate
Butyrate	Fumarate	Formate
Formate	Malate	Glucose
Fructose	Oxalacetate	α-Ketoglutarate
Fumarate	Pyruvate	Lactate
Glucose	Succinate	Mannitol
α-Ketoglutarate	Sulfide	Pyruvate
Lactate	Thiosulfate	Xylose
Malate		
Oxalacetate		
Propionate		
Pyruvate		
Succinate		
Sucrose		
Thiosulfate		

biomass in their feed[14]. Since phototrophic bacteria are able to grow in continuous culture using molecular nitrogen as sole nitrogen source[15], cultures can be grown on media essentially lacking bound nitrogen.

Inhibition of nitrogenase

As mentioned above, photoproduction of hydrogen is catalyzed by nitrogenase. In photosynthetic bacteria the enzyme is quickly inhibited by low concentrations of ammonium salts. This efficient inhibition poses severe problems when waste material containing varying amounts of nitrogen compounds is used as a hydrogen donor. It has been shown however that the glutamate analog L-methionine-DL-sulphoximine (MSO) relieves the repression exerted by exogenous ammonia on nitrogenase activity. Nitrogen fixation or acetylene reduction was shown to operate in the presence of ammonia upon the addition of MSO[16,17]. The inhibition of the photoproduction of hydrogen by ammonia is also released by MSO[18]. Genetically altered bacteria could also be of interest for the photoproduction of hydrogen. Glutamine auxotroph mutants of *Rhodopseudomonas capsulata* have been described, which synthesize nitrogenase and produce hydrogen in the presence of exogenous ammonia[19].

Immobilized cells

In resting cells of *Rhodospirillum rubrum* the rate of photoevolution of hydrogen decreases slowly[6]. As with other organisms it should be possible to prolong the active period by immobilizing the cells on polymer lattices. In many cases the use of immobilized enzymes and cells results in increased production[20]. Hydrogen production by immobilized whole cells of different origin (*Rhodospirillum*[21], *Clostridium*[22], *Anabaena*[23]) has been reported, but so far no systematic survey is available, which compares different techniques, yield, stability etc. This technique may improve solar energy conversation into hydrogen.

We thank the Swiss National Science Foundation (grant 4.048.0.76.04 Nationales Energieprogramm) for its generous support.

1 R.O.D. Dixon, Archs Microbiol. *85*, 193 (1972).
2 E. Siefert and N. Pfennig, Biochemie *60*, 261 (1978).
3 A. Colbeau, B.C. Kelley and P. Vignais, J. Bact. *144*, 141 (1980).
4 H. Gest, J.G. Ormerod and K.S. Ormerod, Archs Biochem. Biophys. *97*, 21 (1962).
5 P. Hillmer and H. Gest, J. Bact. *129*, 732 (1977).
6 H. Zürrer and R. Bachofen, Appl. environ. Microbiol. *37*, 789 (1979).
7 S. Kumazawa and A. Mitsui, in: Handbook of Biosolar Resources, vol. 1, Fundamental Principles. Ed. A. Mitsui and C.C. Black, Jr. CRC Press, Palm Beach, in press.
8 M. Kobayashi, M. Kobayashi and H. Nakanishi, J. Ferment. Technol. *49*, 817 (1971).
9 R.H. Shipman, I.C. Kao and L.T. Fan, Biotechnol. Bioengng *17*, 1561 (1975).
10 H. Sawada, R.C. Parr and P.L. Rogers, J. Ferment. Technol. *55*, 326 (1977).
11 P. Hillmer and H. Gest, J. Bact. *129*, 724 (1977).
12 H. Zürrer, unpublished data.
13 R.H. Shipman, L.T. Fan and I.C. Kao, Appl. Microbiol. *21*, 161 (1977).
14 M. Kobayashi and S. Kurata, Process Biochem. *13*, 27 (1978).
15 T.O. Munson and R.H. Burris, J. Bact. *97*, 1093 (1969).
16 N.M. Weare and K.T. Shanmugam, Archs Microbiol. *110*, 207 (1976).
17 B.L. Jones and K.J. Monty, J. Bact. *139*, 1007 (1979).
18 H. Zürrer and R. Bachofen, Experientia *36*, 1166 (1980).
19 J.D. Wall and H. Gest, J. Bact. *137*, 1459 (1979).
20 G. Durand and J.M. Navarro, Process Biochem. *13*, 14 (1978).
21 M.A. Bennett and H.H. Weetall, J. Solid-Phase Biochem. *1*, 137 (1976).
22 I. Karube, T. Matsunaga, S. Tsuru and S. Suzuki, Biochim. biophys. Acta *444*, 338 (1976).
23 G.R. Lambert, A. Daday and G.D. Smith, FEBS Lett. *101*, 125 (1979).

Conversion of biomass to fuel and chemical raw material

After calculating the potential of biomass production for fuel mainly from agriculture, B.A. Stout reviews the different techniques for converting the rather bulky biomass into a usable form of energy, starting from direct combustion to anaerobic digestion to production of ethanol through alcohol fermentation.
J. Wiegel discusses in detail the direct conversion of cellulose, the most abundant plant product, to ethanol by the single bacterium, *Clostridium thermocellum*, and also by a defined mixed culture.
The problem of degrading cellulose by microorganisms is reviewed by K.E. Eriksson whose main emphasis is on cellulose degrading fungi.
The different ways in which lignins are degraded are summarized by T. Higuchi (aerobic sytems) and J.-P. Kaiser and K. Hanselmann (anaerobic ones). Some possible uses for this second most abundant biomass component are proposed; however, the full and imaginative exploitation of lignin as a raw material remains a challenge for chemists in the future.

Agricultural biomass for fuel

by Bill A. Stout

Department of Agricultural Engineering, Michigan State University, East Lansing (Michigan 48824/USA)

Introduction

The U.S. energy problem. The united States imported about 8 million barrels of oil each day in 1979. At a cost approaching $30 per barrel, the annual cost amounted to about 80 billion dollars! Needless to say, excessive dependence on foreign oil imports has resulted in serious trade deficits. Inflation, fueled in part by energy problems, has been unacceptably high. Thus, the soundness of the U.S. dollar, our economic wellbeing and even our national security are all inextricably bound to the energy problem.
There is no overall energy shortage: for all practical purposes the sun radiates an infinite energy supply, nuclear reactions release huge amounts of energy and coal supplies are extensive. But can we manage these vast energy resources in an economically and environmentally acceptable manner?

Biomass – what is it and how can it be used for fuel?

What is biomass? It's everything that grows – all organic matter except fossil fuels. Examples of biomass available for substitute fuels include traditional agricultural crops and residues, animal manure, forests, aquatic plants, algae and other microorganisms. Biomass contains energy stored from the photosynthetic process – starches, sugars, cellulose, lignin, etc. Dry biomass contains perhaps 16 MJ/kg – more than one-half as much as a pound of coal! Biomass has many uses – as food, fiber, soil organic matter, bedding, structural material, and it may be used for fuels. The latter is not a new concept. Homes and industry in the U.S. were once heated and powered by wood before fossil fuels (coal, oil and natural gas) displaced wood as the major fuel. In the meantime, however, our national policy of cheap oil and strict environmental regulations has gradually led to the present situation where domestic oil no longer satisfies our vast oil appetite. Now we must turn to, and develope, other energy options.
Dry biomass can be burned to produce heat, steam and/or electricity, and it can be converted to liquid or gaseous form for use in mobile vehicles by anaerobic fermentation, alcoholic fermentation, or gasification.
The use of biomass for fuels raises complex and widely diversified issues and its impact must be assessed variously according to specific feedstocks, geographic areas, conversion technology and end-use application. Although some generalizations are possible, few similarities exist among the use of corn to produce alcohol to operate an internal-combustion engine in Indiana, a wood-fired electric generating plant in Vermont or a kelp farm off the California coast.

How large is the potential for biomass production?

Food for humans is the most important use of biomass. Feed for livestock, organic matter for soil conservation and nutrients, bedding, and structural materials are other important biomass uses. The question is – can U.S. agriculture meet all these biomass needs and still produce a surplus for use as a fuel? And can our forests be managed in such a way as to meet the needs for lumber, paper and other forest products and still provide fuels?

Biomass from agricultural residue. According to Stanley Barber, Professor of Agronomy at Purdue University, an estimated 360 million metric tons of residues are produced each year from 10 major crops in the U.S.[1]. Not all residue is collectible with present machinery and some must remain on the land to maintain it within acceptable erosion limits. The above estimate excludes at least 2 t/ha corn and soybean residues and 0.5 t/ha of small grain residues that were likely left in the field. 71 metric tons of collectible 'surplus' residue (usable) might be considered for fuel – with 87% from corn and small grains (table 1).

Growing crops for fuel. Agricultural crops grown under modern management methods are effective multipliers of fossil energy by capturing and converting solar radiation. Table 2 shows yields in tons and net energy as well as the net energy ratio for various crops. Yields averaged over 15 t/ha per year for Napier grass, kenaf and corn. Napier grass provided the greatest net energy return followed by whole corn plants, kenaf, slash pine, alfalfa, and corn kernels. The net energy ratio indicates slash pine returns 26.8 times as much energy as is required to produce the crop; this is followed by alfalfa (15.1), kenaf (13.6) and Napier grass (13.4)[2,3].

Calvin has written extensively about the direct photosynthetic production of hydrocarbons from Euphorbias, Aselepias and other hydrocarbon-containing plants[4].

Lipinsky at the Battelle Memorial Institute has focused on the use of sweet sorghum and sugar crops for fuel or industrial feedstock[5]. The ethanol concept has the potential for a positive net energy return if the byproducts are utilized effectively.

Opinions differ on the availability of land for biomass production. Zeimetz estimates that over 90% of the 190 million ha of U.S. cropland is of sufficient quality to support biomass production. However, conservation measures must be applied to about one half of this land to prevent soil and environmental degradation[6]. An additional 89 million ha of pasture and rangeland have the potential for sustaining biomass crops. Another 65 million ha of forest land might be suitable for growing biomass for energy. Whether or not this land would actually be used for biomass crops depends on price/cost relationships. Much of the land would require investment to bring it to its full production potential. Also, withdrawal of cropland, pasture, range and forest lands for biomass farms or any other use might conflict with the growing demand for food, feed and fiber products.

Larsen et al.[7] emphasize that crop residues on the land are not necessarily surplus. It is difficult to say how much residue can be removed because the answer depends on so many site specific factors – soil type and fertility level, topography, and climate. Posselius and Stout[8] have developed a computer program that determines how much crop residue can be removed from each field considering wind and water erosion, nutrient removal and other factors.

Forages. The present production on pasture and haylands in the U.S. provides feed for the nation's livestock with little surplus. By developing a new market for biomass fuels, millions of tons of additional biomass could be produced from the current pasture and hayland acreage (tables 3 and 4). The 'surplus' in table 4 is 93 million metric tons if 2.5 t/ha is produced above and beyond livestock feed requirements and 186 million metric tons if 5 t/ha are produced. Additional fertilizer would be needed, but Barber[1] estimates a favorable energy output/input ratio of 8:1 for producing biomass on hayland.

The combined output of residues and forages could produce $2–4 \cdot 10^{18}$ J of energy or 15–30 million m^3 of

Table 1. Collectible 'surplus' residues

Crop	Amount (mega tons metric)
Corn	34
Small grains	31
Rice	5
Sorghum	1
Sugarcane	½
Total:	71

Source: Barber[1].

Table 2. Energy potential for various crops

Crop	Yield (t/ha-year)*	Net energy produced (GJ/ha)**	Net energy ratio***
Alfalfa	12.1	202	15.1
Corn, whole	19.3	324	13.0
Corn, kernels	7.7	161	8.6
Kenaf	19.5	309	13.6
Napier grass	50.2	803	13.4
Slash pine	14.5	238	26.8
Wheat, whole	7.4	114	8.0
Wheat, grain	2.9	43	3.4

* t/ha-yr means metric tons/hectare/year;
** GJ = gigajoule = 10^9 joules;
*** $\frac{\text{gross energy produced} - \text{energy input}}{\text{energy input}}$.

Source: Keener and Roller[2].

Table 3. Present pasture and hayland in Eastern U.S.

Region	Hay (million hectares)	Cropland pasture	Non-cropland pasture
Northeast	2	2	1
North Central	7	8	8
South	3	9	10
Total:	12	19	19

Source: Barber[1].

Table 4. 'Surplus' biomass potential from pasture and hayland in Eastern U.S. (yield in addition to livestock needs).

Region	+2.5 t/ha (million metric tons)	+5 t/ha
Northeast	11	22
North Central	43	86
South	39	77
Total:	93	186

Source: Barber[1].

alcohol per year, enough to substitute for 5–9% of the nation's gasoline supply.

DOE estimates of available biomass raw material. The U.S. Department of Energy alcohol fuels policy review commissioned 5 individual studies to assess biomass raw material availability and economics[9]. While the focus was on alcohol fuels, the data assembled give a good overview of biomass availability for any fuel use.

To clarify the meaning of 'available' feedstocks, consider that usually, 'available' refers to what is noncompetitive with the clearly higher values of a particular feedstock. Assuming no new or marginal cropland is brought into production, available grain crops are generally those which can be grown on existing cropland in the absence of any USDA policy of production restriction and which are not needed for projected demands of food, feed or export markets. Food processing wastes or by-products include such things as citrus rind, pulp, and corn starch strains from a corn sweetener plant.

Available crop residues exclude an average of 35% of all residues estimated as the minimum the farmer must leave on the land. The amount of residue that must be returned to the land is highly site specific and depends on the soil type, topography, climatic factors and crops[10].

The maximum available U.S. biomass resources total 700 million dry metric tons annually (table 5). Wood accounts for 61% of this total, agricultural residues 23%, municipal solid waste 10%, grains 5% and food processing wastes 1%. A more conservative estimate of 72.8 million dry metric tons of biomass potentially available from wastes supplemented by grains grown on set-aside lands is given by DOE[1].

Significance of biomass fuels. Clearly, millions of tons of biomass could be available for fuel. Researchers at Purdue University[11] estimated the technical energy potential for direct combustion of biomass (excluding grains) to be $1.7–3.4 \cdot 10^{18}$ J. The technical potential for alcohol production (using residues, forage and grains) would be 42–68 million m^3 per year or 9–15% of the U.S. gasoline consumption. These numbers represent the technical potential only. The Office of Technology Assessment[12] concluded that $11–16 \cdot 10^{18}$ J could be produced from biomass by the year 2000 if biomass fuels were vigorously promoted. Where biomass is available and the technology for using it for fuel in a cost-effective manner exists or can be developed, it seems prudent to do so. For a nation that uses $79 \cdot 10^{18}$ J of energy each year, biomass fuels are likely to provide only a small percentage of our national energy needs. But if biomass fuels meet 1% of our nation's energy needs, this is significant!

Biomass conversion technologies

Many processes or technologies exist for converting biomass to a more useful form for fuel or industrial feedstocks. Most are classified as wet or dry processes (fig. 1). Dry processes include direct combustion and gasification; wet processes include anaerobic and alcohol fermentation. (Methanol production is not discussed because it is not normally made from agricultural feedstocks.) The 2 primary uses of this energy are heat and fuel for mobile vehicles.

Table 5. Projected maximum U.S. biomass resources available

	Quantities in million dry metric tons per year							
	1980		1985		1990		2000	
	Quantity	%	Quantity	%	Quantity	%	Quantity	%
Wood*	453	61	421	56	389	49	498	48
Agricultural residues	175	23	200	26	218	28	252	24
Grains**								
Corn	20	–	18	–	7	–	–	–
Wheat	11	–	14	–	15	–	18	–
Grain sorghum	4	–	3	–	3	–	3	–
Total grains	34	5	35	5	25	3	21	2
Sugars**								
Cane	–	–	3	–	12	–	12	–
Sweet sorghum	–	–	5	–	51	–	144	–
Total sugars	–	–	7	1	63	8	156	15
MSW	78	10	83	11	90	11	105	10
Food processing wastes	6	1	7	1	8	1	10	1
Total	746	100	753	100	793	100	1042	100

* Assumes wood from silvicultural energy farms starting in 1995.
** Estimates for grains and sugars assume an aggressive development program to establish sweet sorghum as a cash crop. This program would divert land from corn in 1990 and 2000 – 1.9 and 2.8 million hectares (4.7 and 7 million acres) respectively.
Source: Department of Energy[9]. The report of the alcohol fuels policy review, p. 48.

If heat is needed and the biomass is relatively dry, direct combustion may be the most efficient and effective process. Heat may be used to produce steam and electricity if desired. Various gasification processes will produce a low or medium Btu gas (gas of low or medium caloric value, about $\frac{1}{4}$ and $\frac{1}{2}$ of natural gas respectively) if that energy form is desired for stationary engines or heat.

If the biomass is wet, e.g., animal manure, anaerobic fermentation will yield a low Btu gas, primarily methane. For mobile vehicles requiring a high energy density fuel (most commonly a liquid), synthetic liquid fuels may be produced from many feedstocks. Alcohol fuels may be produced by direct fermentation of sugar crops or by hydrolysis of starches or cellulosic materials followed by fermentation (fig. 2).

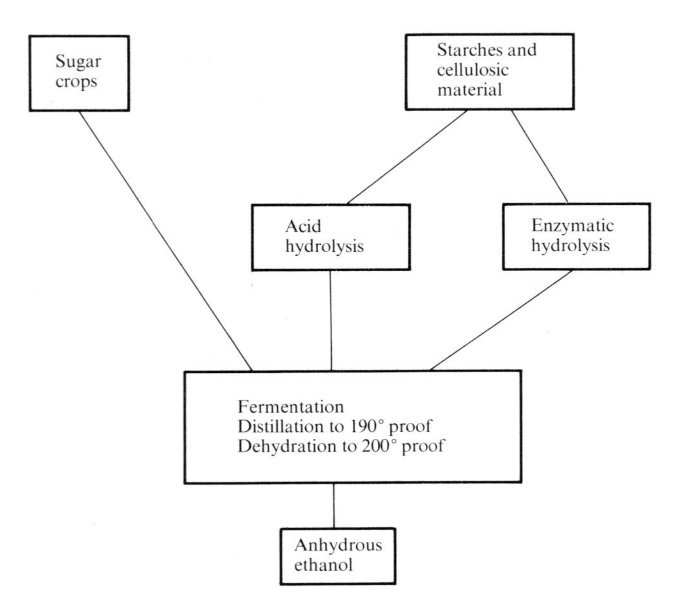

Figure 1. Process alternatives for converting biomass to gaseous, liquid or solid fuels.

Direct combustion. Technology for direct combustion is old and highly developed. It is in wide use commercially, accounting for most of the $1.9 \cdot 10^{18}$ J of energy presently generated from wood. Much research is under way to develop suitable combustion systems for wet biomass and to study optimum particle size, feeding systems, particulate control, biomass mixtures with oil or coal, suspended burning systems, etc.

Buchele[13] and others have conducted research on converting the energy value of cornstalks to useful forms by burning cornstalks as a companion fuel with high sulfur coal in boilers at electric generating plants. Shredded cornstalks were fed to the traveling-grate boiler of the Ames, Iowa, power plant at a rate of 4.5 t/h. The stalks burned well with no special problems.

Burning in an excess of air. One purpose of burning is to eliminate unwanted waste; burning in a open pile and incineration are examples. Some type of furnace is required to collect and distribute the heat generated, and combustion may occur inside or outside tubes. Provisions must be made to: a) introduce the organic particles or shredded material; b) provide an adequate air flow to maintain an excess oxygen supply; c) remove the residue or ash; and d) control particulate emissions. Air flow may be by natural or forced draft.

There are two types of air-suspended combustion systems: a) those which suspend the burning fuel in the gas stream in the combustion enclosure; and b) those which suspend the fuel in the gas stream and in another medium, the fluidized bed. Advantages of flue-gas stream suspension include a more rapid response to automatic control, an initial cost saving due to lack of grate surface and mechanical stoking devices, and the ability to complete combustion with a much smaller percentage of excess air in the furnace.

Fluidized-bed suspension burning systems have all

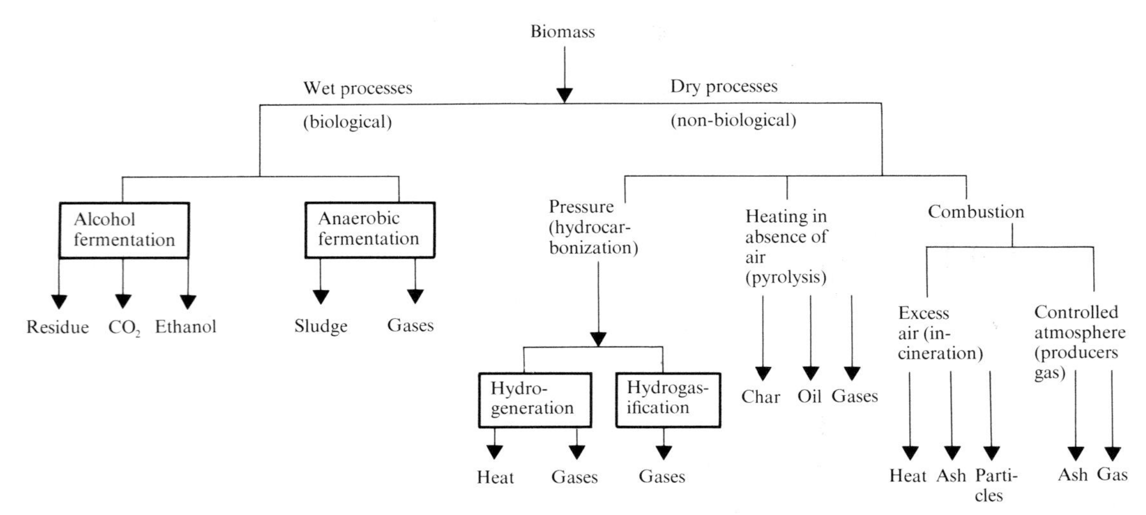

Figure 2. Production of 200° proof anhydrous ethyl alcohol from starch, sugar or cellulosic materials.

the advantages of the fluegas stream suspension, plus one that is important when a system must operate intermittently. Fluidized beds, usually sand, comprise a 'termal flywheel' or large capacity. Once operating temperature is reached, they retain heat over a long period, losing only about 110 °C during an overnight shutdown. The savings of auxiliary fuel for preheat on the next start-up are appreciable.

Burning in a controlled atmosphere. Gasification is the conversion of a solid or liquid to a gas. If the oxygen supply is restricted, incomplete combustion occurs releasing combustible gases such as carbon monoxide, hydrogen and methane. A solid residue or char remains. Gasification is discussed in detail later.

Heating in the absence of air. Pyrolysis is the transformation of an organic material into another form by heating in the absence of air. If heat is applied slowly, the initial products are water vapor and volatile organic compounds. Increased heat leads to recombination of the organic materials into complex hydrocarbons and water. The principal products of pyrolysis are gases, oils and char.

Producer gas generation. A producer gas generator produces a combustible gas from crop residues, wood chips or charcoal. (During World War II, in Europe and Japan, producer gas generators or gasifiers were often used to operate tractors, automobiles and buses, because petroleum was scarce.) The feedstock is heated to 1000 °C and reacts with air, oxygen, steam or various mixtures of these to produce a gas containing about 30% carbon monoxide (CO), 15% hydrogen (H_2) and up to 3% methane (CH_4) (table 6).

There are 2 basic generator designs: the updraft and the downdraft. In an updraft generator, hot gases flow counter to the feedstock. Part of the fuel stock is pyrolyzed and the resulting gas has a high tar content. In the downdraft system, pyrolysis products are broken down as they pass through the reaction zone before combining with the exiting gases. Since downdraft generators have the potential to eliminate tar from gas, they are probably better suited for burning crop residues as a fuel source[15].

Table 6. Gas analysis from an updraft producer gas generator using charcoal

	Percentage by volume
CO	25–30
H_2	10–14
CO_2	5–8
O_2	0.5–1.5
CH_4	0–2.5
N_2 and others	50–53

Source: Posselius et al.[14]. An updraft producer gas generator, p. 3.

Anaerobic digestion

Anaerobic digestion is a conversion process for wet biomass such as animal manure, municipal sewage and certain industrial wastes. Through this process complex organics are converted into methane and other gases. An effluent is also produced which can be used as fertilizer or animal feed. An extensive bibliography[16] on anaerobic digestion was prepared for EPA in 1978.

Anaerobic digestion is a biological process carried out by living microorganisms:

Organic matter + bacteria + water → methane + carbon dioxide + hydrogen sulfide + stabilized effluent.

This process occurs only in the absence of free oxygen. Methane-forming bacteria are sensitive to environmental conditions such as pH (6.6–7.6 optimum), temperature (35 °C and 54 °C are 2 preferred levels), and carbon/nitrogen ratio (30^{-1} optimum).

Man-made digesters, or containers that keep the feedstock isolated from air can be of either the batch or continuous flow type. Advantages of the batch type include:

- feedstock availability is often sporadic and comes in batches;
- daily management is minimal; and
- relatively inexpensive.

Disadvantages of the batch type are:

- much labor is needed to load and unload digester;
- gas production is sporadic; and
- not as efficient as continuous digester.

Table 7. Approximate daily production and heat values for biogas

Livestock (454 kg b.wt)	Approximate biogas production (m^3/day)	Approximate heat value (MJ)*	Approximate equivalents Gasoline (l)**	Diesel fuel (l)**	Natural gas (m^3)**	Propane (kg)**
Beef	0.85	19	0.57	0.53	0.51	0.4
Dairy	1.3	29	0.87	0.76	0.76	0.6
Poultry broilers	2.6	58.3	1.7	1.6	1.6	1.2
Poultry layers	2.0	45.6	1.4	1.2	1.2	1.0
Swine	0.82	18.4	0.57	0.49	0.48	0.4

* Assumes biogas containing 60% methane or heating value 22 MJ/m^3.
** Heating values: gasoline, 5.6 MJ/l; diesel fuel, 37.1 MJ/l; natural gas, 37 MJ/m^3; propane, 49 MJ/kg.

While early digesters were usually of the batch type, the continuous flow type is considered an improvement.

Although many factors affect output, table 7 illustrates the gas production rate and energy output for various feedstocks. To translate energy output into common language, the daily manure from a single 630-kg dairy cow could produce 1.8 m^3 of biogas[17].

Biogas consists of 60–70% methane, 30–40% carbon dioxide and a trace of hydrogen sulfide, ammonia gas and water vapor. Biogas has an energy content around 22 MJ/m^3. Methane or biogas are 'permanent' gases and cannot be liquified at any pressure at commonly occurring temperatures, seriously limiting their use in mobile vehicles.

These gases are better suited for use in high compression (13–14:1) stationary engines designed or modified to operate on methane. In biogas-powered stationary engines, waste heat can be recirculated in the digester coil and gas can be used as it is produced without a compressor storage unit. Full engine power is realized only if carbon dioxide is removed from the biogas mixture to increase the energy content of the gas. Longer engine life is attained if hydrogen sulfide is also eliminated from the gas before use.

Biogas may be used to heat livestock buildings by scrubbing H_2S only, but the 30–40% CO_2 will necessitate additional venting and this requires more heating energy. The CO_2 would not present a problem in greenhouse heating, however.

Digester waste or sludge is an excellent fertilizer containing all the potassium and phosphorus and up to 99% of the nitrogen originally in the manure. In addition, trace elements such as boron, calcium, iron, Mg, S and Zn remain unchanged. Sludge could also be used in livestock rations if mixed with molasses, grains and roughage. Water must be removed by centrifuge to concentrate the protein, but some of the protein will be dissolved in the water and lost.

The relative economy of an anaerobic digester is probably the most important factor in determining its feasibility. However, an accurate economic picture for the anaerobic digester is difficult to project if specific aspects of its implementation remain unknown: a) the cost of energy; b) what will sludge be used for; c) what type of system will be used; d) how much salvaged material will be used; and e) what is the nature of the farm operation. Nevertheless, rough guidelines can be provided. One method of analysis gives the estimated digester construction costs per animal. These estimated costs range from \$200 to \$300/cow and from \$40 to \$120/pig. A second method considers both construction costs and potential economic returns to provide the minimum digester size (in animal units) necessary for economic feasibility. Estimates for the minimum digester size for economic viability range from a 200- to 400-cow digester.

Alcohol fuels

Ethyl alcohol (ethanol) is made by fermentation of sugars. Where grains or other starchy materials are used this step is preceded by enzymatic conversion of starches[18]. Much research is underway to hydrolyze cellulosic materials and then convert them to alcohol.

Ethanol is a premium fuel and can be blended with unleaded gasoline and burned with no modification of today's engines. A 10% blend of ethyl alcohol with 90% unleaded gasoline is presently being marketed as 'Gasohol'.

Ethanol costs more than gasoline today but the technology for producing it is very dynamic. Much controversy surrounds the energy balance, but the latest technology yields a positive net energy return[19]. Furthermore, there is no reason to use petroleum in the production of alcohol.

Alcohol plant sizes range from less than 4 kL/day (classified as a small scale by the Department of Energy)[9] to more than 75 ML per year. Large plants offer economies of scale, but farm level plants may be competitive when collection, storage and transportation costs are minimal[20,21].

The technology and thus the economics of producing alcohol are quite dynamic. Many have taken firm stands against alcohol fuels based on obsolete data. There is every reason to expect that with improved technology – heat recycling, improved distillation methods, membrane separation and integrated systems which permit wet feeding of by-products – the energy balance and economics of alcohol production can be improved. Certainly, it is in our national interest to attempt to make alcohol an attractive extender of scarce gasoline.

Ethanol is an excellent fuel for spark-ignition engines and may be considered for diesel engines, gas turbines, fuel cells and petrochemical feedstocks[18,19]. Many reports on alcohol fuel applications are available. For example, the American Petroleum Institute Task Force EF-18 reviewed the properties of ethanol and methanol and their suitability as automotive fuels[22]. It concluded that since alcohol fuels are more expensive than gasoline, their best use would be in premium fuel applications where their clean burning and low nitrogen oxide formation characteristics could be used advantageously. Straight alcohol fuels, if used in engines specifically designed for optimum use of their properties, offer potential advantages that could outweigh the disadvantages in certain situations[23].

Major issues

Biomass for fuels is a complex subject involving the growth, collection, densification, transport, conversion and utilization of organic material. Often, biomass for fuels must compete with important alternative uses. The impact of biomass for fuels on food, feed and fiber prices is not fully known. And the need to return organic material to the soil for erosion control and organic matter maintenance continues to be of concern. Also competition between food crops and fuel production from biomass is an unresolved issue and will need a great deal more attention[24]. Certainly, a net energy gain from biomass fuels relative to the petroleum input is essential for a successful biomass fuels program. However, an overall net energy gain may not be necessary in the short run if a low quality bulky fuel is upgraded to a high quality clean burning fuel, especially a high energy density fuel to power existing mobile vehicles.

Finally, a word on the autonomy of a fuels program. Of course, economics drives our free enterprise system. Although our economic system is already highly distorted by regulations, subsidies and tax incentives, a biomass fuels program should eventually stand on its own. Temporary subsidies and incentives may be justifiable to promote development of such a program due to the high risks and uncertainties involved.

Bio-Energy Directory, 2nd edn, 1979 Bio-Energy Council, 1625 Eye Street, N.W., Washington, DC 20005.

1 St. Barber, Energy resource base for agricultural residues and forage crops. Mid-American Biomass Energy Workshop, Purdue University, May 21, 1979.
2 H.M. Keener and W.L. Roller, Energy production by field crops. ASAE paper No. 75-3021, ASAE, St. Joseph, MI 49085, 1975.
3 W.L. Roller et al., Grown organic matter as a fuel raw material source. Ohio Agricultural Research and Development Center. Report to NASA, October 1, 1975.
4 M. Calvin, Hydrocarbons via photosynthesis, Energy Res. *1*, 299–327 (1977).
5 E.S. Lipinsky, Fuels from biomass-integration with food and materials system. Science *199*, 644–651 (1978).
6 K.A. Zeimetz, Growing energy. USDA Agricultural Economic Report No. 425, June 1979.
7 W.E. Larsen et al., Effects of tillage and crop residue removal on erosion, runoff, and plant nutrient. Special Publication No. 25, Soil Conservation Society of America, 1979.
8 J. Posselius and B. Stout, Crop residue availability for fuel. AEIS No. 440, File 18.8. Cooperative Extension Service, Michigan State University, East Lansing, August 1980.
9 DOE report. Report of the alcohol fuels policy review, US Department of Energy, Washington, DC 20585, 1979.
10 J.R. Goss, Food, forest wastes = low Btu fuel. Agric. Engng *59*, 30–33 (1978).
11 W.E. Tyner and J.C. Bottum, Agricultural energy production: Economic and policy issues. Bull. No. 240, Department of Agricultural Economics, Purdue University, September 1979.
12 Office of Technology Assessment: Energy from Biological Processes. Congress of the United States, Washington, DC 20006, July 1980.
13 W.F. Buchele, Direct combustion of crop residues in boiler furnace. Proc. Conf. Production of Biomass from Grains, Crop Residues, Forages and Grasses for Conversion to Fuels and Chemicals, 1977, p. 312–331.
14 J. Posselius, C. Myers, B. Stout and J. Sakai, An updraft producer gas generator. AEIS No. 394. Michigan State University, March 1979.
15 R.H. Hodam and R.O. Williams, Small-scale gasification of biomass to produce a low Btu gas. Proc. Symposium on Energy from Biomass, 1978.
16 T.P. Abeles et al., Energy and economic assessment of anaerobic digesters and biofuels for rural waste management. OASIS 2000. University of Wisconsis Center, Barron County, Rice Lake, Wisconsin, June 1978.
17 D.L. Van Dyne and C.B. Gilbertson, Estimating U.S. livestock and poultry manure and nutrient production. USDA-ESCS Bulletin No. 12, 1978.
18 R. Ofoli and B. Stout, Making ethanol for fuel on the farm. AEIS No. 421. Cooperative Extension Service, Michigan State University, East Lansing, February 1980.
19 R. Ofoli and B. Stout, Ethyl alcohol production for fuel: Energy balance. ASAE Energy Symposium, Kansas City, Miss., September/October 1980.
20 Solar Energy Research Institute. Fuel from farms. A guide to small-scale ethanol production. SERI/SP-451-519 UC-61. Technical Information Center, US Department of Energy, Oak Ridge, Tenn. 37830, February 1980.
21 United States Department of Agriculture. Small-scale fuel alcohol production. US Government Printing Office, Washington, DC 20402, March 1980.
22 American Petroleum Institute. Alcohols – a technical assessment of their application as fuels. API Publication No. 4261, July 1976.
23 A. Rotz, M. Cruz, R. Wilkinson and B. Stout, Utilization of alcohol in spark-ignition and diesel engines. Extension Bulletin E-1426. Cooperative Extension Service, Michigan State University, East Lansing, July 1980.
24 Food and Agriculture Organization. FAO expert consultation on energy dropping versus food production. FAO, Rome, June 1980.

Ethanol from cellulose

by Jürgen Wiegel*

Institut für Mikrobiologie der Gesellschaft für Strahlen- und Umweltforschung, Grisebachstrasse 8, D–3400 Göttingen (Federal Republic of Germany)

Summary. An excess of organic waste, containing up to 60% cellulose and hemicellulose is produced worldwide. The conversion of this cellulosic material to ethanol is discussed: The two-step process consisting of a hydrolysis step to glucose and the subsequent fermentation by yeasts; and the one-step process, a fermentation of the cellulose by the anaerobic thermophile *Clostridium thermocellum,* or by a thermophilic, anaerobic, defined mixed culture. The use of the latter seems to be very feasible. To achieve an economic process, it is suggested to combine this approach with a thermophilic fermentation of the effluent and/or stillage obtained to produce methane.

Ethanol for technical and industrial purposes has been in use for only 100 years. Presently, there is an increasing demand for ethanol for fuel and feedstock chemicals. The petrochemical sources are very limited, and thus, the cost of oil is increasing continuously. After 50 years of producing ethanol mainly from petrochemical sources, the conversion of biomass to ethanol has become interesting again. Cellulose and hemicellulose are potentially important substrates for such processes. This is mainly due to the fact that cellulose is abundant, renewable and, at present, poorly utilized[1–7].

It has been calculated[1–3], that a total of 85×10^9 tons of cellulose and hemicellulose are produced annually in the world; of this figure, land plants account for 20 tons produced per capita each year. Many microorganisms degrade these polymers aerobically or anaerobically. Human beings and higher animals cannot degrade cellulose, except in commensalism with microorganisms, e.g. bacteria in the rumen of cattle or in the gut of termites[8]. Only small amounts, about 2%, of the annual cellulose production is decomposed by human beings through burning or industrial processes. Most of the harvested cellulose becomes waste or parts of agricultural and food wastes, municipal and industrial wastes and urban refuses containing more than 40% of paper and paper products. The amount of the various kinds of waste produced is increasing world-wide.

About 22% of the landmass of the globe is covered by large forests. With present wood harvesting methods, about 40% of the organic material is left as waste in the forest[9] and normally is decomposed by microorganisms. Most of it is aerobically mineralized to carbon dioxide and water, and a smaller amount is anaerobically degraded to alcohols, fatty acids, carbon dioxide and molecular hydrogen. In addition, on a world-wide basis, about 1.3×10^9 tons of cellulose and hemicellulose are produced annually as waste from grain (straw), cotton, bamboo etc. About the same amount of cellulosic waste results from printed paper and paper products. The cellulose content of the waste produced annually in the USA (to take an example from an industrialized country) is summarized in the table. According to statistics, an American citizen produces about 2.2 tons of liquid and solid cellulosic waste in 1 year. The following calculation may illustrate the theoretical potential of a bioconversion of waste to ethanol: assuming that at least 1 mole ethanol per mole glucose equivalent of the cellulose can be formed from the 2.2 tons of cellulosic waste, the tremendous amount of 630 kg or 768 l absolute ethanol per citizen and year could be produced. The cost of raw materials is still a controversial point when analyzing the problems in the bioconversion of cellulosic material to ethanol but the long term availability of large quantities of cellulosic biomass, required for the successful production of liquid fuel, should not be a substantial problem. Many of the potential raw materials for ethanol fermentation can be obtained at practically zero cost: lowgrade wood, waste from processing of wood and pulp, agricultural waste from corn, grain and sorghum, or used newspapers and governmental papers out of date. Waste materials have no significant value; however, they probably will receive new values when suitable methods to convert them into useful products are applied. Considerable expense is involved in the collection and in the transportation of the wastes or the biomass. Consequently, industrial companies and communities will be better off if they are able to treat their waste at the location of production. Fermentation processes are now required that work economically in small units without high investment costs and without highly trained man-power. Unfortunately, realization of such projects is being strongly hindered by alcohol legislation and in those industrialized countries which are producing such copious amounts of waste. It seems easier and cheaper to dump the waste.

And yet, the primary goal at present should be the conversion of cellulosic waste into useful products in order to stop pollution of our environment. Profit-making should be only a secondary concern. The conversion of waste to ethanol can be an effective way in fighting pollution since ethanol is a clean energy source and our present level of biotechnology should now enable us to intelligently utilize cellulose, hemicellulose and waste material.

The bioconversion to ethanol has not yet been studied with normal waste under technical conditions as has been done for the conversion to methane or SCP. Under laboratory conditions, new processes have been developed and new prospective bacteria have been isolated. Now the conversion of normal waste resources has to be studied in pilot plants from the point of view of economy.

Presently, there are 2 major ways of producing ethanol from cellulose.

Annual production of solid cellulosic waste in the USA

Waste type	Waste per year (tons $\times 10^6$)	Assumed cellulose content (%)	(tons $\times 10^6$)
Agricultural and food wastes	400	60	240
Manure	200	50	100
Urban refuse	150	45	67.5
Logging and other wood wastes	60	55	33
Industrial wastes	45	33.3	15
Municipal sewage solids	15	33.3	5
Miscellaneous organic waste	70	25	17.5
Total:	940		478

[a] Based on values from Humphrey et al.[10].

a) The two-step process: cellulose is converted enzymatically or by a treatment with chemicals to glucose which in the 2nd step is fermented to ethanol by yeasts.
b) The one-step process: cellulose is degraded anaerobically to ethanol by the cellulolytic thermophilic *Clostridium thermocellum* or by defined anaerobic thermophilic recombined cultures consisting of cellulolytic and glycolytic bacteria.
Both methods can be applied to the various cellulose sources, to the cellulosic wastes, or to cellulosic material specially grown for the bioconversion into methane and/or ethanol.

The two-step process

Presently, *Saccharomyces* strains are being used in the conventional ethanol fermentation. These yeasts are not able to hydrolyze cellulose; consequently, the biopolymers have first to be hydrolyzed to glucose. Several processes are known using either a treatment with chemicals or with cellulase preparations.
The chemical processes require first a milling to obtain suitable particle sizes. The hydrolysis is performed either with alcali, acids, or organic solvents. For the alcaline treatment the cellulose is incubated with 1–2% sodium hydroxide or with NH_3[11] for 24 h at room temperature or with 4% sodium hydroxide at 80–100 °C and 200–300 atm for some minutes[12,13]. A short incubation time with 0.5 N sulfuric acid at 120 °C leads to the hydrolysis of the hemicellulose only, whereas the hydrolysis of the cellulose requires a treatment over several hours[14]. None of these methods gives a quantitative reaction, and concomitantly a partial decomposition of the sugars obtained occurs. An almost quantitative hydrolysis is obtained with a treatment of a mixture of 5% cadmiumoxide with 28% ethylendiamine in H_2O[15]. This and the other

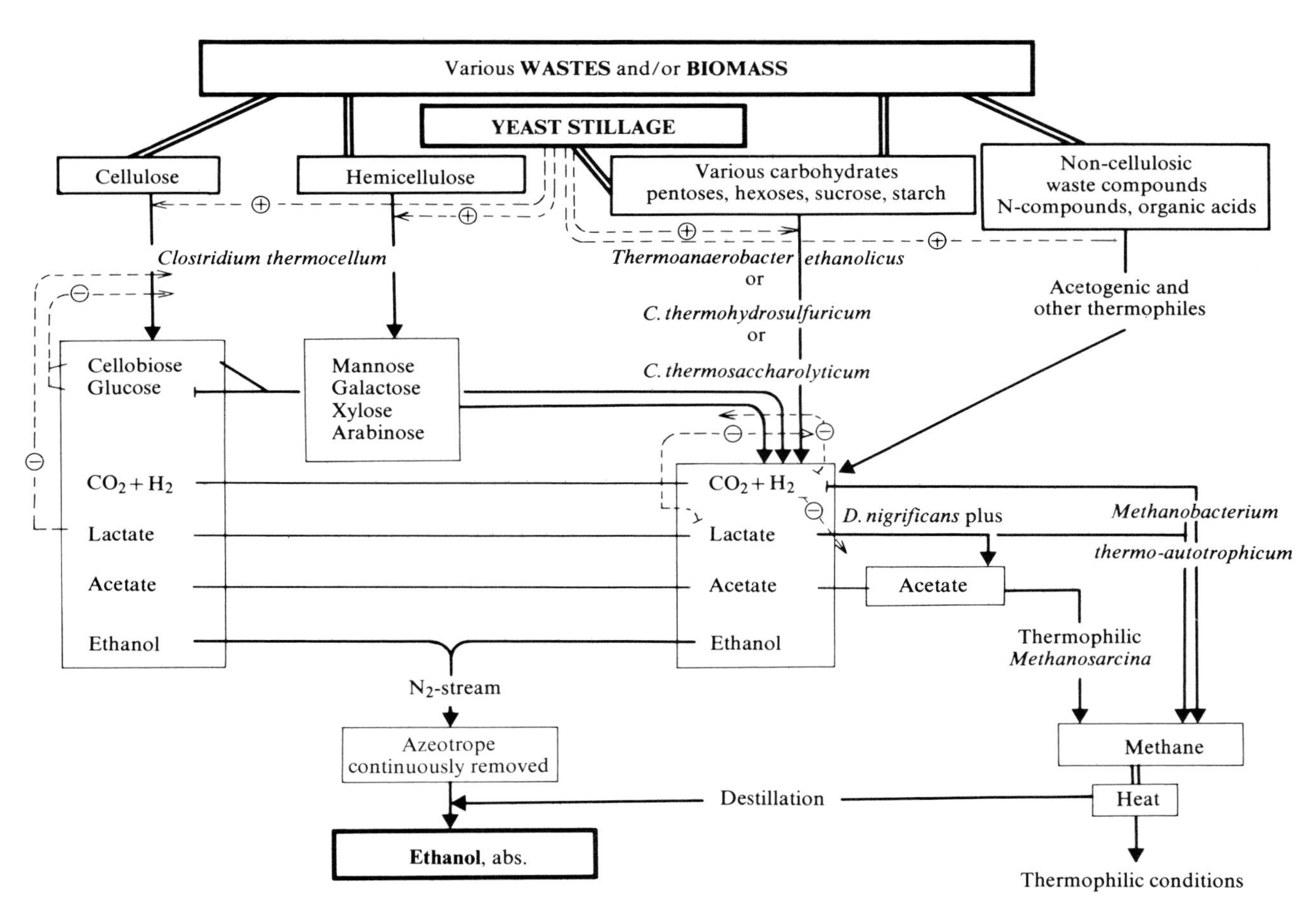

Ethanol production from cellulosic wastes and biomass by thermophilic anaerobes. The various wastes and the biomass contain cellulose, hemicellulose, various carbohydrates as sugars, cellobiose, starch etc. and many N-compounds, fatty acids. etc. These different components are degraded by the various bacteria as indicated. The stillage of yeast ethanol fermentation is added to promote the degradation (– – ⊕ – → –). Several products obtained during the conversion are inhibitory (– – ⊖ – → –) and thus they have to be kept low in the fermentation broth through utilization by other bacteria. Some of the by-products (e.g. acetate and H_2/CO_2) can be directly converted to methane, whereas lactate has to be degraded to the methanogenic substrate acetate. This reaction is carried out by *Desulfotomaculum nigrificans* producing acetate, carbondioxide and H_2S. In the absence of high concentrations of sulfate and in the presence of *M. thermoautotrophicum* CH_4 is formed instead of H_2S. The combining of the cellulose degradation to ethanol with the methane fermentation helps a) to avoid product inhibition as far as possible, b) to diminish the overall fermentation residue and c) to obtain a cheap and clean energy source for heat production required for the elevated temperature of the fermentation vessel and for the ethanol distillation. Ethanol is removed from the fermentation vessel by a low stream of oxygen-free N_2.

methods, however, have the disadvantage that the employed chemicals are strong pollutants or even strong poisons. Therefore they have to be removed from the sugars before the fermentation process can be started; thus, the conversion of the organic wastes leads to anorganic wastes and pollutants. This is unfortunate as the pollutive effect of these chemicals is less reversible than that of the organic material.
A better method is the enzymatic hydrolysis. Although this process is more expensive, the cellulose is almost quantitatively hydrolyzed to glucose and no caramelization products are formed. Recently, Reese and Mandels at the Army Natick Research and Development Command (USA) obtained suitable enzyme preparations with high catalytic activities[16,17]. However, there is still a demand for cheaper enzymes of a better quality. The desired enzymes should have a lower product inhibition and a higher stability. Especially the enzyme preparations for the conversion of cellulose dispersed in other wastes need a high stability against inactivation through proteases, heavy metals and elevated temperatures. The stability of the single components of the cellulase complex varies considerably under hydrolysis conditions e.g. pH 4.8, 50 °C and 24 h[16]. For instance, merthiolate and other Hg-compounds are extremely potent inhibitors of the cellobiohydrolase (over 60% inhibition) whereas most of the various endoglucanases are less affected. Shaking and mixing generally decrease the cellulose degradation rates and also the enzyme recovery[16]. The use of *Trichoderma viride, T. reesi* or similar cellulases should be economical for processes using relatively pure cellulose. From waste with a low cellulose content, the enzyme is difficult to recover or is not reusable due to its inactivation. Another main disadvantage of this two-step process is the inability of the ethanol-producing yeasts to utilize the pentoses derived from hemicellulose[18]. For the enzymatic hydrolysis of straw and of waste consisting of wood and bark, pre-treatment with pressurized water steam[35] seems very promising. This method does not lead to severe pollution; the hemicellulosic part is extracted with water. This steaming technique and the organo-solvent process[36] are gaining more and more importance for various strategies converting material containing lignocellulose. The links between lignin and (hemi)cellulose have to be destroyed, otherwise the enzymes have only limited access to the cellulosic part. Such pre-treatment of lignocellulosic biomass is also necessary for the one-step process.

The one-step process

There are several bacteria which can hydrolyze cellulose. None of the mesophilic cellulolytic anaerobes known produce ethanol as the main fermentation product. The ability to produce ethanol from various sugars is widespread among bacteria; however, the production of ethanol as the sole fermentation product is relatively rare (for distribution of ethanol production among microorganisms see Wiegel[19] and Lorry[20]). Among the ethanol producing organisms there are several thermophilic (T_{opt} above 42 °C and T_{max} above 50 °C) and extreme thermophilic (T_{opt} above 65 °C and T_{max} above 70 °C) anaerobes[19]. Thermophilic processes are much more suited for the industrial production of ethanol than are the mesophilic ones[19,21]. Some of the advantages are: fast degradation rates, a relatively cheap fermentation process since heating a fermentation vessel is easier than cooling it, less danger of contamination, no growth of pathogenic microorganisms – most of them are destroyed by the elevated temperature, cooling is necessary for mesophilic processes due to the production of heat through microbial degradation of the biomass and mixing the fermentation broth. Several thermophilic microorganisms can degrade cellulose: *Actinomyces, Sporocytophaga* species, fungi and clostridia. Examples of the thermophilic cellulolytic fungi are *Chaetomium thermophile var. disstum* and *Talarmyces mersonii.* Some of the strains exhibit a very high cellulase activity[22,23]. However, there are several disadvantages to the fungi: they grow slowly, many produce antibiotics or substances which are poisonous for animals and humans, and ethanol – if it is at all produced – is only a minor product. As is the case with the fungi, the potential of the thermophilic actinomyces *Thermomonospore curvata* and related species is very low for the ethanol production. Recently, Belamy[24] described thermophilic *Sporocytophaga* species with a high cellulase activity. They seem obligate syntrophic with other glycolytic bacteria, similar to an extreme thermophilic *Clostridium (C. caldocellum)*[25]. All these thermophiles might be useful in developing stable enzyme preparations to be used in the two-step procedure, but they do not produce high amounts of ethanol. Only the thermophilic *Clostridium thermocellum (C. thermocellulaseum)* produces up to 1 mole ethanol per mole glucose equivalent of the cellulose degraded[18,19].

Thus, the direct conversion of cellulose to ethanol is possible and has been subject to study by several groups[18,26,27]. Although very useful mutants of *C. thermocellum* with low product inhibition have been obtained[18,26], the yield has not exceeded significantly 1 mole ethanol per mole glucose equivalent. Recently Wang and co-workers[18,26] started to use co-cultures with *Clostridium thermosaccharolyticum.* This glucolytic bacterium has some useful properties. Contrary to the cellulose degrader, it utilizes starch directly and converts pentoses to ethanol, too. Pentoses will always be present in the fermentation broth of cellulosic material due to the hydrolysis of hemicellulose by the cellulase of *C. thermocellum,* which utilizes only slowly the pentoses. However, *C. thermosaccharolyticum* produces ethanol only under special conditions, vegetative cells normally form butyrate[28-31]. The yield of ethanol is not higher than 1 mole per mole of glucose utilized, neither in pure nor in co-culture with *C. thermocellum*, so far.

Two other bacteria, extremely thermophilic, seem more suitable for such a co-culture: *Clostridium thermohydrosulfuricum* and the recently described *Thermoanaerobacter ethanolicus*[30,32]. *C. thermohydrosulfuricum* produces up to 1.6 mole ethanol per mole glucose, if the pH-value of the culture shifts from about pH 7.5 to below 6.9 during the fermentation[31]. *T. ethanolicus* ferments glucose and pentoses up to 1.9 mole ethanol per mole sugar. In addition, both organisms utilize starch, cellobiose, various hexoses and the various pentoses derived from hemicellulose hydrolysis. With all the substrates, ethanol is the main fermentation product. The cellobiose concentration in the cellulolytic co-culture may play an important role due to regulatory effects on the cellulase activity. A pH-shift is not required for high ethanol production with *T. ethanolicus*. Thus this bacterium presently seems to be an ideal organism for co-cultures with thermophilic, cellulose degrading bacteria[19,30,32]. It produces ethanol between pH 4.4 and 9.8; the pH optimum for growth and the ethanol production rate is between 5.8 and 8.5. This unusually broad pH-optimum suits it for an industrial application especially in combinations with cellulolytic co-cultures since in these cultures the pH-value always drops due to the concomitant production of lactate and acetate. The temperature range for growth and ethanol production is from 38 to 78 °C. The ethanol yield does not change drastically with the fermentation temperature. Strain JW 200 has been proved to adapt easily to higher ethanol (8%) concentrations. More than 1.4 mole ethanol per mole glucose equivalent of the cellulose degraded was obtained in a co-culture with *C. thermocellum* JW 20[33]. Both organisms need yeast extract for growth and for a high ethanol yielding fermentation. This requirement can easily be fulfilled by the addition of the stillage from yeast ethanol fermentation. Presently this stillage is used as animal feed, either directly or after an enrichment with protein, using pentose utilizing yeasts; the distillers' solubles are often converted to fodder by concentration. Since *Thermoanaerobacter* converts pentoses and starch directly to ethanol, the additional stillage would also increase the fermentable C-source and thus the ethanol yield.

The application of the extreme thermophilic cellulolytic and glycolytic bacteria in co-cultures makes possible a continuous fermentation process from cellulose to ethanol. Both organisms have a temperature optimum of about 68 °C, but still exhibit a rapid metabolic activity at 72/74 °C. The boiling point of the ethanol-water azeotrope is 78.2 °C. Subsequently, only an oxygenfree stream of N_2 or a very low vacuum is necessary to separate the ethanol from the fermentation broth. The removal of ethanol from the fermentation broth through extraction procedures with other solvents to save destillation energy, as proposed by Dellweg and Misselhorn[34], does not seem feasible for thermophilic processes using waste materials. The fermentation broth is already at an elevated temperature and other interfering materials, possibly poisonous, might also be extracted from the wastes. Many of the energy cost tied to the thermophilic ethanol fermentation process could be recovered by producing biogas from the remaining residues (distillers' wastes).

Even at high conversion ratios of the cellulose and other additional carbohydrates to ethanol, the outflow or the stillage still has a high content of proteins, volatile fatty acids, lactate and other organic compounds. Many of these are methanogenic substrates or could be converted into the same by other thermophiles. One example is the production of methane and acetate by a co-culture of *Desulfotomaculum nigrificans* (lactate to acetate, CO_2 and 2 H_2) and *Methanobacterium thermoautotrophicum* (CO_2 and 4 H_2 to methane). The second product of this co-culture, acetate, can be converted to methane by a thermophilic *Methanosarcina* spec. as isolated by Schobert (Jülich, West Germany; pers. commun.). From the higher volatile fatty acids, methanogenic acetate can be produced by acetogenic bacteria. In addition, many amino acids can be converted to acetate, CO_2 and H_2 by not yet identified thermophiles (unpublished results). The weight yield of methane generation can be about 0.23 g per g total solids in a conventional ethanol stillage, which contains up to 20% solids[5]. A direct thermophilic fermentation of the distillation residues (outflow of continous culture) to methane, would lead to a clean energy source and would diminish the waste production. The obtained methane could be used for heating the fermentation vessels, if neccessary, for the distillation process or for drying the distillation waste. If the latter one contains higher amounts of lignin, the dried residues may also be used for heating processes. (Dry lignin has up to 3500 kJ/kg).

The combined process of the ethanol and methane fermentation by thermophilic organisms is summarized in the figure. It should be possible to increase the net energy conversion rate of cellulose and biomass far above 50%[5,3].

Conclusions. Cellulose, obtained either as waste or as specially grown biomass, can be efficiently converted to ethanol by thermophilic, anaerobic co-cultures *(C. thermocellum* and *T. ethanolicus).* In addition to defined mixed cultures containing acetogenic and methanogenic thermophiles, a minimum of residual waste can be obtained. The direct microbial conversion of cellulose to ethanol has many advantages over the two-step process consisting of a yeast fermentation after the cellulose is hydrolyzed by chemical treatments or by fungal cellulase preparations. Much more work has to be done, including studies with actual waste under technical conditions, before a final judgement can be made. But the direct process seems one of the most promising alternatives for ethanol production from cellulosic waste.

* Acknowledgment. Part of this work was supported by Energy and Research Development Administration contract number EY-76-509-0888-M003, and by the Deutsche Forschungsgemeinschaft.

1 B. Finnerty, in: Microbial Energy Conversion, p. 83. Ed. H.G. Schlegel and J. Barnea. Erich Golze KG, Göttingen 1976.
2 B. Berg, Archs Microbiol. *118*, 61 (1978).
3 L.A. Spano, in: Microbial Energy Conversion, p. 157. Ed. H.G. Schlegel and J. Barnea. Erich Goltze KG, Göttingen 1976.
4 H. Sahm, in: Rothenburger Symposium, p. 75. Braun AG, Melsungen 1978.
5 The National Biomass Program, 3rd Annual Biomass Energy System Conference Proceedings, SERI/TP 33–285 (1979).
6 G. Halliwell, Prog. ind. Microbiol. *15*, 1 (1979).
7 T.K. Ghose, in: Bioconversion of Cellulosic Substances into Energy, Chemicals and Microbial Protein, p. 599. New Delhi 1977.
8 R.E. Hungate, The Rumen and its Microbes. Academic Press, New York 1967.
9 M. Linko, in: Microbiology applied to Biotechnology; Dechema Monographie No. 83, p. 209. Verlag Chemie, Weinheim/New York 1979.
10 A.E. Humphrey, A. Moreira, W. Armiger and D. Zabriskie, Biotech. Bioengng Symp. *7*, 45 (1977).
11 D.S. Chaha, J.E. Swan and M. Moo-Young, Devs ind. Microbiol. *18*, 433 (1977).
12 T.C. Rexen, Animal Fd Sci. Technol. *1*, 73 (1976).
13 Y.W. Han and C.D. Callihan, Appl. Microbiol. *27*, 159 (1974).
14 G.H. Grant, Y.W. Han and A.W. Anderson, Appl. environ. Microbiol. *35*, 549 (1978).
15 M.R. Ladisch, C.M. Ladisch and G.T. Tsao, Science *201*, 743 (1978).
16 E.T. Reese and M. Mandels, Biotechnol. Bioengng *22*, 323 (1980).
17 R.F. Gomez, in: Proc. Colloque Cellulolyse Microbienne, p. 177, Marseille 1980.
18 I.C. Wang, I. Biocic, H.-Y. Fang and S.-D. Wang, in: Proc. 3rd Annual Biomass Energy System Conference, SERI/TP 33–285 (1979).
19 J. Wiegel, Experientia *36*, 1434 (1980).
20 J.E.L. Corry, J. Bact. *44*, 1 (1978).
21 J.G. Zeikus, Env. Microbiol. Tech. *1*, 243 (1979).
22 M. Tansey, ASM-News *45*, 417 (1979).
23 S.L. Rosenberg, Mycologia *70*, 1 (1978).
24 W.D. Belamy, ASM-News *45*, 326 (1979).
25 J. Wiegel, in preparation.
26 C.L. Cooney, D.I.C.Wang, S.D. Wang, I. Gordon and M. Jiminez, Biotechnol. Bioengng Symp. *8*, 103 (1979).
27 D.V. Garcia-Martinez, A. Shinmyo, A. Madia and A.L. Demain, Eur. J. appl. Microbiol. *9*, 189 (1980).
28 N.D. Sjolander, J. Bact. *34*, 419 (1937).
29 E.J. Hsu and Z.J. Ordal, J. Bact. *102*, 369 (1970).
30 J. Wiegel and L.G. Ljungdahl, in: Technische Mikrobiologie, p. 117. Ed. H. Dellweg. Verlag Versuchs- und Lehranstalt für Spiritusfabrikation und Fermentationstechnologie im Institut für Gärungsgewebe und Biotechnologie, Berlin 1979.
31 J. Wiegel, L.G. Ljungdahl and J.R. Rawson, J. Bact. *139*, 800 (1979).
32 J. Wiegel and L.G. Ljungdahl, Archs Microbiol. *128*, 343 (1981).
33 L.G. Ljungdahl and J. Wiegel, USA patents 4.292.406 and 4.292.407 (1981).
34 H. Dellweg and K. Misselhorn, in: Microbiology applied to Biotechnology; Dechema Monographie No. 83, p. 35. Verlag Chemie, Weinheim/New York 1979.
35 H.H. Dietrichs, Holzforschung *32*, 193 (1978).
36 S.I. Aronovsky and R.A. Gortner, Indian Engng Chem. *28*, 1270 (1936).

Degradation of cellulose

by Karl-Erik Eriksson

Swedish Forest Products Research Laboratory, Box 5604, S-11486 Stockholm (Sweden)

Microorganisms degrading cellulosic materials

One of nature's most important biological processes is the degradation of lignocellulosic materials into carbon dioxide, water and humic substances. Different kinds of microorganisms are involved in the process of degrading woody materials, but it is mainly a task for fungi. Bacteria are considered to have only a limited capability of wood degradation. The strong wood-degrading effect that fungi have has to do, in part, with the organization of their hyphae which gives the organisms a capacity for penetration.

Different types of fungi give rise to different types of wood rot. One normally distinguishes between soft-rot, brown-rot and white-rot fungi[1]. Fungi from the first 2 groups mainly attack the polysaccharides of wood and other lignocellulosic materials while the white-rotters also are capable of a substantial attack on the lignin. The degradation of the different compounds in lignocellulosic materials is catalyzed by enzymes produced by the respective microorganisms. Knowledge of these reactions may be of importance for the conversion of biomass into chemicals and fuels.

The enzyme mechanisms for cellulose degradation by fungi are known in great detail and will be summarized below. The corresponding enzyme mechanisms for lignin degradation are less known and will not be subject to description in this article.

Enzyme mechanisms involved in cellulose degradation

The enzyme mechanisms involved in cellulose degradation have been particularly well studied in 2 fungi, namely the white-rot fungus *Sporotrichum pulverulentum*[2] and the mold *Trichoderma reesei*[3] (the fungus *T. viride* QM6a and strains derived from it are now referred to as *T. reesei*).

The fungus *S. pulverulentum* produces 3 different types of hydrolytic enzymes, namely a) 5 different endo-1,4-β-glucanases which attack at random the 1,4-β-linkages along the cellulose chain; b) 1 exo-1,4-β-glucanase which splits off cellobiose or glucose units from the non-reducing end of the cellulose; c) 2 1,4-β-glucosidases which hydrolyze cellobiose and water-soluble cellodextrins to glucose and cellobionic acid to glucose and gluconolactone[2].

It has been generally accepted that essentially the same picture is also true for cellulose hydrolyses by *T. reesei*[4]. However, a few differences have been recognized such as the number of the various hydrolytic enzymes, the degree to which the β-glucosidase activity is bound to the fungal cell wall, etc. The action of the exo-glucanase in *S. pulverulentum* differs from the action of the corresponding enzymes in *T. reesei* in that the exo-glucanase from *S. pulverulentum* splits off both glucose and cellobiose, while the exo-glucanases from *T. reesei* only split off cellobiose[3]. However, to degrade crystalline cellulose a synergistic action between endo-1,4-β- and exo-1,4-β-glucanases seems necessary for both fungi. Crystalline cellulose is not attacked by one of these types of enzymes alone. Amorphous cellulose is, however, degraded by both types of enzymes separately[2,3]. In a recent paper by Gritzali and Brown[5] a much simpler enzymic pattern of the fungus *T. reesei* QM9414 has been suggested compared with that previously found[4]. Using a different cultivation technique, only 1 endo-glucanase and 2 exo-glucanases were obtained[5]. Instead of being grown on cellulose, glucose was used as a carbon source. After washing, the glucose grown cells were transferred to a buffer where the hydrolytic cellulose degrading enzymes were induced by the addition of sophorose. Polyacrylamide gel electrophoresis of the concentrated culture solution revealed that the separation pattern of cellulases induced by sophorose is much simpler than the separation pattern of enzymes obtained after several days of cultivation on cellulose. It seems therefore likely that the apparent multiplicity of cellulases in cellulose culture is due to protein modification by proteases. Indeed, Nakayama et al.[6] reported that partial proteolysis of an endo-glucanase from *T. viride* yielded enzymes with a changed substrate specificity and protein structure. In *S. pulverulentum* the 5 endo-glucanases are very similar in molecular weight, amino acid composition, etc. However, they differ somewhat in function[7]. Recent investigations of culture solutions after growth of *S. pulverulentum* on cellulose have demonstrated the existence of 2 different proteases, 1 of carboxy-peptidase and the other of chymotrypsin type. These enzymes seem to influence the release of endo-1,4-β-glucanases from the fungal cell wall and also appear to modify the fungal cell wall[8,9]. Whether or not these enzymes are responsible for the multiplicity of endo-glucanases in *S. pulverulentum* is not known. However, the recent finding of Gritzali and Brown[4] concerning the very simple enzyme picture in *T. reesei* QM 9414 when the cellulases are induced by sophorose in a short term culture points to this possibility. An investigation into the effect of similar cultivation conditions on the endo-glucanase pattern in *S. pulverulentum* will be undertaken.

In *S. pulverulentum* an oxidative enzyme of importance for cellulose degradation has been discovered in addition to the hydrolytic enzymes described above[10]. The enzyme has been purified and characterized and found to be a cellobiose oxidase, which oxidizes cellobiose and higher dextrins to their corresponding onic acids thereby using molecular oxygen. The enzyme is a hemoprotein and also contains a FAD group. It is not yet known whether this enzyme also oxidizes the reducing end group formed when 1,4-β-glucosidic bonds are split through the action of endo-glucanases. It was recently reported by Vaheri[11] that cultures of *T. reesei* grown on cellulose also contained gluconolactone, cellobionolactone and cellobionic acid. These findings indicate that *T. reesei* also produces an oxidative enzyme involved in cellulose degradation. However, further confirmation is necessary in this case.

The fungus *S. pulverulentum* has 3 means of converting cellobiose. The first is hydrolysis by the 1,4-β-glucosidases. The second is through the already described enzyme cellobiose oxidase and the third is through the enzyme cellobiose: quinone oxidoreductase[12,13]. This enzyme is of importance for the degradation of both cellulose and lignin. Although the enzyme seems to be involved in both lignin and cellulose degradation, the highest yields of the enzyme were obtained when cellulose powder was used as a carbon source. In *S. pulverulentum* development of cellobiose: quinone oxidoreductase activity and cellulolytic activity occurred simultaneously. The enzyme is relatively specific for its disaccharide substrate, while the requirements on the quinone structure are less specific and the enzyme is able to reduce both ortho- and para-quinones. A reaction scheme for the enzyme is presented by Eriksson[2] where also a total reaction scheme for cellulose degradation in *S. pulverulentum* is given.

Regulation of endo-1,4-β-glucanases in the white-rot fungus *S. pulverulentum* has recently been investigated using a newly developed sensitive method[14]. The method is based upon the viscosity lowering effect of endo-1,4-β-glucanases on solutions of carboxy-methyl cellulose (CMC). The effect of inducers and repressors can be determined with the method as well as whether the enzymes are localized on cell wall surfaces or actively released into the surrounding medium. The results show that cellobiose causes induction of endo-1,4-β-glucanases at concentrations as low as 1 mg/l. It was also shown that glucose causes catabolite repression of enzyme formation at concentrations as low as 50 mg/l. Mixtures of inducer and repressor give rise to a delayed enzyme production compared with solutions of inducer only.

Studies of the mold *T. reesei* QM6a using the same technique show that cellobiose under our conditions was not an efficient inducer. However, sophorose causes induction of endo-1,4-β-glucanases at a concentration of 1 mg/l. The comparison between the regulation of endo-1,4-β-glucanase production in the 2 fungi also demonstrates several other important differences. Thus, a solution of CMC alone induces enzyme formation in *S. pulverulentum* but not in the *T. reesei* strain. Under our experimental conditions no endo-1,4-β-glucanases were actively excreted into the solution by *T. reesei.* The enzymes were bound to the cell wall. However, *S. pulverulentum* released the enzymes into the medium although they first appeared bound to the cell wall. It was recently shown by Gritzali and Brown[5] that sophorose gives rise to active excretion of endo-1,4-β-glucanases into the culture solution of *T. reesei* QM 9414. The differences in the results of the two studies must be due either to the differences in the fungal strains or in cultivation conditions.

Production of fuels and chemicals from cellulosic materials

The increasing pressure upon the fossil fuel resources has given rise to a world-wide interest in the production of liquid fuels and chemicals from renewable resources. An area which at present attracts very special attention is the production of ethanol by fermentation of sugar from lignocellulosic materials. Cellulose can be hydrolyzed to soluble sugars by acids or enzymes. Drawbacks of the acid processes include low yields due to degradation of the products, production of inhibitors affecting the ethanol formation by the yeast, corrosion of the equipment and high capital costs. It therefore seems as if, in the future, an enzymic hydrolysis process would be preferred. If this will be the case, it is an absolute requirement that the lignocellulosic material first be delignified. Different methods exists to do so and one example is the so called IOTECH process[15].

A prerequisite for enzymatic saccharification of cellulosic materials is that cellulose and hemicellulose degrading enzymes be cheaply produced. Production can be achieved by cultivation of fungi as well as of bacteria. Enzyme productivity can be enhanced by several means. Primarily, it can be brought about by selection. Genetic manipulation can give rise to strains of microorganisms hyperproducing with respect to one or several of the 3 different hydrolytic enzymes involved in cellulose degradation. It also seems reasonable to assume that DNA-hybridization techniques will be used in the future for large-scale production of these saccharification enzymes.

The development that hitherto has taken place in the area 'hyperproduction of cellulases' has been primarily in the United States[16] and also in Finland[17]. Mutations to create hyper-producing strains have been carried out mainly with the fungus *T. reesei.* DNA hybridization techniques for hyper-production of cellulases have been instituted at the Massachusetts Institute of Technology (MIT) in the United States and in the Biotechnical Laboratory at the Technical Research Center of Finland in Helsinki.

In addition to the above described enzymatic degradation of cellulose for conversion of the produced glucose to ethanol, another development is taking place. At MIT, cellulosic materials are directly, in 1 fermentation step, converted to ethanol. For this purpose, the thermophilic bacterium *Clostridium thermocellum* is used. *C. thermocellum* grows at 60 °C and degrades cellulose to glucose which, by the same organism is converted to ethanol. Another interesting bacterium, also studied at MIT, is *Clostridium thermosaccharolyticum.* As is clear from the name, also this bacterium is thermophilic but instead of using cellulose as substrate, it can convert xylan to ethanol. The ethanol concentrations obtained with these 2 bacterial strains are around 5%. Still higher concentrations are, however, expected to be reached[18].

Another interesting project concerning ethanol production from lignocellulosic materials is taking place at the University of Georgia in Athens, Georgia. Also here the work is carried out with 2 different *Clostridium* strains, namely *C. thermohydrosulphuricum* and another bacterium which is not yet named. This bacterium has been isolated from hot wells in Yellowstone National Park in USA. Both of these bacteria are extremely thermophilic and function up to temperatures of 78 °C, i.e. in the immediate vicinity of the boiling point of ethanol (78.6 °C).

C. thermohydrosulphuricum can ferment glucose, mannose, galactose, ribose, raffinose, xylose, cellobiose, sucrose, starch and pectin. However, cellulose or xylan cannot be utilized by this strain. The products formed are ethanol, lactic acid and acetic acid. The yields of ethanol are as high as 1.5 moles ethanol/mole sugar if the fermentation is carried out at 72 °C. The thermophilic bacterium which has been isolated from hot wells in Yellowstone differs from *C. thermohydrosulphuricum* by not forming spores and by producing practically the theoretical amount of ethanol, i.e. almost 2 moles/mole fermented glucose. Lactic acid, acetic acid and hydrogen are produced as by-products. By using a mixed culture of, for instance, the bacterium *C. thermocellum* and one of the thermophilic ethanol producing bacteria, ethanol can be produced directly from cellulose. Mixed cultures have given rise to 1.4 moles ethanol/mole glucose[18].

Other biotechnical processes based on cellulosic materials

The utilization of cellulosic materials in biotechnical processes naturally depends upon the development of economically feasible processes. 2 fermentation processes for the production of fodder protein based on cellulosic waste materials from forest industries are already in use and may serve as examples. The first process is founded on the yeast *Candida utilis*, and the other on the fungus *Paecilomyces varioti*. Both of these processes have in common that the substrate is mainly monosaccharides in spent sulfite liquor. Disaccharides and higher oligosaccharides are utilized only to a very limited extent. However, on the basis of the knowledge gained in the studies of the enzymatic degradation of cellulose a fermentation process based on the white-rot fungus *S. pulverulentum*[19] has been developed. The process allows fermentation of solid as well as dissolved lignocellulosic waste and gives, as products, purified water and protein. It seems possible to use the developed technique for the total closure of the white-water system in a newsprint paper mill or in a fiberboard mill. The process has recently been studied in pilot plant scale. The results show no build-up of organic matter taking place and, thus, that all dissolved substances of lignocellulosic origin dissolved from wood in mechanical grinding can be utilized by the white-rot fungus. The fungal mycelium produced in the process can be used either as animal feed[20] or be added to the paper. Both possibilites have been tested and found feasible.

With the present escalation of oil prices it can be foreseen that renewable resources, mainly cellulosic materials, will substitute petroleum as raw material although the extent to which, cannot be surveyed at present. Knowledge of the metabolic pathways of microbial cellulose degradation will, with certainty, play an important role in this evolution.

1 A.A. Käärik, Decomposition of Wood, in: Biology of Plant Litter Decomposition, p. 129. Ed. C.H. Dickinson and G.J.F. Pugh. Academic Press, London 1974.
2 K.-E. Eriksson, Biotechnol. Bioengng *70*, 317 (1978).
3 D.D.Y. Ryu and M. Mandels, Enzyme microb. Technol. *2*, 91 (1980).
4 G.H. Emert, E.K. Gum, Jr, J.A. Lang, T.H. Ling and R.D. Brown, Jr, Adv. Chem. Ser. *136*, 76 (1974).
5 M. Grizali and R.D. Brown, Jr, Adv. Chem. Ser. *181*, 237 (1979).
6 M. Nakayama, Y. Tomita, H. Suzubi and K. Nisizawa, J. Biochem. *79*, 955 61976).
7 M. Streamer, K.-E. Eriksson and B. Pettersson, Eur. J. Biochem. *59*, 607 (1975).
8 K.-E. Eriksson and B. Pettersson, Int. Symp. Wood Pulping Chem. Stockholm *3*, 60 (1981).
9 K.-E. Eriksson, A. von Hofsten and B. Pettersson, to be published.
10 A.R. Ayers, S.B. Ayers and K.-E. Eriksson, Eur. J. Biochem. *90*, 171 (1978).
11 M. Vaheri, Nordforsk's Workshop, May 6–7, 1980. VTT:S Biotechnical Laboratory, Otaniemi, Helsinki.
12 U. Westermark and K.-E. Eriksson, Acta chem. scand. *B28*, 204 (1974).
13 U. Westermark and K.-E. Eriksson, Acta chem. scand. *B28*, 209 (1974).
14 K.-E. Eriksson and S.G. Hamp., Eur. J. Biochem. *90*, 183 (1978).
15 R.H. Marchessault and J. St. Pierre, Chemrawn Conf., Toronto 1978.
16 B.S. Montenecourt and D.E. Eveleigh, Proc. 2nd Annual Symp. on Fuels from Biomass, Troy, N.Y., 1978. TP 613–625.
17 K.M.H. Nevalainen, E.G. Palva and M.I. Bailey, Enzym. microb. Technol. *2*, 59 (1980).
18 Chem. Engng News, Sept. 17, 1979; p. 27.
19 M. Ek and K.-E. Eriksson, Biotechnol. Bioengng *22*, 2273 (1980).
20 S. Thomke, M. Rundgren and S. Eriksson, Biotechnol. Bioengng *22*, 2285 (1980).

Biodegradation of lignin: Biochemistry and potential applications

By Takayoshi Higuchi

Wood Research Institute, Kyoto University, Uji, Kyoto (Japan)

Introduction

An estimated 65% of our biomass is produced on land. Of that biomass, lignin is the most abundant natural polymer next to cellulose and is an important renewable source of aromatic carbon on earth. Since lignin and cellulose, together with the hemicelluloses, are the structural components of the vascular tissues of higher land plants, biodegradation of the vascular tissues is the key process in the recycling of terrestrial biosynthetic carbon.

However, lignin, which is a heterogeneous aromatic polymer containing various biologically stable carbon-to-carbon and ether linkages, is interspersed with hemicelluloses surrounding cellulose microfibrils, resulting in an organic composite material protected from the degradative enzymes of microorganisms. Therefore, elucidation of the lignin biodegradation process is essential for understanding the circumstances of the recycling of carbon on earth, for establishing technology for bioconversion of plant residues and waste lignins to useful materials, and for protecting the environment from lignin-related pollutants.

Biochemistry of lignin biodegradation

Lignins have been shown to be dehydrogenative polymers of p-hydroxycinnamyl alcohols such as p-coumaryl (**1**), coniferyl (**2**) and sinapyl (**3**) alcohols:

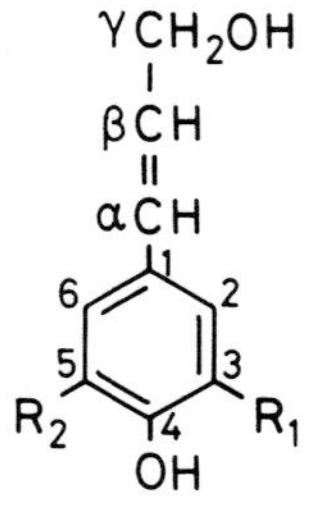

(1) $R_1 = R_2 = H$
(2) $R_1 = OCH_3$, $R_2 = H$
(3) $R_1 = R_2 = OCH_3$

gymnosperm lignin is formed from coniferyl alcohol; angiosperm lignin from mixtures of coniferyl and sinapyl alcohol; and grass lignin from mixtures of coniferyl, sinapyl and p-coumaryl alcohols. These p-hydroxycinnamyl alcohols are dehydrogenatively polymerized by the mediation of peroxidases via radical coupling and subsequent nucleophilic reactions of various nucleophiles to quinonemethide intermediates, which results in *β-β′*, *β-5′*, *β-1′*, *β-O-4′*, 5-5′ and 3-O-4′ linkages between phenyl-propanoid units[1] (fig. 1).

Investigations have indicated that white-rot fungi (Basidiomycetes) and closely related litter-decomposing fungi are dominant lignin-degraders; the brown-rot fungi (Basidiomycetes) cause limited degradation of lignin[2,3]; and soft-rot fungi (Fungi imperfecti), various soil Ascomycetes and Fungi imperfecti *(Fusarium, Aspergillus)* as well as some bacteria *(Nocardia, Streptomyces)* partly degrade lignin. Kirk et al.[4] recently found that the addition of cellulose or glucose is necessary for the decomposition of synthetic [ring-^{14}C]-lignin (DHP) to $^{14}CO_2$ by *Phanerochaete chrysosporium* (a white-rot fungus). Lignin cannot serve as a growth substrate and has no influence on synthesis of the ligninolytic system, which simply appears as part of secondary metabolism in the fungus incubated in a medium of excess carbon source (glucose or cellulose) with a limiting amount of nitrogen source (ammonium salts and some amino acids). The oxygen concentration strongly influences the rate and extent of lignin degradation: stationary cultures (10 ml/125-ml Erlenmeyer flask) at 37–39 °C, pH 4.5 in an atmosphere of approximately 60% O_2 in N_2 gave maximum rates of lignin degradation (ligninolytic culture). Culture agitation resulting in pellet formation greatly suppressed lignin degradation at all O_2 levels. These investigations of the physiological parameters of lignin catabolism have greatly contributed to establishing a sensitive and reproducible assay method in the study of lignin biodegradation.

Chemistry of degraded lignin by white-rot fungi

Analytical comparisons[5] of sound and white-rotted lignins extracted with various solvents have shown that the degradation process is heavily oxidative:

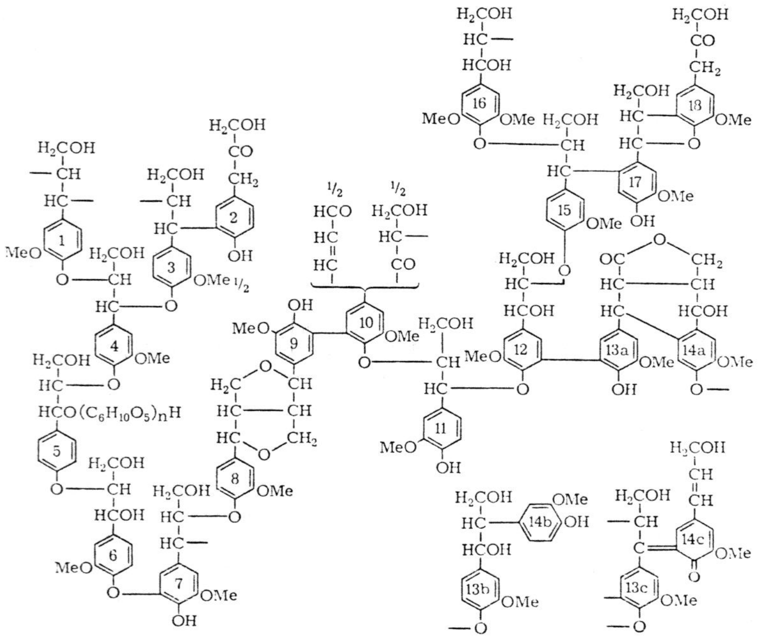

Figure 1. Schematic constitution of spruce lignin (Freudenberg)[1].

degradation of lignin polymer occurs in the side chains, which are oxidized with the formation of α-carbonyl and α-carboxyl groups, and in the aromatic nuclei, which are oxidatively cleaved following demethylation and the introduction of hydroxyl groups in phenolic units to give 2,3- and/or 3,4-dihydroxyphenyl moieties (table). The degraded lignin contained α,β-unsaturated carboxyl groups which were not derived from side chains but presumably from aromatic rings: relative intensity of the 1515 cm^{-1} band due to the aromatic ring in the IR-spectrum was considerably low in decayed lignins in accord with ring cleavage in the lignin polymer. The degradation study of ^{13}C-DHP by various lignin degraders[6] supports the findings on the degradation of aromatic nuclei in the lignin polymer. The low molecular weight fractions extracted from the decayed spruce wood by *P. chrysosporium* contained several aromatic acids (**4–10**), among which vanillic acid (**4**) was by far the most abundant. All of these acids obviously involved $C\alpha$-$C\beta$ cleavage, which resulted either in the direct formation of the aromatic acids, or was followed by oxidation to then yield the acids. Traces of several compounds, each containing an intact ring attached to an aliphatic residue which clearly was formed via oxidative cleavage of a 2nd aromatic ring, have been tentatively identified by GC-MS (fig. 2 (**9**), (**10**))[7].

These results seem to suggest that lignin is degraded by white-rot fungi primarily via a few key reactions; a) demethylation of methoxyl groups, b) hydroxylation at C-2 in aromatic rings as found in the decayed lignin by brown-rotted lignin, c) aromatic ring cleavages in lignin polymers (intradiol 2,3- and 3,4-), with the subsequent release of aliphatic products, and d) $C\alpha$-$C\beta$ cleavage, which is perhaps a major reaction releasing both aliphatic and aromatic products. This picture suggests that various lignin structural units are degraded by the mediation of extracellular enzymes which attack both low molecular weight and polymeric substrates via many intermediate products along several different pathways.

Degradation of dilignols

The use of low molecular weight compounds of a specific substructure (2-unit segments containing specific types of interunit linkage) in lignin is indispensable in elucidating the details of the reactions involved in the degradation of various linkages in lignin macromolecules. Thus, investigations have been done on the degradation of dilignols by *Fusarium solani* M-13-1 which was isolated by an enrichment technique using DHP as the sole carbon source and by *P. chrysosporium*. On being added to the shaking culture of *Fusarium*, guaiacylglycerol-β-coniferyl ether (β-O-4 substructure model, 40–65% in lignins) (**11**) was degraded to (**15**) via compounds (**12**), (**13**) and (**14**). The compound (**15**) was then degraded to (**16**) which was converted to (**17**) and/or (**18**). 2,6-Dimethoxy-p-benzoquinone (**19′**) was isolated when syringylglycerol-β-vanillic acid ether (**15′**) was used as a substrate, which indicates that the β-vanillic acid ethers are degraded via alkylphenyl cleavage as shown in figure 3[8-10]. However, cleavage of the β-O-4 dilignol linkage to β-hydroxypropiovanillone and

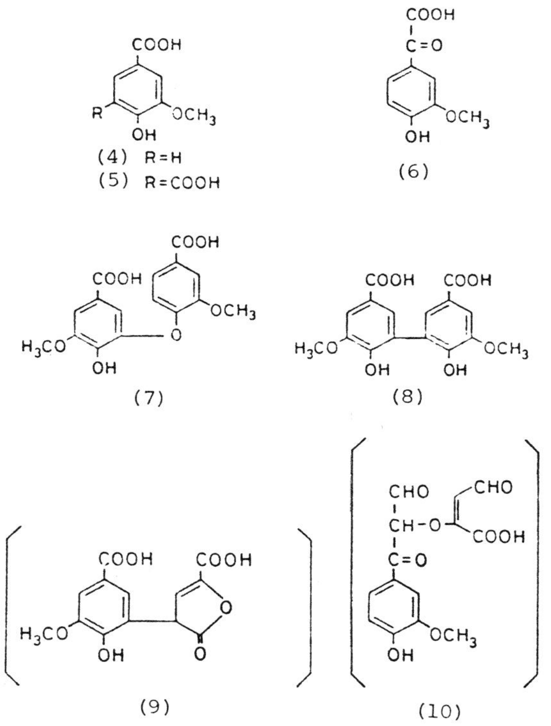

Figure 2. Aromatic acids identified in extracts of white-rotted spruce wood.

Analytical properties of lignin isolated and purified from spruce wood before and after decay by white-rot fungi (Chang et al.)[5]

Lignin	Formula for average C_9-unit	Functional groups (moles/C_9-unit) Conjugated carbonyl	Total carboxyl	Hydroxyl Phenolic	Aliphatic	Total
Sound	$C_9H_{8.66}O_{2.75}[OCH_3]_{0.92}$	0.07	0.10	0.24	0.92	1.16
Decayed by						
Polyporus anceps	$C_9H_{7.70}O_{3.80}[OCH_3]_{0.72}$	0.16	0.59 (0.17)*	0.10	0.77	0.87
Coriolus versicolor	$C_9H_{7.26}O_{3.95}[OCH_3]_{0.74}$	0.17	0.55	0.11	–	–

* Aromatic acid as estimated by ^1H-NMR.

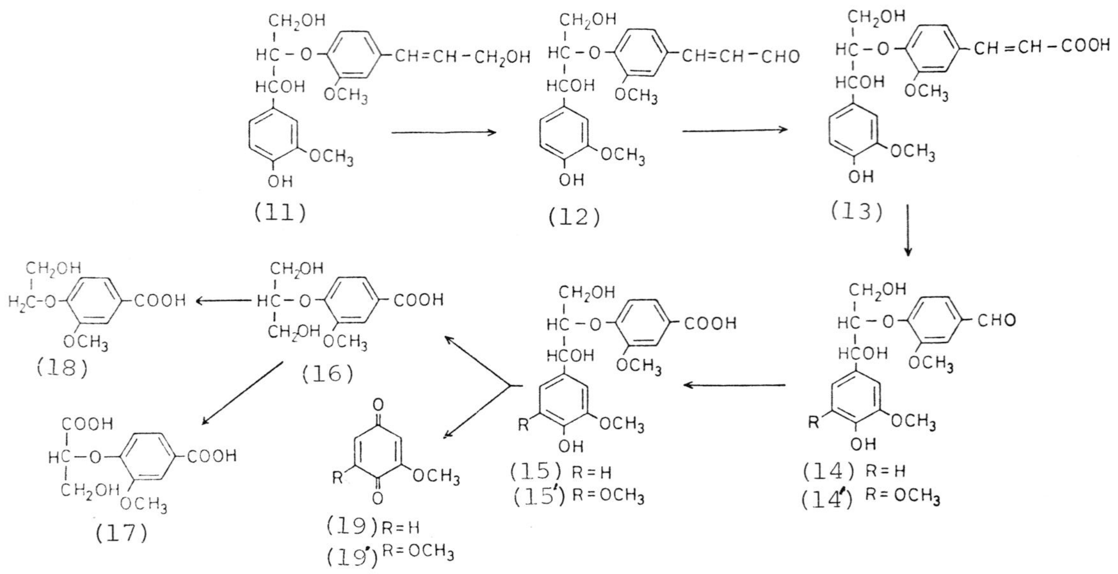

Figure 3. Degradation pathway of guaiacylglycerol-β-coniferyl ether by *Fusarium solani* M-13-1.

coniferyl alcohol was found in *Pseudomonas putida*[11,12].

On investigating *P. chrysosporium*[13] the non-phenolic β-O-4 compound **(20)** was converted to **(21)** which was then converted to **(22)** and **(23)**. The vanillic acid ether **(23′)** was converted to **(24)** via Cα-Cβ cleavage instead of alkyl-phenyl cleavage in phenolic β-O-4 model (fig. 4).

Recently, cleavage of the Cα-Cβ and/or alkyl-phenyl linkages in veratrylglycerol-β-guaiacyl ether, which is a β-O-4 model but not a true degrading intermediate, was also reported in white-rot fungus *(Phanerochaete chrysosporium)*[14] and in bacteria[15].

The β-5′ (phenylcoumaran) substructure model, dehydrodiconiferyl alcohol **(26)** was degraded by *F. solani* M-13-1 to **(29)** via **(27)** and **(28)**. The compound **(29)** was further oxidized to the compounds **(30)** and **(31)**[16]: phenylcoumaran-α′-aldehyde with syringyl group **(29′)** was converted to phenylcoumarone **(32)**[17]. The results suggest the participation of a dioxygenase in the cleavage of the coumarone ring to 5-acetylvanillyl alcohol **(30)** and vanillic or syringic acids (fig. 5). Enzymes which oxidize allyl alcohol groups in side chains of β-O-4 and β-5′ dilignols as well as of the lignin polymer have been found in the culture filtrate of *Fusarium*[18] and *Nocardia*[19], respectively.

On the other hand, in ligninolytic cultures of *P. chrysosporium*[20] the cinnamyl alcohol side chain of 4-O-methyl dehydrodiconiferyl alcohol **(33)** was degraded to **(35)** via the compound with the glycerol side chain

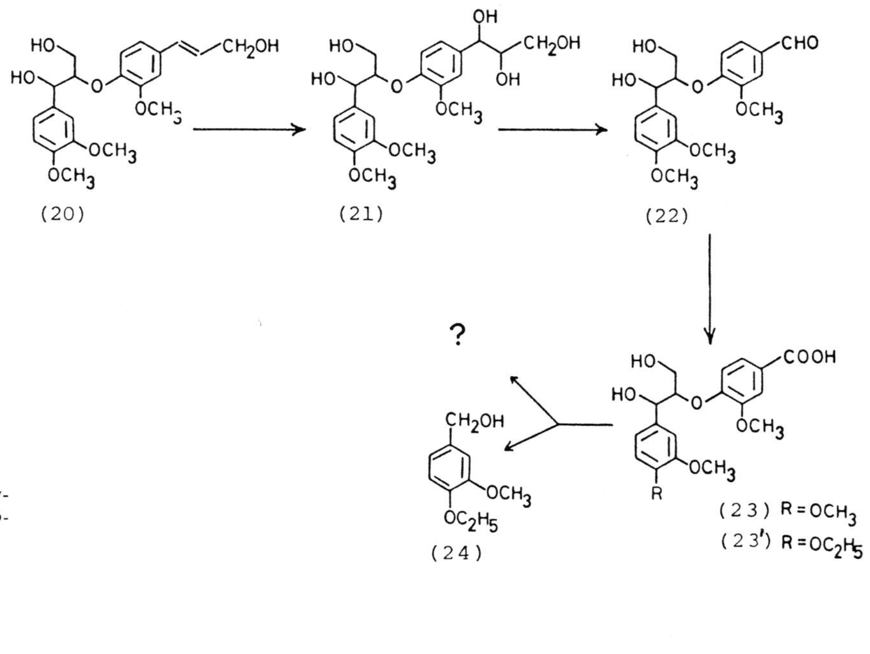

Figure 4. Degradation pathway of arylglycerol-β-aryl ether by *Phanerochaete chrysosporium*.

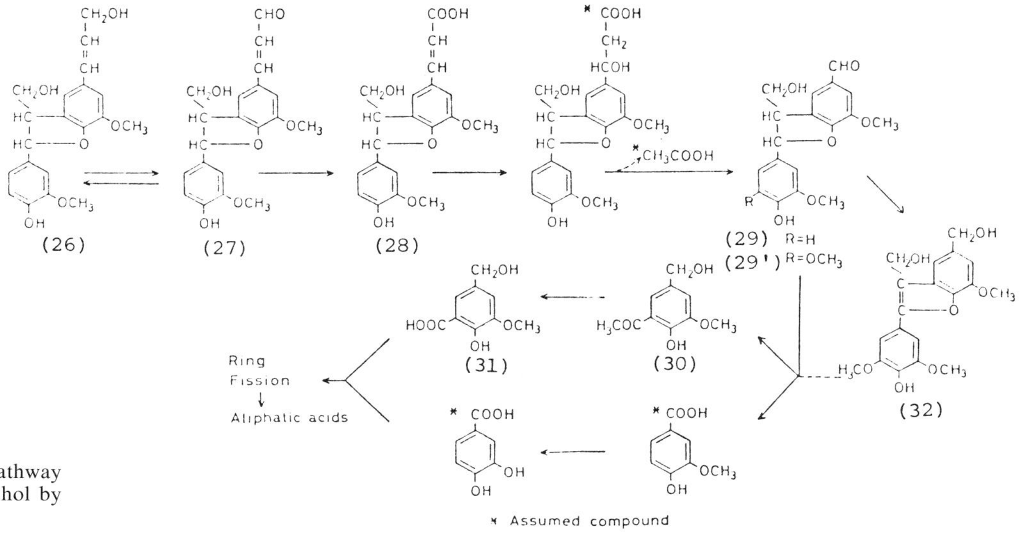

Figure 5. Degradation pathway of dehydrodiconiferyl alcohol by *Fusarium solani* M-13-1.

(**34**) as in the degradation of β-O-4 dilignol, and then (**35**) was degraded to 3-methoxy-4-ethoxybenzoic acid via phenylcoumarone (**36**) (fig. 6). Phenylcoumaran-α'-aldehyde with the syringyl group (**35'**) was converted to phenylcoumarones (**36'**) and α-hydroxyphenylcoumaran (**37**) which was degraded to 2,6-dimethoxy-p-benzoquinone (**19'**) and/or syringaldehyde[21]. Oxygenases seem to be involved in the formation of the glycerol side chain by this fungus.

The β-β' substructure model, syringaresinol monobenzyl ether (**38**) was converted by *F. solani* M-13-1[22] to (**39**), (**40**), (**41**) and 2,6-dimethoxy-p-benzoquinone via alkyl-phenyl cleavage (fig. 7). Dibenzyl and dimethyl ethers of syringaresinol were not degraded, which suggests that both phenolic hydroxyl groups are indispensable for complete degradation of the resinol by this fungus. Diguaiacylpropane-1,3-diol (β-1') substructure model (**42**) was degraded to (**43**), and (**44**), and from 1,2-disyringylpropane-1,3-diol (**42'**), (**43'**), 2,6-dimethoxy-p-benzoquinone (**19'**), (**45**) and syringaldehyde (**46**) were obtained, which again indicates that the compounds were degraded via alkyl-phenyl cleavage as illustrated in figure 8[23].

The degradation products of the main dilignols by *F. solani* M-13-1 and *Phanerochaete* showed the occurrence of the following degradation reactions[24]: 1. cinnamyl alcohol groups are oxidized to the corresponding cinnamic acids via cinnamaldehyde by *Fusarium* but are converted to glycerol groups by *Phanerochaete*, 2. both cinnamic and glycerol side chains are converted to the C6-C1 acid side chains via C6-Cl aldehydes (fig. 9), 3. α-hydroxydilignols such as arylglycerol-β-aryl ethers and diarylpropanes are degraded via cyclohexadienone radical derivatives to the corresponding hydroquinones and glyceraldehydes by alkyl-phenyl cleavage, 4. α-ether (alkoxy and phenoxy) dilignols such as pinoresinol and phenylcoumaran are first converted to α-hydroxy

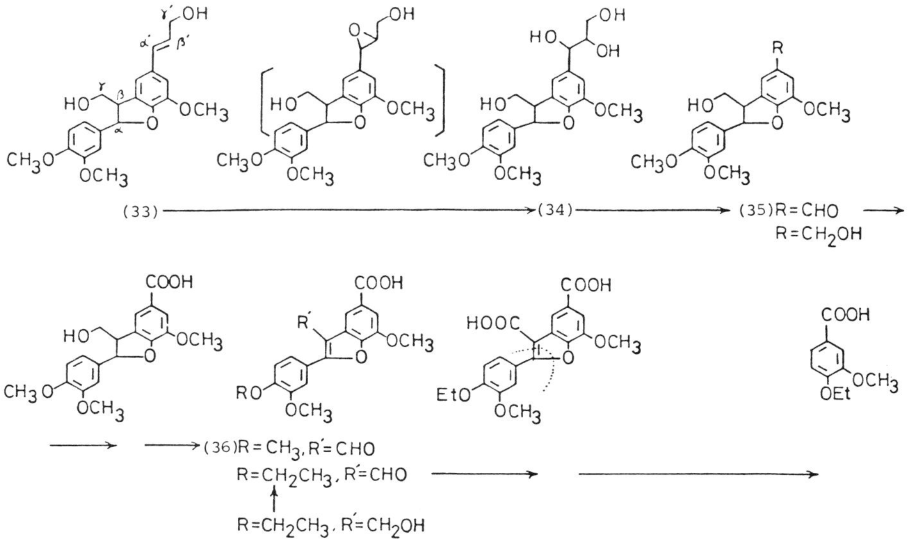

Figure 6. Degradation pathway of 4-O-methyl ether of dehydrodiconiferyl alcohol by *Phanerochaete chrysosporium*.

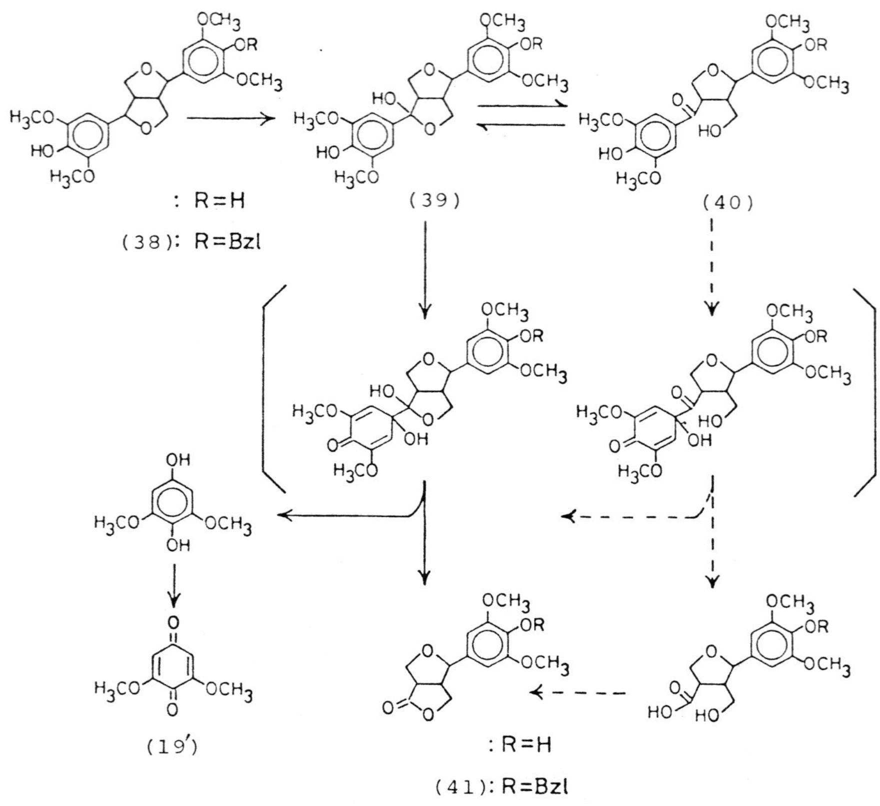

Figure 7. Degradation of syringaresinol by *Fusarium solani* M-13-1.

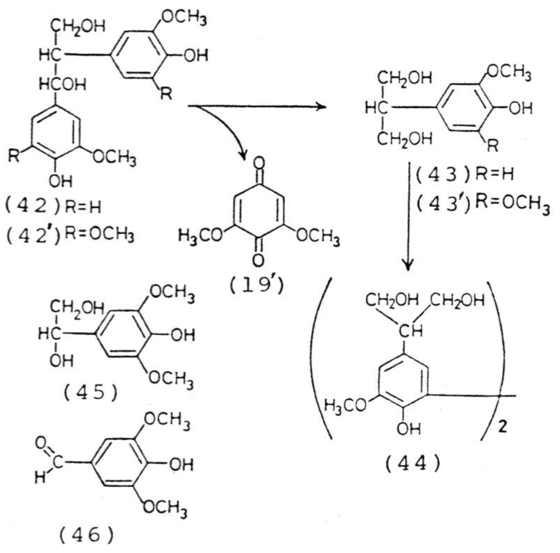

Figure 8. Degradation of 1,2-diarylpropane-1,3-diols by *Fusarium solani* M-13-1.

ethers via quinonemethide intermediates and then α-hydroxy ethers are degraded to the corresponding hydroquinones and lactone or ester derivatives by both fungi (fig. 10), 5. non-phenolic dilignols are not degraded by *Fusarium* but cleaved mainly between Cα and Cβ by *Phanerochaete*. Degradation processes of the phenolic dilignols with *Fusarium* and other white-rot fungi seem to be similar. Even in *Phanerochaete* phenolic dilignols are degraded much faster than non-phenolic dilignols.

Concluding remarks

Chemical analyses of degraded lignins and the degradation of dilignols lead to the conclusion that lignin undergoes simultaneous oxidative degradation of both aliphatic side chains and the aromatic nuclei from the surface of the lignin macromolecule. Aromatic moieties with free phenolic hydroxyl groups

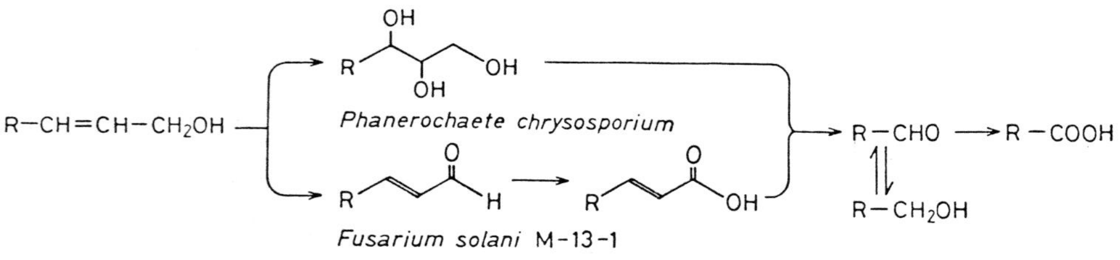

Figure 9. Oxidative degradation of allyl alcohol side chains of dilignols by *F. solani* M-13-1 and *P. chrysosporium*.

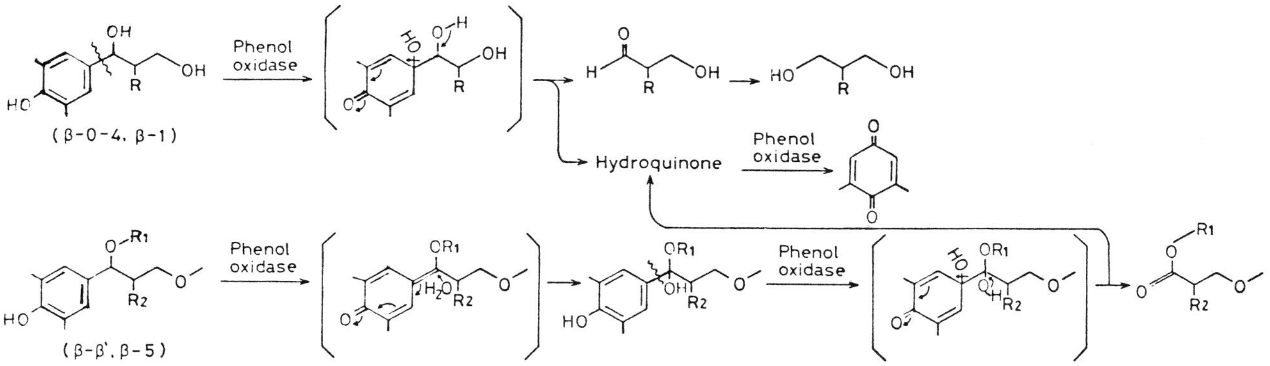

Figure 10. Alkyl-phenyl cleavage of phenolic dilignols by *F. solani* M-13-1 and *P. chrysosporium*.

(20–30% in lignin) would be preferentially attacked via alkyl-phenyl cleavage by the phenol oxidizing enzymes, giving moieties with newly formed phenolic groups which could again be oxidized by the enzymes. Non-phenolic aromatic moieties are mainly cleaved between Cα and Cβ and the aromatic products formed are ring cleaved in polymers. Aromatic alcohol dehydrogenases and monooxygenases are involved as key enzymes in the oxidative degradation of side chains, and dioxygenases are indispensable for cleavage of aromatic rings of lignin. Vanillic and syringic acids and hydroquinones usually found as degradation products of lignin could be metabolized following the decarboxylation of the acids, hydroxylations, and ring cleavage processes as found by Ander et al.[25] (fig. 11), and via humus polymers in soil[26]. We expect that the use of trilignols and tetralignols as a substrate will provide missing links in the degradation processes between dilignols and lignin polymers[24].

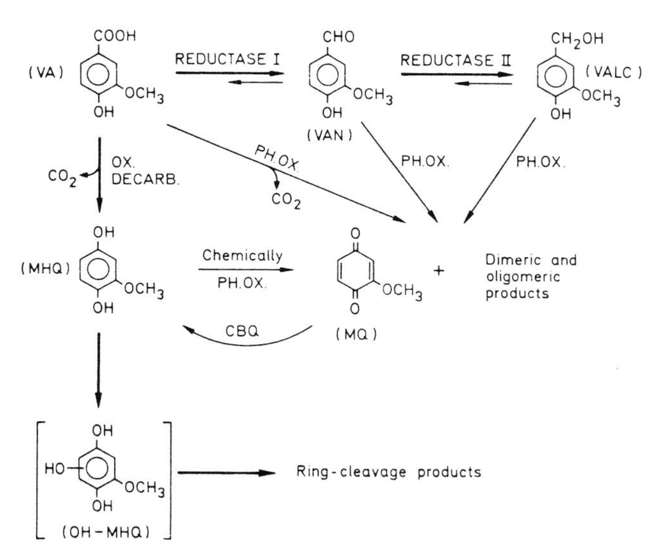

Figure 11. Metabolic pathway of vanillic acid by *Sporotrichum pulverlentum* (Ander et al.)[25].

Applications

Lignin biodegradation research is still at a basic level. However, a considerable potential is expected for the future development of lignin bioconversion processes. Kraft lignins are mostly utilized as fuel to recover chemicals for pulp digestion, whereas lignosulfonates are mainly used as dispersants, stabilizers, binders and for vanillin production. However, a large amount of waste lignin remains still unutilized. Lignin, an aromatic polymer would be the predominant source of low molecular weight chemicals and of high polymers in place of those derived from petroleum. In these fields, lignin bioconversion could be successfully applied with several advantages: conversions with less energy consumption; selected and specific alterations mediated by specific enzymes and/or microorganisms etc. If genetic manipulation is applied to strains of lignin degrading microbes these possibilities could be greatly enhanced[27,28]. Demethylation, hydroxylation, side chain shortening and ring cleavage of lignin could satisfactorily alter polymeric lignin for chemical modifications. Preferential removal of lignin from wood by lignin degraders is useful for biological pulping, and for pretreatment in the production of ethanol and livestock feed from wood polysaccharides. Eriksson and Vallander[29] obtained a cellulase-less mutant (Cel 44) of *Sporotoricum pulverlentum (* = Phanerochaete chrysosporium) by irradiation of a spore suspension with UV-light. 10 days' pretreatment of wood chips by Cel 44 significantly decreased the energy consumption in the production of mechanical pulp. Upon treating thermomechanical pulp[30] and unbleached Kraft pulp[31] with ligninolytic fungi, the maximum rate of lignin degradation was 3% per day over a 2-week incubation; the unbleached Kraft pulp was partially delignified on incubation with ligninolytic fungi, reducing the necessity for bleaching chemicals. Another approach is fungal decolorization of Kraft bleach plant effluents. About 60–70% color reduction was achieved by fungal decolorization of

bleach plant effluent with *P. chrysosporium*[32] and *Tinctoporia borbonica*[33] within 2-4 days.
These kinds of potential applications obviously require an improved understanding of the chemistry and biochemistry of biodegradation, of the nature and properties of the products, and of the microorganisms which could be used. Although such applications seem to be a long way off in the future, research is rapidly providing the fundamental knowledge needed to realize the applications. Recent reviews[2,34-36] and two recent books[3,37] should be consulted for more detailed information on lignin biodegradation.

1 K. Freudenberg, Lignin: its constitution and formation from p-hydroxycinnamyl alcohols. Science *148*, 595-600 (1965).
2 T.K. Kirk, W.J. Connors and J.G. Zeikus, Advances in understanding the microbial degradation of lignin, in: Recent Advances in Phytochemistry, vol. 11, p. 369-394. Ed. F.A. Loewus and V.C. Runeckles. Plenum Press, New York 1977.
3 T.K. Kirk, T. Higuchi and H.-M. Chang, Lignin Biodegradation: Microbiology, Chemistry, and Potential Applications, vol. 1, p. 236. Ed. T.J. Kirk, T. Higuchi and H.-M. Chang. CRC Press, Boca Raton, Fl., 1980.
4 T.K. Kirk, E. Schulz, W.J. Connors, L.F. Lorenz and J.G. Zeikus, Influence of culture parameters on lignin metabolism by *Phanerochaete chrysosporium.* Archs Microbiol. *117*, 277-285 (1978).
5 H.-M. Chang, C.L. Chen and T.K. Kirk, Chemistry of lignin degradation by white-rot fungi, in: Lignin Biodegradation; Microbiology, Chemistry, and Potential Applications, vol. 1, p. 215-230. CRC Press, Boca Raton, Fl., 1980.
6 P.C. Ellwardt, K. Haider and L. Ernst, Untersuchung des mikrobiellen Ligninabbaus durch ^{13}C-NMR Spektroskopie an spezifisch ^{13}C-angereichertem DHP-Lignin aus Coniferylalkohol. Holzforschung *35*, 103-109 (1981).
7 C.L. Chen, H.-M. Chang and T.K. Kirk, Lignin degradation products from spruce wood decayed by *Phanerochaete chrysosporium.* Abstracts ACS/CSJ Chemistry Congress, Honolulu, Hawaii, 1979.
8 T. Higuchi, Microbial degradation of dilignols as lignin model, in: Lignin Biodegradation: Microbiology, Chemistry, and Potential Applications, vol. 1, p. 171-192. CRC Press, Boca Raton, Fl., 1980.
9 T. Katayama, F. Nakatsubo and T. Higuchi, Initial degradation in the fungal degradation of guaiacylglycerol-β-coniferyl ether, a lignin substructure model. Archs Microbiol. *126*, 127-132 (1980).
10 T. Katayama, F. Nakatsubo and T. Higuchi, Degradation of arylglycerol-β-aryl ethers, lignin substructure models by *Fusarium solani.* Archs Microbiol., in press (1981).
11 T. Fukuzumi, Microbial metabolism of lignin-related aromatics, in: Lignin Biodegradation: Microbiology, Chemistry, and Potential Applications, vol. 2, p. 73-93. CRC Press, Boca Raton, Fl., 1980.
12 T. Fukuzumi and Y. Katayama, Bacterial degradation of dimer relating to structure of lignin. I. Mokuzai Gakkaishi *23*, 214 (1977).
13 F. Nakatsubo, T. Umezawa and T. Higuchi, unpublished data, 1981.
14 A. Enoki, G.P. Goldsby and M.H. Gold, Metabolism of the lignin model compounds veratrylglycerol-β-guaiacyl ether and 4-ethoxy-3-methoxyphenylglycerol-β-guaiacyl ether by *Phanerochaete chrysosporium.* Archs Microbiol. *125*, 227-232 (1980).
15 H.G. Rast, G. Engelhart, W. Ziegler and P.R. Wallnöfer, Bacterial degradation of model compounds for lignin and chlorophenol derived lignin bound residues. FEMS Microbiol. Lett. *8*, 259-263 (1980).
16 M. Ohta, T. Higuchi and S. Iwahara, Microbial degradation of dehydrodiconiferyl alcohol. Archs Microbiol. *121*, 23-28 (1979).
17 T. Katayama and T. Higuchi, Biodegradation of β-5′ dilignol by *Fusarium solani* M-13-1. Abstracts 25th Symp. Lignin Chemistry, Fukuoka, Japan, 1980.
18 S. Iwahara, T. Nishihira, T. Jomori, M. Kuwahara and T. Higuchi, Enzymic oxidation of α,β-unsaturated alcohols in the side chains of lignin-related aromatic compounds. J. Ferment. Technol. *58*, 183-188 (1980).
19 L. Eggeling and H. Sahm, Degradation of coniferyl alcohol and other lignin-related aromatic compounds by *Nocardia* sp. DSM 1069. Archs Microbiol. *126*, 141-148 (1980).
20 F. Nakatsubo, T.K. Kirk, M. Shimada and T. Higuchi, Metabolism of a phenylcoumaran substructure model compound in ligninolytic culture of *Phanerochaete chrysosporium.* Archs Microbiol. *128*, 416-420 (1981).
21 T. Umezawa, F. Nakatsubo, T. Higuchi and T.K. Kirk, Degradation of phenylcoumaran dilignol by *Phanerochaete chrysosporium.* Abstracts 25th Symp. Lignin Chemistry, Fukuoka, Japan, 1980.
22 Y. Kamaya, F. Nakatsubo, T. Higuchi and S. Iwahara, Degradation of d,l-syringaresinol, a β-β' linked lignin model compound by *Fusarium solani* M-13-1. Archs Microbiol. *129*, 305-309 (1981).
23 H. Namba, F. Nakatsubo and T. Higuchi, Degradation of 1,2-diarylpropane-1,2-diols by *Fusarium solani.* Abstracts 25th Symp. Lignin Chemistry, Fukuoka, Japan, 1980.
24 T. Higuchi and F. Nakatsubo, Synthesis and biodegradation of oligolignols. Kemia-Kemi *9*, 481-488 (1980).
25 P. Ander, A. Hatakka and K.E. Eriksson, Vanillic acid metabolism by the white-rot fungus *Sporotrichum pulverulentum.* Archs Microbiol. *125*, 189-202 (1980).
26 J. Martin and K. Haider, Microbial degradation and stabilization of ^{14}C-labeled lignins, phenols, and phenolic polymers in relation to soil humus formation, in: Lignin Biodegradation: Microbiology, Chemistry, and Potential Applications, vol. 2, p. 77-100. CRC Press, Boca Raton, Fl., 1980.
27 M. Gold, T. Cheng, K. Krisnangkura, M. Mayfield and L. Smith, Genetic and biochemical studies on *P. chrysosporium* and their relation to lignin degradation, in: Lignin Biodegradation: Microbiology, Chemistry, and Potential Applications, vol. 2, p. 65-71. CRC Press, Boca Raton, Fl., 1980.
28 S.M. Salonen and V. Sunham, Regulation and genetics of the biodegradation of lignin derivatives in pulp mill effluents, in: Lignin Biodegradation; Microbiology, Chemistry, and Potential Applications, vol. 2, p. 179-198. CRC Press, Boca Raton, Fl., 1980.
29 K.E. Eriksson and L. Vallander, Biomechanical, pulping, in: Lignin Biodegradation: Microbiology, Chemistry, and Potential Applications, vol. 2, p. 213-224. CRC Press, Boca Raton, Fl., 1980.
30 H.H. Yang, M.J. Effland and T.K. Kirk, Factors influencing fungal degradation of lignin in a representative lignocellulosic thermomechanical pulp. Biotechnol. Bioengng *22*, 65-77 (1980).
31 T.K. Kirk and H.H. Yang, Partial delignification of unbleached Kraft pulp with ligninolytic fungi. Biotechnol. Lett. *1*, 347-352 (1979).
32 D. Eaton, H.-M. Chang and T.K. Kirk, Fungal decolorization of Kraft bleach plant effluents. Proc. Tappi Annual Meeting 1980.
33 T. Fukuzumi, Microbial decolorization and deforming of pulp waste liquors in lignin biodegradation, in: Lignin Biodegradation: Microbiology, Chemistry, and Potential Applications, vol. 2, p. 161-177. CRC Press, Boca Raton, Fl., 1980.
34 P. Ander and K.E. Eriksson, Lignin degradation and utilization by microorganisms. Prog. ind. Microbiol. *14*, 1-58 (1978).
35 D.L. Crawford and R.L. Crawford, Microbial degradation of lignin. Enzyme Microb. Technol. *2*, 11-22 (1980).
36 T. Higuchi, T.K. Kirk, M. Shimada and F. Nakatsubo, Some recent advances in lignin biodegradation research as related to potential applications. Proc. 2nd int. Symp. Bioconversion and Biochemical Engineering, IIT Delhi, India, in press.
37 T.K. Kirk, Degradation of lignin, in: Biochemistry of Microbial Degradation. Ed. D.T. Gibson. Marcel Dekker, New York, in press (1981).

Aromatic chemicals through anaerobic microbial conversion of lignin monomers

by J.-P. Kaiser and K.W. Hanselmann[1,2]

Institute of Plant Biology, University of Zürich, Department of Microbiology, Zollikerstrasse 107, CH-8008 Zürich (Switzerland)

Summary. Large efforts are directed towards production of ethanol from cellulosic biomass in order to reduce our dependence on petroleum based ethylene. No satisfactory process exists to date, however, which would make the aromatic molecules present in wood available to economic exploitation. A combination of physicochemical pretreatment of lignocellulose and selective microbial conversion of the mixture of aromatic monomers into a few phenolic products is outlined. Anaerobic microbial communities are employed since they offer thermodynamic and physiological characteristics necessary for efficient conversion. Under anaerobic conditions most of the carbon and energy initially present in the substrate can be recovered as useful products; oxidative losses as CO_2 and H_2O are minimized. The 3,4-disubstituted aromatic lignin monomers are converted to catechol while 3,4,5-trisubstituted monomers are mineralized to CH_4 and CO_2. Further studies are directed towards an understanding of the physiological functions of the populations participating in the conversion process, the reason for catechol recalcitrance and the tolerance of the community towards phenolic endproducts.

The rate at which world oil reserves are presently being consumed is about 3.4% per year which means that consumption of proven reserves may be complete in 29 years[3]. This period of time could be extended by a few years if consumption would decrease further. But clearly within less than half a century we will have to find ways to change petroleum consuming civilizations to those that use alternative sources for chemicals and fuels. Although it might not be our first option, biomass will certainly be the only alternative in the long run[4].

We consider in this presentation some microbiological aspects of producing chemicals from biomass, in particular the production of aromatic compounds from lignin by means of anaerobic microbial processes. Using microbial communities to produce valuable synthesis chemicals from substrate mixtures is to be encouraged. Fermentation processes in which aromatic products derived from lignocellulose would accumulate could become interesting economically for wood-rich countries. Furthermore, economically competitive production of lignochemicals could lessen our present dependence on petrochemicals.

Microbial metabolism of aromatics

To provide a useful source of aromatic synthesis chemicals, enzymatic or chemical processes that monomerize lignin must leave intact the aromatic nucleus and ensure microbial conversion of the resulting mixture of phenolics into a few separable compounds. In nature, degradation of lignin takes place through the sequential action of different microbial populations. This process and some of the organisms involved in the aerobic pathway have been widely studied[5-9]. Monomeric compounds found during these investigations in decaying wood and in culture media of fungi grown with isolated lignins include vanillic-, ferulic-, syringic-, p-hydroxybenzoic-, p-hydroxycinnamic and 3-methoxy-4-hydroxyphenylpyruvic acids as well as some of their aldehydes. These products are converted to the corresponding diphenols by a sequence of 4 reactions[6]:

1. Elimination of methylgroups by O-demethylase or laccase forming formaldehyde and methanol respectively.
2. Oxidation of the C-3 side chain.
3. Oxidation of the alcohol- and aldehyd functions at C-1.
4. Hydroxylation of monophenols to diphenols.

Catechol, protocatechuic acid and gallic acid are the corresponding products for ring fission. Pathways for ring cleavage in the presence of oxygen are well established and their regulation has been studied in detail[10,11]. Oxygenases play the major role in initial ring fission.

Based on the original report by Tarvin and Buswell (1934)[12] that certain aromatic compounds could be fermented to methane, pioneering work has been carried out on reductive ring fission under anaerobic conditions[13-17]. Anaerobic dissimilation of the aromatic ring is now known to be promoted in at least 4 different situations[18,19]: through anaerobic photometabolism by certain Rhodospirillaceae; by *Pseudomonas* spp., a *Bacillus* sp. and *Moraxella* spp. and mixed microbial populations in presence of nitrate as electron acceptor; by microbial communities through fermentation to methane; and by certain sulfate reducing bacteria.

The ring is cleaved by reductive transformation of the aromatic nucleus into the corresponding cyclohexane followed by oxidative ring fission[18]. This mechanism is outlined by the postulated pathways for the degradation of syringic acid (fig. 1).

Interactions in microbial communities

Ferry and Wolfe[20] have made a microbiological ana-

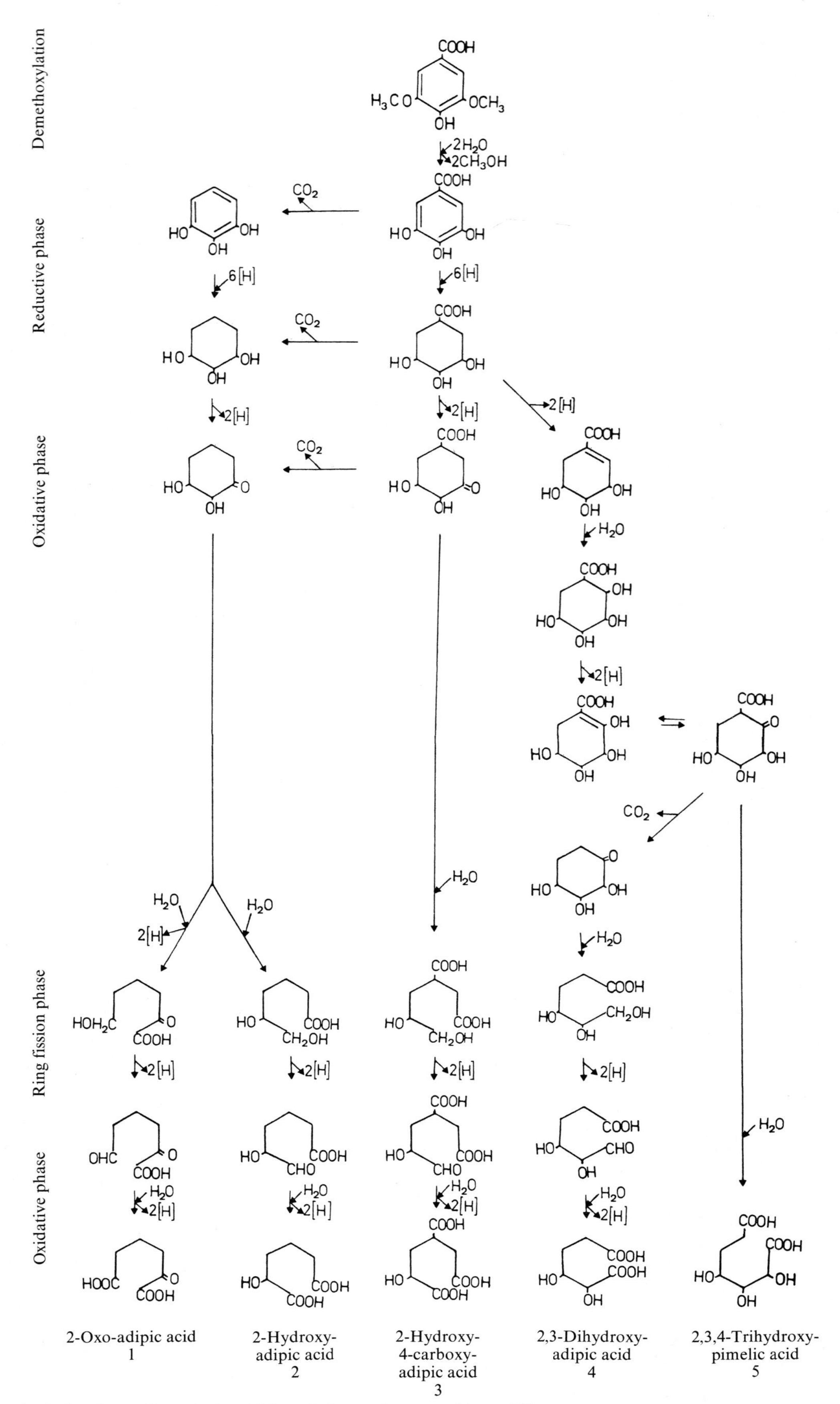

Figure 1. Hypothetical pathways for syringic acid degradation under anaerobic conditions.

lysis of a benzoate degrading methanogenic consortium and postulated the following sequence of reactions:

$$(1)\ 4\,C_6H_5COO^- + 24\,H_2O \rightarrow \mathit{4\,HCOO^-} + \mathit{12\,CH_3COO^-} + 12\,H^+ + \mathit{8\,H_2}$$

$$\downarrow (2) \qquad \downarrow (3) \qquad \downarrow (4)$$

Reaction (1) is carried out by a gram negativ, facultatively anaerobic organism. It catalyzes the reaction only in the presence of symbiotic populations that continually remove the products (reactions 2–4) thereby maintaining favorable thermodynamic conditions for ring fission to proceed.

$$(2)\ \text{a)}\ 4\,HCOO^- + 4\,H_2O \rightarrow 4\,HCO_3^- + 8\,[H]$$
$$\text{b)}\ 1\,HCO_3^- + 8\,[H] + H^+ \rightarrow CH_4 + 3\,H_2O$$
$$\overline{\mathit{4\,HCOO^-} + H_2O + H^+ \rightarrow CH_4 + 3\,HCO_3^-}$$

$$(3)\ \mathit{12\,CH_3COO^-} + 12\,H_2O \rightarrow 12\,CH_4 + 12\,HCO_3^-$$

$$(4)\ 2\,HCO_3^- + \mathit{8\,H_2} + 2\,H^+ \rightarrow 2\,CH_4 + 6\,H_2O$$

Reactions (2) and (4) are assigned to *Methanobacterium formicium* and *Methanospirillum hugatei* respectively which were isolated from the community. The acetate utilizing methanogenic organism could not be obtained in axenic culture. Complete mineralization of the aromatic ring (reaction 5) under anaerobic conditions, therefore, requires synthropically interacting populations.

$$(5)\ 4\,C_6H_5COO^- + 31\,H_2O \rightarrow 15\,CH_4 + 13\,HCO_3^- + 9\,H^+$$

Healy and Young[21–23] have applied methanogenic communities to the degradation of lignin monomers which could be produced during alkaline heat treatment of wood. All of the 11 aromatic model compounds tested were degraded and methane-produced. Their communities also consisted of several different populations which were necessary to make the ring fission reaction exergonic. Ring fission could thus be prevented through regulation of certain populations of the community.

Conversion and mineralization of lignin derived aromatics

Enrichments from anaerobic freshwater lake sediments which we have obtained with syringic acid as the sole carbon- and energy source, mineralize syringic acid (fig. 2a) to CH_4 and CO_2 with temporary accumulation of acetate (fig. 3). The benzene ring is cleaved rapidly while the further catabolism of degradation intermediates is presumably limited by the size and activity of the appropriate companion populations. Vanillic acid and other 3,4-substituted aromatic compounds are converted by our community to catechol (fig. 2b).

Syringic acid adapted cultures were fed with vanillic acid repeatedly and the products were allowed to accumulate. From figure 2b it may be seen that the absorption band at 251 nm, characteristic for vanillic acid, decreases while the new compound with absorption at 275 nm accumulates. Separation of acidified ether extracts with HPLC (fig. 4) shows that catechol

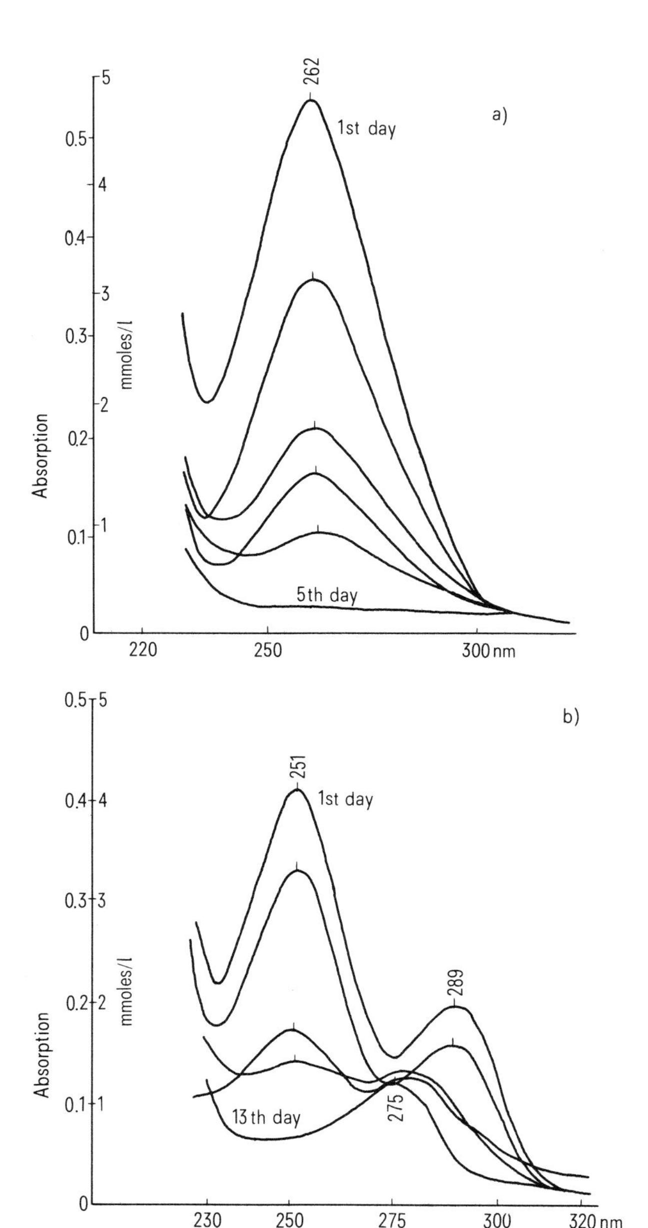

Figure 2. *a* Spectral changes during syringic acid mineralization. Syringic acid adapted cultures were fed 5 mmoles/l syringic acid. Within 5 days all of the substrate is converted as indicated by the disappearance of the absorption band at 262 nm. Samples for absorbancy measurements were diluted 80-fold. *b* Spectral changes during vanillic acid conversion. Syringic acid adapted cultures were fed 5 mmoles/l vanillic acid twice within 1 week. The substrate from the 1st feeding was degraded; the substrate from the 2nd one is converted into catechol (absorption band at 275 nm). Samples for absorbancy measurements were diluted 80-fold.

is formed from vanillic acid via protocatechuic acid as an intermediate. All of the syringic acid analogs with substituents in the 3,4 and 5 position of the ring were completely mineralized (fig. 5), while 3,4-substituted substrates tested were converted to catechol only (fig. 6). Ring fission of this intermediate did not take place with our communities under the strictly anaerobic conditions maintained. The refractory nature of catechol has been documented for aerobic[9] and anaerobic conditions by other investigators[14,24]. Healy and Young[21] however, found complete mineralization. At present, we do not know why 3 consecutive substituents are necessary for mineralization to occur. The knowledge that microbial communities can metabolize one class of monomers while only converting another could become a useful tool in reducing the complex mixture of aromatic monomers obtained in lignin hydrolysis to a few recoverable products.

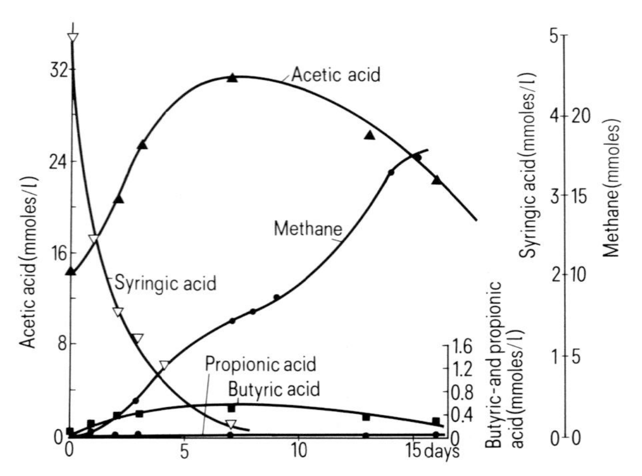

Figure 3. Intermediary products and methane formed by a syringic acid mineralizing culture. Syringic acid is converted within 1 week. Acetic acid accumulates and serves as a long lasting source for methane production.

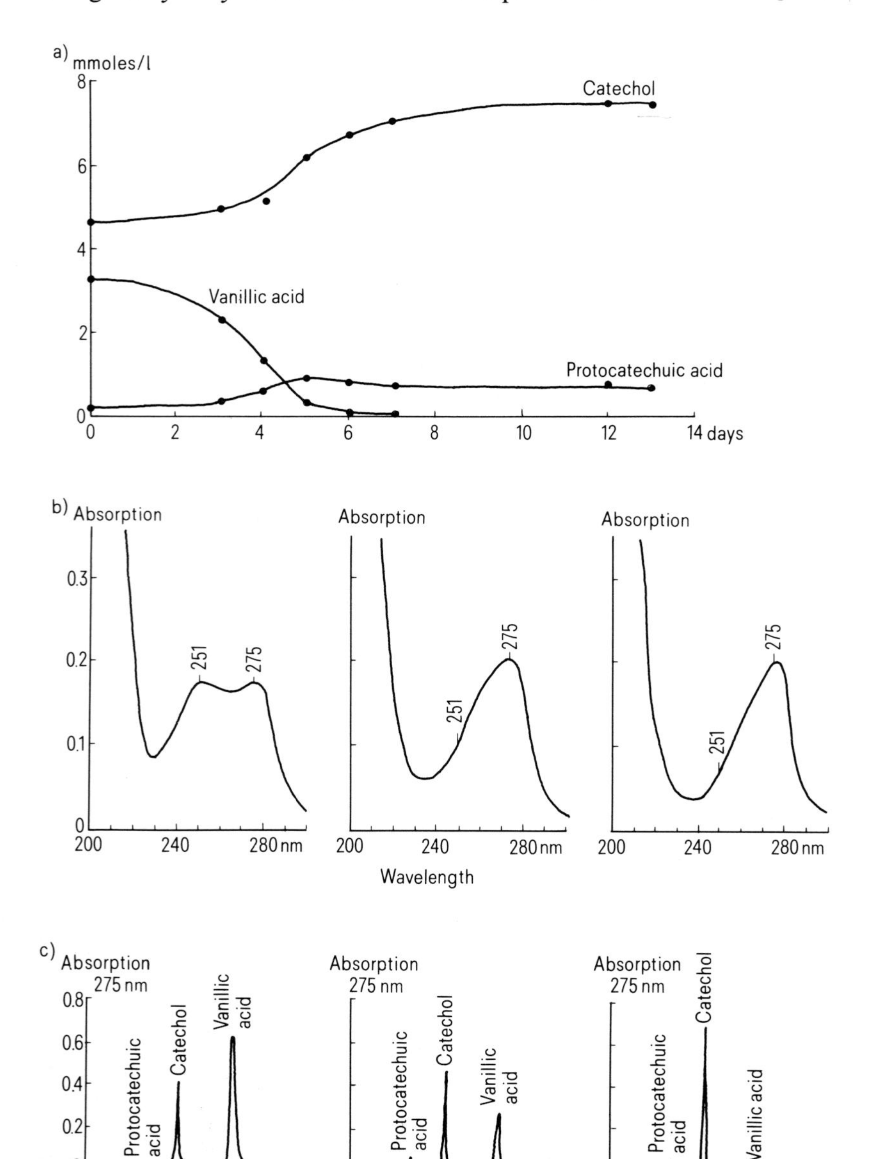

Figure 4. Time course (a), spectral changes (b) and product separation patterns (c) from a vanillic acid converting batchculture. Catechol was allowed to accumulate during previous conversions of vanillic acid. Protocatechiuc acid appears as intermediate. Spectral changes (b) and separation patterns (c) are representative for days 1, 3 and 7.

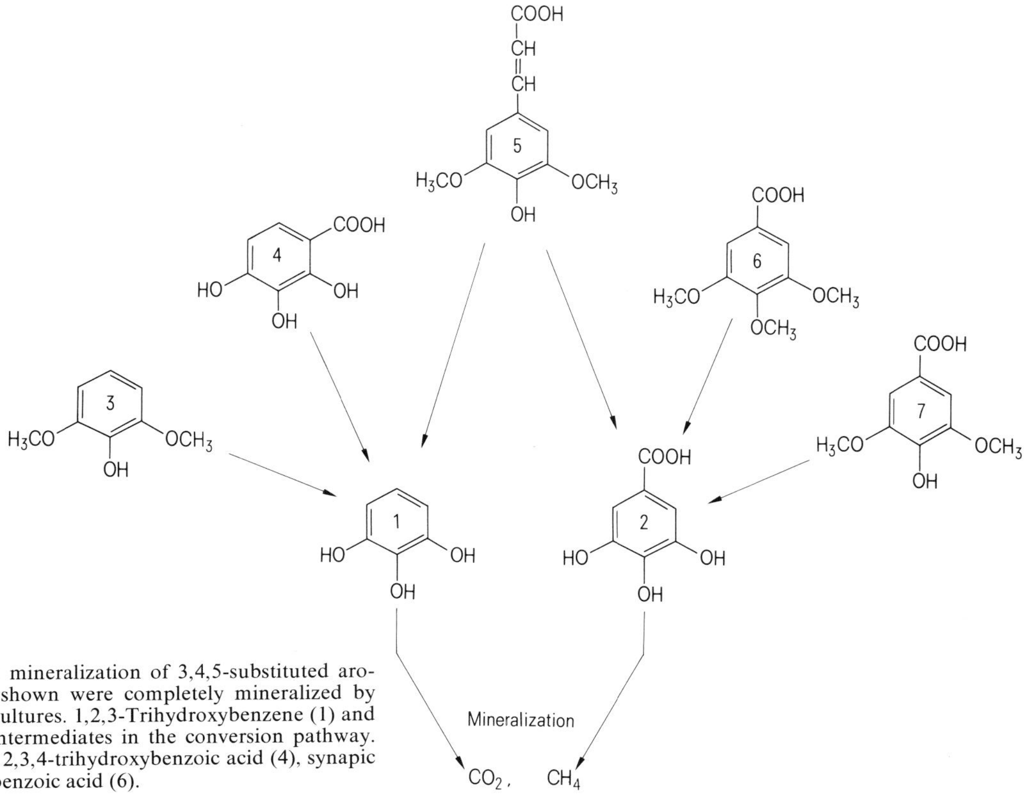

Figure 5. Conversion and mineralization of 3,4,5-substituted aromatics. The 7 substrates shown were completely mineralized by syringic acid (7) adapted cultures. 1,2,3-Trihydroxybenzene (1) and gallic acid (2) might be intermediates in the conversion pathway. 2,6-Dimethoxyphenol (3), 2,3,4-trihydroxybenzoic acid (4), synapic acid (5), 3,4,5-trimethoxybenzoic acid (6).

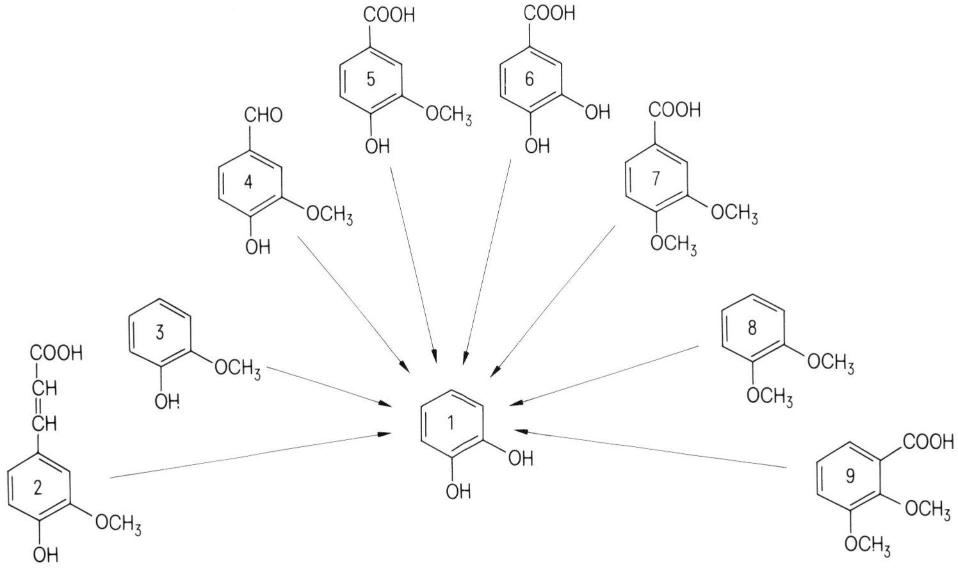

Figure 6. Conversion of 3,4-substituted aromatics into catechol (1) by syringic acid adapted cultures. No or only small amounts of methane were formed during metabolism of the 9 substrates listed. The aromatic ring was not cleaved. 4-Hydroxy-3-methoxycinnamic acid (2), 2-methoxyphenol (3), vanillin (4), vanillic acid (5), protocatechuic acid (6), veratric acid (7), 1,2-dimethoxybenzene (8), 2,3-dimethoxybenzoic acid (9).

Microbiological feasibility

Lignin fractions and phenolic compounds are used in herbicides, insecticides, fungicides, and as wood preservatives. Besides these man-made applications which are directed towards limiting the life span of certain organisms, many naturally occurring aromatic substances act similarly in natural ecosystems. Since their toxicity contributes to their recalcitrance, they tend to accumulate as humic- and fulvic acids in soil, peat and sediments and, through the years, have been conserved under extreme environmental conditions as coal, oil and natural gas[25].

The continuous presence of these toxic substances, on the other hand, has led to selection of microorganisms capable of coping with them metabolically. We therefore find microbial communities today with abilities to convert and metabolize lignoaromatics anaerobically.

Three main characteristics distinguish naturally occurring degradative systems from those that are preferably employed industrially: substrates available in nature occur as complex mixtures; degradation is carried out by microbial communities; and degradation ultimately leads to mineralization in most cases. Industrially interesting processes require high substrate-to-product conversion from well-defined substrates by stable cultures, preferably consisting of one axenic population that can be dependably regulated. Degradation should stop at the level when a useful product is achieved that still contains most of the matter and energy present in the original substrate. Thus, there is still a large gap between the recognition of an ecologically functional pathway and its adaptation to an industrially interesting process.

Many organisms are able to oxidize completely organic polymers aerobically. No single organism has been found, however, that can achieve the same thing under anaerobic conditions. Only interacting microbial populations, each with limited metabolic abilities, can achieve together complete mineralization under anaerobic conditions. Symbiotic interactions are therefore responsible for the regulation and energetic efficiency of an overall process. They determine both, the kind and the amount of metabolic end products, and the conversion rates.

Although there are organisms with broad substrate utilization abilities, no single species exists in nature that can degrade all naturally occurring substances. Diversity in metabolic pathways reflects the diversity found in biomass composition. In an ecosystem, one expects a larger number of different populations as the substrate mixture increases in complexity. At the first levels of degradation, population diversity depends on the kind of intramolecular bonds that have to be hydrolyzed and the classes of monomers derived for fermentation. Glycosidic-, amin- and ester linkages are hydrolyzed easily while the different C-C bonds and the ether bonds in lignin require extremely oxidative conditions. Pretreatment is required, therefore, to make this biopolymer accessible to industrial fermentation.

Nature allows populations to become dominant for the conversion of a particular substrate and later to be replaced by other populations which attack other substrates. Thus, degradation of a complex substrate mixture is achieved through oscillating population activities in syntrophic associations. Microbial communities that show phases of adaptation and population fluctuations are more difficult to handle and seem less suited for industrial applications. We have tried to overcome some of the difficulties mentioned with our communities that degrade lignin monomers.

A stable community, able to mineralize 3,4,5-trisubstituted aromatic monomers and to convert 3,4-disubstituted aromatics into catechol will continue to do so in the presence of a mixture of syringic- and vanillic acid (fig. 7). Conversion to catechol seems to be stimulated by low levels of syringic acid. At higher concentrations vanillic acid conversion is inhibited until the concentration of syringic acid has dropped below a certain level. This experiment indicates that the great variety of monomeric substrates obtained through pretreatment of lignin could possibly be reduced to a few products through microbial fermentation. Certain classes of substrates would be converted into one common aromatic product which could be harvested more easily while certain companion substrates would be degraded further into products that would not interfere with the extraction process (e.g. acetic-, butyric acid) or which would escape as gases (CH_4; fig. 8). The principles of selective fermentation are presently employed, for example, in the recovery of pentoses from hardwood hydrolysates. The hexoses are fermented selectively to ethanol while xylose and arabinose remain unaltered and can be concentrated and recovered by crystallization.

Our cultures contain 4 to 5 different populations (fig. 9) which have been maintained through many transfers for almost 2 years as a balanced community. The interactions within this community stabilize the metabolic functions both thermodynamically and physiologically and maintain the environmental conditions for optimal fermentation. Although interactions lead to highly stable processes, failure of only one population disrupts the entire community. This fragility coupled with our limited knowledge about the behavior of mixed populations in general have prevented successful industrial applications (with the exception of the fermentation of certain foods). Research directed towards a better understanding of microbial communities could lead one day to economic exploitation of many more of the anaerobic processes which offer advantages over aerobic conversions and single population processes.

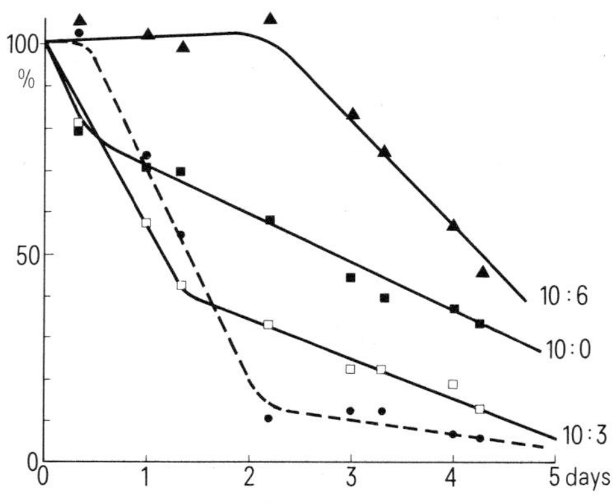

Figure 7. Microbial conversion of aromatics in substrate mixtures. Syringic acid adapted cultures were fed vanillic acid (10 mmoles/l) and 0 (■), 3 (□) or 6 (▲) mmoles/l syringic acid respectively. 1 culture was fed syringic acid only (5 mmoles/l, ●). Solid lines: disappearance of vanillic acid; dashed line: disappearance of syringic acid.

Multicomponent mixture of fermentable and refractory substrates (e.g. phenolic lignin hydrolysis products).

A B C D E F G H I

Selective microbial conversion of certain classes of chemical substances into a few intermediary products

S T

Further conversion of S is inhibited

Further conversion and mineralization of T is encouraged

X + Y

Final mineralization products from T (e.g. CH_4, CO_2, H_2O).

D + H + S

Final product mixture

Chemical separation of product mixture

Byproducts

D, H

S

Desired end product

Figure 8. Major steps in selective fermentation with pure cultures or microbial communities.

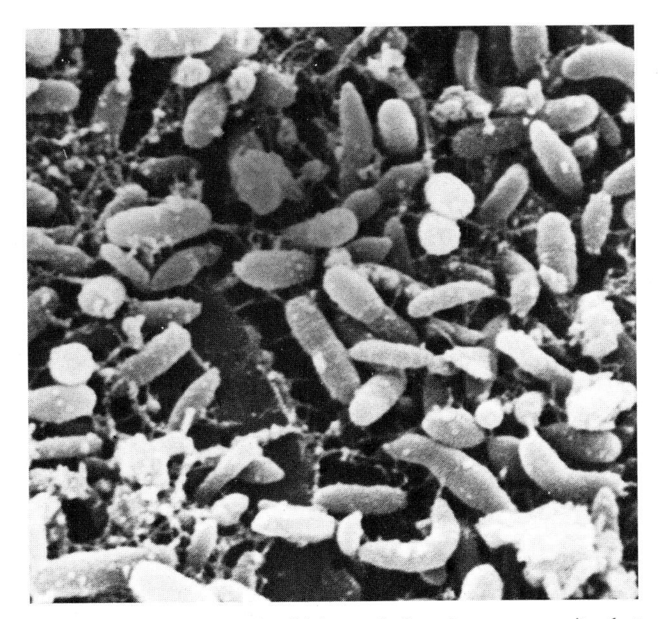

Figure 9. Diversity of microbial populations in a community that degrades syringic acid fermentatively.

Many anaerobes produce large amounts of reduced end products. Product composition depends on the energy metabolism of the organisms involved and can be altered by adjusting growth conditions, by changing the environment, and by culturing together with other organisms. For example in the absence of an inorganic electron acceptor (NO_3^-, $SO_4^=$, CO_2), reducing equivalents are deposited in organic acceptors formed from the substrates offered. This leads to the production of reduced organic compounds which might be useful synthesis chemicals (table 1). Diversion of electrons to an acceptor other than the one that leads to the desired product results in decreased yields and product mixtures. The homolactic and the homoacetic fermentations where carbon and electrons

Table 1. Primary synthesis chemicals produced by anaerobic microbial fermentations

Chemical	Example of producing microorganisms
Ethanol	*Clostridium thermocellum*, yeasts
Butanol	*Clostridium acetobutylicum*
Isopropanol	*Clostridium butylicum*
Acetone	*Clostridium acetobutylicum*
Butanediol	*Enterobacter aerogenes*
Formic acid	*Enterobacter punctata*
Acetic acid	*Clostridium formicoaceticum*
Propionic acid	*Clostridium propionicum*
Butyric acid	*Clostridium butyricum*
Caproic acid	*Clostridium kluyveri*
Lactic acid	*Lactobacillus casei*
Succinic acid	*Ruminococcus flavofaciens*
Methane	*Methanobacterium thermoautotrophicum*
Hydrogen	*Rhodospirillum rubrum*, *Clostridium butyricum*
Carbon dioxide	*Clostridium butyricum* and many others

The microbially produced chemicals are present in fermentation broths. Their concentration depends on the substrate offered, the microbial species employed, the fermentation conditions and the tolerance of the population towards toxic effects of the accumulating endproducts.

are preserved in 2 moles of lactate (e.g. *Lactobacillus casei*) or 3 moles of acetate (e.g. *Clostridium thermoaceticum*) belong to the most efficient substrate-product conversion processes. Diversion of electrons to inorganic electron acceptors might be employed to regulate product formation or to prevent the buildup of undesired organic by-products. However, resulting reduced inorganic products (N_2, HS^-, CH_4, H_2O) might constitute an energy loss or inhibit certain populations. Thus, proper consideration should also be given to the composition of the biomass as a substrate and its possible influence on directing metabolic routes if minimal carbon- and energy losses and maximal product yields are expected. Wood pulping waste waters, for example, have been proposed as substrates for the production of chemicals[26,27]. Feeding this resource (table 2) after chemical hydrolysis and adjustment of an appropriate C:N:P ratio to microbial cultures could create problems and unwanted side effects. The extremely high content of sulfur present in the sulfonate groups at the C-3 side chain would presumably be liberated during chain oxidation. In its oxidized state it could serve as an electron acceptor, thereby diverting energy into an inorganic, inhibitory and noxious end product and reducing the yield of the desired reduced organic products.

Further developments

Discussions have been limited to naturally occurring populations that can be selected for particular purposes from appropriate ecosystems. Mutation and selection of strains with improved conversion rates and higher tolerance towards toxic products, cloning of hyperproductive, metabolically deregulated strains and the incorporation of plasmid associated metabolic functions into appropriate strains are possibilities that could widen applicability of anaerobic conversion processes[28]. Also from studies about the ecology of the gut of termites and other wood eating insects we can learn more about microbial communities involved in anaerobic lignin degradation. One day it might become possible to synthesize microbial communities from defined strains and make industrial fermentation processes with mixed cultures more feasible.

Table 2. Composition of calciumlignosulfonates[a] (as dry powder)[h]

Dry substances (DS)[b]	92–96%
Lignin[c]	50–70% of DS
Other organic substances[d]	6–15% of DS
Inorganic salts[e]	8–14% of DS
Nitrogen as N[f]	0.80% of DS
Sulfur as S[g]	4–7% of DS
Methoxy groups ($-OCH_3$)	6–10% of DS
C:N[f]	52.5
C:H	8.4
C:S[g]	6–10.5

[a] Values calculated from analysis data supplied by Cellulose Attisholz AG., Switzerland;
[b] dryed at 103 °C;
[c] polymers with mol. wts of 2000–30,000 daltons, corresponding to polymers with 11–170 phenolic monomers with an average mol. wt of 180 daltons;
[d] reducing substances, hexoses, pentoses, small fibers, yeast residues;
[e] after combustion at 800 °C; mostly calciumsulfate, calciumcarbonate and oxides of Ca, Mg, Na, K and Fe.
[f] The high nitrogen content and the variability in the concentration of 'other organic substances' are due to addition of ammonia to the sulfite liquor from gymnosperms and its use for the production of ethanol and yeast biomass.
[g] The high sulfur content stems from the acid used in the sulphite pulping process.
[h] The powder is soluble in water. The pH of a 50% (w/v) aqueous solution varies between 3.5 and 5.

Conclusions and outlook

Since the beginning of this century, microorganisms have been employed to produce chemicals, antibiotics, enzymes, vitamins and food and feed biomass. During times of inexpensive and seemingly unlimited supplies of fossil biomass, microbial processes have not remained competitive. Thus the study of many microbes with industrial potential, their physiology, the regulation of their metabolism, genetics and interactions with other organisms have been neglected. The encouraging success of genetic engineering in a time of concern about the dwindling resources of fossil biomass have resulted in a renewed interest in the biosynthetic and catabolic abilities of predominantly anaerobic organisms. Anaerobes are more difficult to handle since they must be maintained in an oxygen-free environment. They offer features, however, that make them attractive catalysts for efficient substrate-to-product conversions with high yields. Some of them have an energy metabolism which is most efficient in the presence of syntrophic companion organisms. Some have been found in extreme environments: in hot springs, at low pH and at high salt concentrations. These organisms might become particularly well suited for fermentations with high product accumulation, substrate compositions with low water activities or elevated process temperatures[29].

The decision whether to make a certain product through microbial fermentation, or through chemical synthesis, is primarily based on economic considerations. The price and the availability of raw materials, investments, energy costs, production rates and yields, recoverability of the products, their versatility and ecological impacts influence the decision. Two factors are worth emphasizing here. 1. The efficiency of anaerobic microbial processes depends largely on the energetic efficiency with which the organisms can dispose of reducing power. 2. Not all of the plant-derived raw materials are economically and politically suited as substrates. It would be unwise to base a large fermentation industry on a raw material which predominantly serves to feed mankind. Where the food- and the chemical industries compete for the same

plant product the former must be given priority. A fermentation-based chemicals industry should therefore concentrate on raw materials that are less important in the human food chain. Even raw materials for which an established use already exists should not be diverted to the production of chemicals without due evaluation of the influence on other aspects of our living habits and on the economics of other industrial branches. Plant raw materials which have been underutilized in the past and those that can be grown for a particular new application are the most promising candidates to support a fermentation-based chemical industry. Lignocellulose from wood, the most abundant renewable biomass component produced by terrestrial ecosystems, has a good potential as raw material to support production of many chemicals[30]. At first, research should be directed towards the production of lignochemicals with structures that cannot be obtained as petrochemicals. This minimizes economic competition for the same product while processing competence in handling lignin microbiologically can increase.
Today's efforts are directed towards efficient fermentation of the cellulose component of wood into chemical feedstocks[31]. Lignin is obtained as a by-product and supplies the energy needed to operate the pretreatment, the fermentation and the extraction processes[32,33]. Energetically, this may lead to an economic utilization of cellulose; it does not make use, however, of the synthetic work of plants which is stored in the chemically advanced structures of lignin monomers. Energy expenditure during resynthesis of the aromatic structures should also be considered before lignin is used as fuel only. Unfortunately, the microbial fermentation- and conversion processes for aromatic lignin monomers are not yet so highly developed that they already can be applied industrially.
A renewed effort is being made at present to search for new organisms and to study metabolic pathways and biochemical abilities of organisms and communities potentially useful for the production of chemicals. The field has been neglected during times of fossil biomass surplus and some time will be needed to reactivate it.

Note added in proof. After completion of this manuscript, Aftring and Taylor[34] published evidence for anaerobic degradation of phthalate (1,2-benzene dicarboxylate) with nitrate as electron acceptor by a pure culture of a *Bacillus* sp., Bache and Pfennig[35] identified the organism responsible for anaerobic demethoxylation of lignin monomers as *Acetobacterium woodii.* In the presence of bicarbonate the methoxygroups are oxidized to acetate. Our data support these observations.

1 Acknowledgment. We thank Ms H. Müller and Dr G. Hanselmann-Mason for their help in preparing the manuscript. Part of this work has been presented at the annual meeting of the Swiss Society for Microbiology 1981.
2 Author for correspondence and reprint requests.
3 Union pétrolière suisse. Rapport annuel, Zürich 1980, 52 pp.
4 R.S. Wishart, Industrial energy in transition: a petrochemical perspective. Science *199*, 614–618 (1978).
5 P. Ander and K.-E. Eriksson, Lignin degradation and utilization by microorganisms. Prog. ind. Microbiol. *14*, 1–58 (1978).
6 R.B. Cain, The uptake and catabolism of lignin-related aromatic compounds and their regulation in microorganisms, in: Lignin biodegradation: Microbiology, Chemistry and Potential Applications, vol. 1, p. 21–60. Ed. T.K. Kirk, T. Higuchi, and H. Chang. CRC Press, Boca Raton, FL 1980.
7 T. Fukuzumi, Microbial metabolism of lignin-related aromatics, in: Lignin Biodegradation: Microbiology, Chemistry and Potential Applications, vol. 2, p. 73–94. Ed. T.K. Kirk, T. Higuchi and H. Chang. CRC Press, Boca Raton, FL 1980.
8 H. Kawakami, Degradation of lignin-related aromatics and lignins by several pseudomonads, in: Lignin Biodegradation: Microbiology, Chemistry and Potential Applications, vol. 2, p. 103–105. Ed. T.K. Kirk, T. Higuchi and H. Chang. CRC Press, Boca Raton, FL 1980.
9 M. Kuwahara, Metabolism of lignin-related compounds by bacteria, in: Lignin Biodegradation: Microbiology, Chemistry and Potential Applications, vol. 2, p. 127–146. Ed. T.K. Kirk, T. Higuchi and H. Chang. CRC Press, Boca Raton, FL 1980.
10 S. Dagley, New pathways in the oxidative metabolism of aromatic compounds by microorganisms. Nature *188*, 560–566 (1960).
11 R.Y. Stanier and L.N. Ornston, The β-ketoadipate pathway. Adv. microbial Physiol. *9*, 89–151 (1973).
12 D. Tarvin and A.M. Buswell, The methane fermentation of organic acids and carbohydrates. J. Am. chem. Soc. *56*, 1751–1755 (1934).
13 M.T. Balba and W.C. Evans, The methanogenic fermentation of aromatic substrates. Biochem. Soc. Transactions *5*, 302–304 (1977).
14 F.M. Clark and L.R. Fina, The anaerobic decomposition of benzoic acid during methane fermentation. Archs Biochem. *36*, 26–32 (1952).
15 M. Guyer and G. Hegeman, Evidence for a reductive pathway for the anaerobic metabolism of benzoate. J. Bact. *99*, 906–907 (1969).
16 C.L. Keith, R.L. Bridges, L.R. Fina, K.L. Inverson and J.A. Cloran, The anaerobic decomposition of benzoic acid during methane fermentation. IV. Archs Microbiol. *118*, 173–176 (1978).
17 P.M. Nottingham and R.E. Hungate, Methanogenic fermentation of benzoate. J. Bact. *98*, 1170–1172 (1969).
18 W.C. Evans, Biochemistry of the bacterial catabolism of aromatic compounds in anaerobic environments. Nature *270*, 17–22 (1977).
19 F. Widdel, Anaerober Abbau von Fettsäuren und Benzoesäure durch neu isolierte Arten sulfat-reduzierender Bakterien. Thesis, University of Göttingen, FRG, 1980.
20 J.G. Ferry and R.S. Wolfe, Anaerobic degradation of benzoate to methane by a microbial consortium. Archs Microbiol. *107*, 33–40 (1976).
21 J.B. Healy and L.Y. Young, Catechol and phenol degradation by a methanogenic population of bacteria. Appl. environm. Microbiol. *35*, 216–218 (1978).
22 J.B. Healy and L.Y. Young, Anaerobic biodegradation of eleven aromatic compounds to methane. Appl. environm. Microbiol. *38*, 84–89 (1979).
23 J.B. Healy, L.Y. Young and M. Reinhard, Methanogenic decomposition of ferulic acid, a model lignin derivative. Appl. environ. Microbiol. *39*, 436–444 (1980).
24 P.L. McCarthy, L.Y. Young, J.M. Gossett, D.C. Stuckey and J.B. Healy, Heat treatment for increasing methane yields from organic materials, in: Microbial Energy Conversion, p. 179–

199. Ed. H.G. Schlegel and J. Barnes. Pergamon Press, Oxford 1977.
25 R.F. Christman and R.T. Oglesby, Microbial degradation and the formation of humus, in: Lignins, occurrence, formation, structure and reactions, p. 769–796. Ed. K.V. Sarkanen and C.H. Ludwig. Wiley Intersci., New York 1971.
26 W.C. Browning, The lignosulfonate challenge. Appl. Polymer Symp. *28*, 109–124 (1975).
27 A.L. Compere and W.L. Griffith, Industrial chemicals and chemical feedstocks from wood pulping wastewaters. Tappi *63/2*, 101–104 (1980).
28 D.E. Eveleigh, The microbial production of industrial chemicals. Scient. Am. *245*, 120–130 (1981).
29 J.G. Zeikus, Chemical and fuel production by anaerobic bacteria. A. Rev. Microbiol. *34*, 423–464 (1980).
30 I.S. Goldstein, ed., Organic chemicals from biomass, to be published by CRC Press, Boca Raton, FL.
31 S. Rosenberg and C.R. Wilke, Lignin biodegradation and the production of ethylalcohol from cellulose, in: Lignin Biodegradation; Microbiology, Chemistry and Potential Applications, vol. 2, p. 199–212. Ed. T.K. Kirk, T. Higuchi and H. Chang. CRC Press, Boca Raton, FL 1980.
32 Inventa, Ethanol process by wood saccharification, process description 81-D1. Inventa, Donat-Ems 1981.
33 P. Wettstein and B. Domeisen, Production of ethanol from wood. 1st Int. Energy Agency (IEA) Conf. on new energy conservation technologies and their commercialization, 1981.
34 R.P. Aftring and B.F. Taylor, Aerobic and anaerobic catabolism of phthalic acid by a nitrate-respiring bacterium. Archs Microbiol. *130*, 101–104 (1981).
35 R. Bache and N. Pfennig, Selective isolation of *Acetobacterium woodii* on methoxylated aromatic acids and determination of growth yields. Archs Microbiol. *130*, 255–261 (1981).

Lignochemicals

by K.W. Hanselmann[1]

University of Zürich, Institute of Plant Biology, Department of Microbiology, Zollikerstrasse 107, CH-8008 Zürich (Switzerland)

Introduction

Only 6–8% of the petroleum produced worldwide serves, at present, as raw material for the production of basic chemicals. But since petrochemicals supply 98% of the basic feedstocks used by the chemical industry, its dependence is obvious[2]. Economical production of chemicals from natural oil and gas has been possible because of their relative purity and their continued availability at low cost. Highly developed technologies have evolved based on the well known and specific hydrocarbon composition of petroleum and natural gas. Changing from these sources to biomass, would require the development and adaptation of new process technologies. Wishart[3] envisions 3 phases necessary to convert today's petroleum-based chemical industry to one based on biomass. During phase 1, chemicals will still be based on oil and gas resources, but they will be produced at a higher yield with improved processes and technologies. Phase 2 will be characterized by increased production of synthesis gas from heavy oil and coal to produce bulk chemical feedstocks such as ethylene and methanol. Production of chemicals from biomass will take place in phase 3 which is beginning to be developed now and which will be operative economically and with commercial significance later in the 21st century. A renewed interest in converting biomass components into organic chemical feedstocks is expressed by the increasing number of publications dealing with the subject[4–12]. I would like to discuss here some aspects of physico-chemical pretreatment and microbial conversion for the production of aromatic synthesis chemicals from wood.

Development of microbial processes

The conversion of lignin monomers to catechol with microbial communities under anaerobic conditions is outlined in the preceding article (Kaiser and Hanselmann). According to Wishart's plan, the microbial conversion processes which are being studied and improved today will not have an immediate applicability. Therefore, the decision for, or against, their development should not turn solely on how economical the processes are at present. A proper evaluation should be based on a consideration of the changing raw material resources and the time required to develop industrially interesting microbial fermentation processes, as well as an appreciation of microbial abilities to convert biomass efficiently. Although the production of chemical feedstocks via microbial processes is at present not economically competitive with the production of petrochemicals, the development of these processes should no longer be postponed. We know little about many of the potentially interesting biochemical pathways in microbes and even less about their suitability to be employed in biotechnology. The spectacular hopes invested in genetic engineering can only become reality if we gain enough knowledge about the physiological basis of the metabolic sequences that look useful for biomass conversion.

Wood as a biomass resource

Wood represents a stored form of renewable solar energy and materials (fig. 1). It is the most abundant

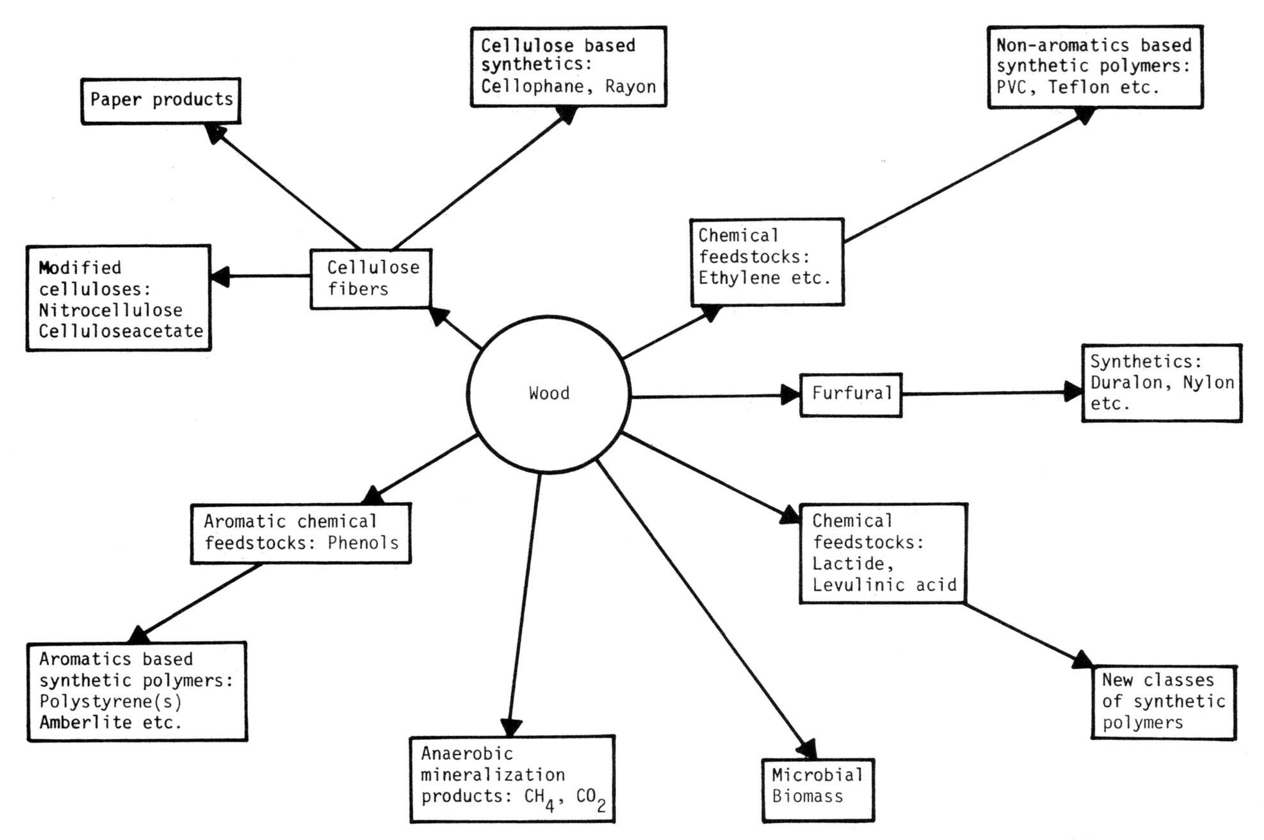

Figure 1. Integrated use of wood for the production of chemicals and paper. Through proper combination of physical, chemical or enzymatic pretreatment of wood and its components with microbial conversion processes, it is possible to obtain wood derived replacements for petrochemical based products.

product of terrestrial plants synthesized as an extremely stable, easily handled crop with a relatively high energy- and mass density (see Schwarzenbach and Hegetschweiler, this issue). Lignocelluloses form extremely stable structures which are very resistant to microbial attack under many environmental conditions. These chemical and physical properties have made wood a desirable building material and energy source for thousands of years. Only during the last few decades have coal, followed by natural gas and oil, begun to fulfill functions originally assigned to wood.

Before addressing the question of how wood might reassume its role as the basic raw material for conversion into fuels and chemicals, we should remind ourselves that nature supplies us with a raw material which, in itself, is many-sided. Wood polymers can often be utilized as – or even more – effectively in the natural form, without prior hydrolysis and degradation, to yield products for resynthesis into materials which will serve the same purpose. Appreciation of the beautiful structural characteristics of wood and its great versatility could also promote economy in the use of energy and materials. Turning to wood as a resource for chemicals will, therefore, be economically more interesting if some use can be made of the synthetic work performed by the plant during photosynthetic CO_2-fixation. Wood serves best if the molecules derived from it are those with the more complex structures: Cellulose, pentoses, hexoses and phenolic compounds. A process in which plant polymers are degraded to yield small molecules which serve as starting material for the synthesis of more complex ones is certainly less energy-efficient. It can be justified only if it produces feedstocks appropriate for existing chemical technologies. Wood can, for example, be converted to carbon monoxide and hydrogen through gasification at elevated temperatures and pressures[13]. The resulting gas mixture has a high H_2/CO ratio and is practically free of sulfur contaminants as opposed to gas originating from coal. This synthesis gas can serve as raw material for the chemical production of methanol, or possibly, for the production of acetate by *Butyribacterium methylotrophicum* (J.G. Zeikus, personal communication). Production and use of synthesis gas is hampered, economically, by the costs of the large amount of process energy required and, technologically, by the state of development of synthesis processes based on it[13].

Structurally more complex molecules might have to be isolated and purified from a mixture of very similar molecular species present in wood. Low yields

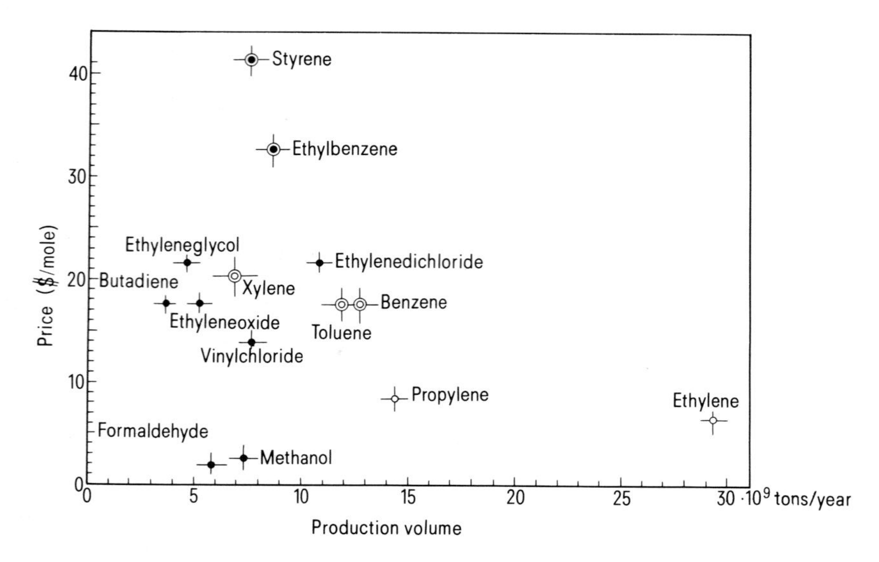

Figure 2. Unit prices and production volume of petrochemicals produced in the U.S. in 1979[34]. Vinylchloride, ethylenedichloride, ethyleneoxide, butadiene and ethyleneglycol are synthesized fromethylene. Ethylbenzene and styrene are based on benzene and ethylene. The high prices for the BTXes make lignin derived aromatics commercially interesting substitution products.

and processing difficulties might make the biomass approach less desirable. If the complex mixture of substrates can be converted microbiologically or enzymatically into a few smaller molecules (preferably belonging to one chemical substance class), however, the yield will increase and separation will become a minor problem (see Kaiser and Hanselmann, this issue).

Synthesis chemicals from wood

The synthetics industry has the largest demand for basic petrochemicals. In Western Europe, for example, 77% of the petrochemicals are used to synthesize polymers: 63% for plastics, 8% for fibers and 6% for elastomers. Solvents (13%), detergents (3%) and various other products (7%) account for approximately one fourth of the market volume[14]. Petroleum-based chemicals with the largest production volume in the U.S. are listed in figure 2. Table 1 shows possible means for deriving them – or replacement products – from biomass. Olefins can be obtained through fermentation of carbohydrates to ethanol and isopropanol and subsequent dehydration to ethylene and propylene. Many of the other primary chemical intermediates can be obtained through chemical modifica-

Table 1. Current primary chemical feedstocks and possible routes to plant chemical replacements

Rank*	Primary chemicals and intermediates	Current source**	Route to plant chemical replacement
1	Ethylene	P (cracking), NG	Dehydration of ethanol derived from anaerobic fermentation of carbohydrates
2	Propylene	P (cracking), NG; byproduct of ethylene production	Dehydration of isopropanol derived from anaerobic fermentation of carbohydrates
3	Benzene	P, C	?
4	Toluene	P, C	?
5	Ethylenedichloride	Synthesis from ethylene	Synthesis from ethanol
6	Ethylbenzene	Synthesis from ethylene and benzene	Synthesis from ethanol and ?
7	Vinylchloride	Synthesis from ethylene	Synthesis from ethanol
8	Styrene	Synthesis from ethylene and benzene	Synthesis from ethanol and ?
9	Methanol	NG, C	Byproduct of hemicellulose fermentation; microbial oxidation of methane
10	Terephthalic acid	P, C	?
11	Xylene	P, C	?
12	Formaldehyde	NG, P	Reduction of formate derived from anaerobic fermentation of carbohydrates
13	Ethyleneoxide	Synthesis from ethylene	Synthesis from ethanol
14	Ethyleneglycol	Synthesis from ethylene	Synthesis from ethanol
15	Butadiene	Synthesis from ethylene	Dehydration of butanediol derived from anaerobic fermentation or synthesis from ethanol

* Listed according to rank of 1979 U.S. production volume[34].
** P = petroleum, C = coal, NG = natural gas liquids (ethane, propane, butane).
? = biomass source and microbial processes are not known.

tion and synthesis based on a few fermentation products (fig. 3). For the moment, there are no satisfactory chemical or biological technologies available to supply aromatic hydrocarbons from biomass to replace the BTXs (benzene, toluene, xylene), although aromatic structures are present in abundance in lignin and other plant constituents.

The production from biomass of two entities would be of particular commercial interest: ethylene because of its large production volume and its importance as a basic chemical, and the aromatics because of their high market value (fig. 2). The development of chemical and microbiological techniques affording access to the valuable aromatics in the lignin polymer is hampered, however, by problems in separating monomers from a complex mixture of lignin hydrolysis products. The toxic effect of phenolic monomers on microbes and the recalcitrance of the lignin polymer to microbial hydrolysis under anaerobic conditions present further challenges.

Besides producing feed stock chemicals for synthetic purposes, one might search more intensively for enzymes capable of catalyzing conversion reactions at higher rates with better yields under less energy consuming conditions. Enzymatic conversion of propylene to propyleneoxide, for example, has been suggested by S.L. Neidleman (cited by Eveleigh[15]) with 3 enzymes (a pyranose-2-oxidase, a chloroperoxidase and a chlorohydrin epoxidase) derived from *Ondemansiella mucida, Caldariomyces fumago* and a *Flavobacterium sp.* respectively.

Qualitative availability of biomass

A great number of low molecular weight substances and polymeric products have been extracted from wood and bark (table 2). These are mostly specialty compounds with small production volumes but they demonstrate the great variety of plant chemicals that are available for particular purposes[16–18]. While purity and source-dependent constancy in composition are characteristic for petroleum resources, compositional flexibility and economic adaptability are properties attributed to plant biomass. Species with a high content of defined components can be grown according to needs for food, chemical feedstocks, synthesis chemicals, pharmaceuticals or fibers (fig. 4). Strains with the desired biomass composition (e.g. sunflowers with high seed oil content, grains with high content of amylose and very little amylopectin) can be bred or species carrying valuable traits can be selected from a large gene pool of the wild plants.

Thus the quality and quantity of biomass production can partially be controlled and be grown according to

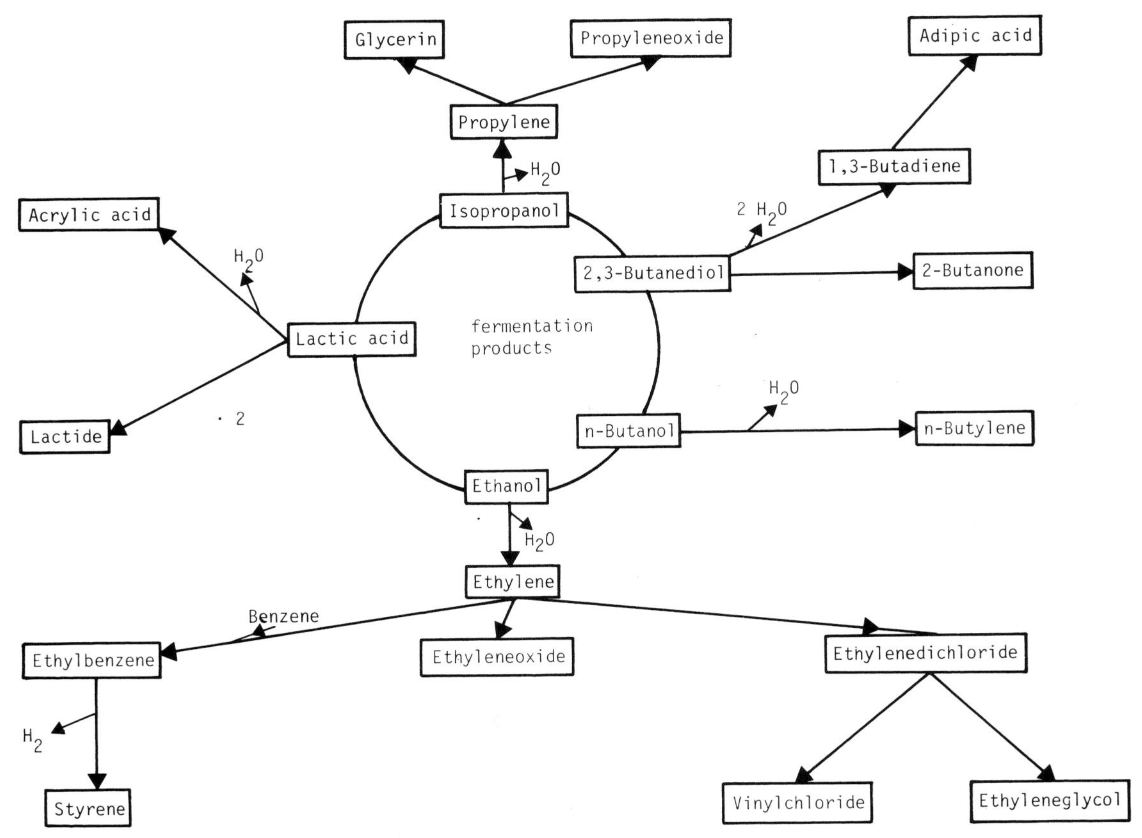

Figure 3. Fermentation products as feedstocks for synthesis chemicals. With the exception of the aromatics the most frequently used petrochemicals could be derived from a few products obtainable through anaerobic microbial fermentations and dehydration.

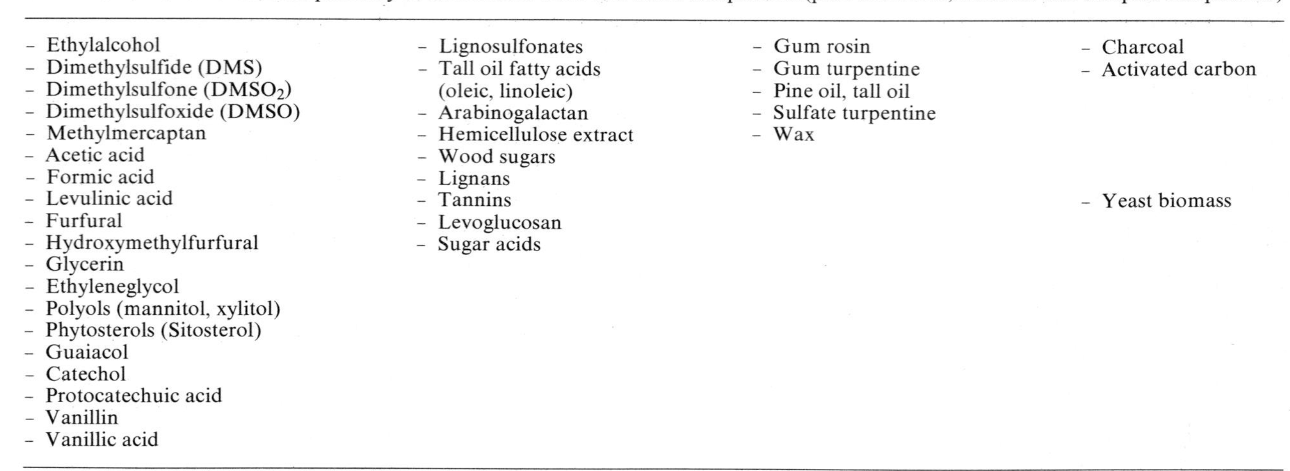

Table 2. Non-fibrous chemicals presently obtained from wood and wood components (pure chemicals, mixtures and complex components)

– Ethylalcohol	– Lignosulfonates	– Gum rosin	– Charcoal
– Dimethylsulfide (DMS)	– Tall oil fatty acids (oleic, linoleic)	– Gum turpentine	– Activated carbon
– Dimethylsulfone ($DMSO_2$)	– Arabinogalactan	– Pine oil, tall oil	
– Dimethylsulfoxide (DMSO)	– Hemicellulose extract	– Sulfate turpentine	
– Methylmercaptan	– Wood sugars	– Wax	
– Acetic acid	– Lignans		
– Formic acid	– Tannins		– Yeast biomass
– Levulinic acid	– Levoglucosan		
– Furfural	– Sugar acids		
– Hydroxymethylfurfural			
– Glycerin			
– Ethyleneglycol			
– Polyols (mannitol, xylitol)			
– Phytosterols (Sitosterol)			
– Guaiacol			
– Catechol			
– Protocatechuic acid			
– Vanillin			
– Vanillic acid			

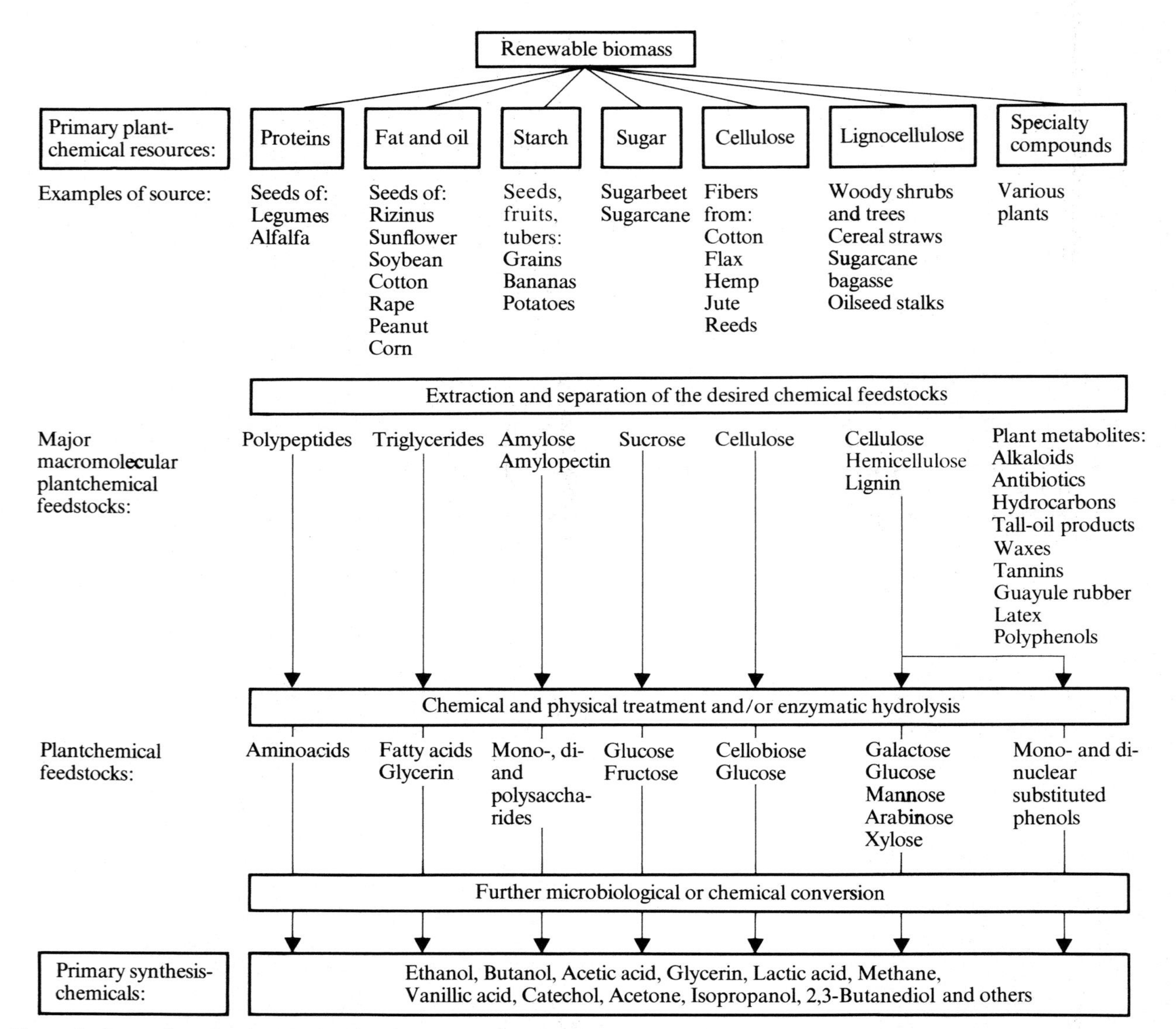

Figure 4. Conversion of primary plantchemical resources into primary synthesis chemicals. The great diversity of plant constituents combined with one of the many microbial fermentation pathways leads to high flexibility in the production of various synthesis chemicals.

needs. However, seasonal growth differences in temperate regions, limited species selection for different climatic zones, slow growth (particulary of certain desirable wood species), the vulnerability of plant monocultures to infectious diseases and insect pests, the relatively low ratio of biomass per unit area, the lack of appropriate and economic conversion facilities and technologies, and the competition with alternative uses, are disadvantages for a biomass-based chemical producing industry. To date, little effort has been expended on breeding woody plants to improve cellulose and lignin content. But some use can be made of the ecological diversity in the plant kingdom. The natural abundance of different species adapted to various environments offers a large pool of possibly desirable lignocellulose producers[19]. Sjöström[20] has compiled data on the composition of 20 tree species (table 3). Lignocellulose content varies between 60% and 76% of the dry weight. Softwood species have a higher lignin to cellulose ratio than hardwoods. The major aromatic building blocks of softwood lignin are the coumaryl-, coniferyl- and sinapyl precursors. However, the differences between carbohydrate and lignin content and the kind of aromatic monomers become less important if an integrated use of the whole biomass is considered.

Integrated use of wood for the production of chemicals

The great variety of anaerobic metabolic routes taken by different microorganisms allow utilization of any of the many forms of biomass (figs 3 and 4). This leads to great versatility in the kind of substrates that can be processed[21] and a flexible adaptability to products that are in great demand (e.g. fig. 2). Polymers with non-aromatic and aromatic building blocks could be synthesized, for example, if the microbial conversion processes suggested in the flow diagrams (fig. 5, a-e) proved to be technologically applicable and economically feasible.

After the 2nd World War, wood distilleries which produced many organic solvents through cellulose fermentation and other processes were closed because competition from inexpensive petroleum based chemicals could not be met. As a consequence, no research priority was given to the development of new manufacturing techniques for chemicals from alternative raw materials. Reversal of this trend is opportune today and should be fostered.

Many processes that use wood today do not use the lignin as raw material but as fuel. If the 'waste' produced in cellulose production is burned, for example, it supplies some of the process energy; if it has to be disposed of, it constitutes a burden to the environment. Cellulose production would become uneconomical if the full costs of disposal had to be included in manufacturing costs. But it can remain economical if lignin is converted into high priced chemicals. Thus the development of technologies that achieve better use of all the components present in a biomass resource should be encouraged.

Table 3. Lignocellulose content of some wood species*

	Lignocellulose (% of dry wood weight)	Lignin: cellulose ratio	Species
Softwoods (average of 9 species)	67.3	42:58	
1	69.1	40:60	*Picea abies*
2	68.2	39:61	*Larix sibirica*
3	68.2	45:55	*Tsuga canadensis*
4	68.1	43:57	*Pseudotsuga menziesii*
5	67.9	43:57	*Abies balsamea*
6	67.7	41:59	*Pinus silvestris*
7	67.0	41:59	*Picea glauca*
8	65.1	49:51	*Juniperus communis*
9	64.6	42:58	*Pinus radiata*
Hardwoods (average of 11 species)	67.3	36:64	
1	76.3	41:59	*Eucalyptus camaldulensis*
2	73.4	36:64	*Gmelina arborea*
3	73.2	30:70	*Eucalyptus globulus*
4	69.2	31:69	*Ochroma lagopus*
5	67.4	38:62	*Acer rubrum*
6	65.9	38:62	*Acer saccharum*
7	64.2	39:61	*Fagus silvatica*
8	63.7	33:67	*Acacia mollissima*
9	63.1	39:61	*Alnus incana*
10	63.0	35:65	*Betula verrucosa*
11	60.8	35:65	*Betula papyrifera*

* Calculated from data given by Sjöström[20].

Convertibility of biomass

Plants store photosynthesis products in special cellular organelles and in other structural components as polymers with large molecular weights or as hydrophobic aggregates[22]. Reactivity and osmotic activity of the monomers are decreased through polymerization, making stored plant reserves chemically inert substances. Our ability to harvest and store plant reserves is based on these characteristics. With the exception of lignin and cellulose in lignified tissues most plant polymers can be hydrolyzed easily with the aid of enzymes. The hydrolysis of wood-cellulose into fermentable sugars, however, represents a rate limiting step in the cellulose to ethanol conversion process. Thus large research efforts have been directed toward improving microbial productivity of cellulases. The lignin polymer with its non-reactive intermonomer C-C and C-O linkages is very slowly degraded by some fungi and bacteria under aerobic conditions (see Higuchi, this issue). In anaerobic environments lignin is recalcitrant[35]. Lignin incrustations make cellulose in vascular and supportive tissues inaccessible to the cellulases of many organisms. For practical purposes, therefore, mechanical or chemical dissolution of the lignocellulose complex has to be employed. Today, steam explosion of wood is regarded as the most efficient separation process[11]. This process yields cellulose digestable to microorganisms, partially hydrolyzed hemicellulose, and low molecular weight lignin oligomers consisting of 3–10 aromatic units. Further hydrolysis yields a mixture of phenolic compounds whose composition and yield depend on the cracking procedure employed and the source of the lignin. Alkaline oxidation at elevated pressures and temperatures in the presence of a catalyst, for example, yields

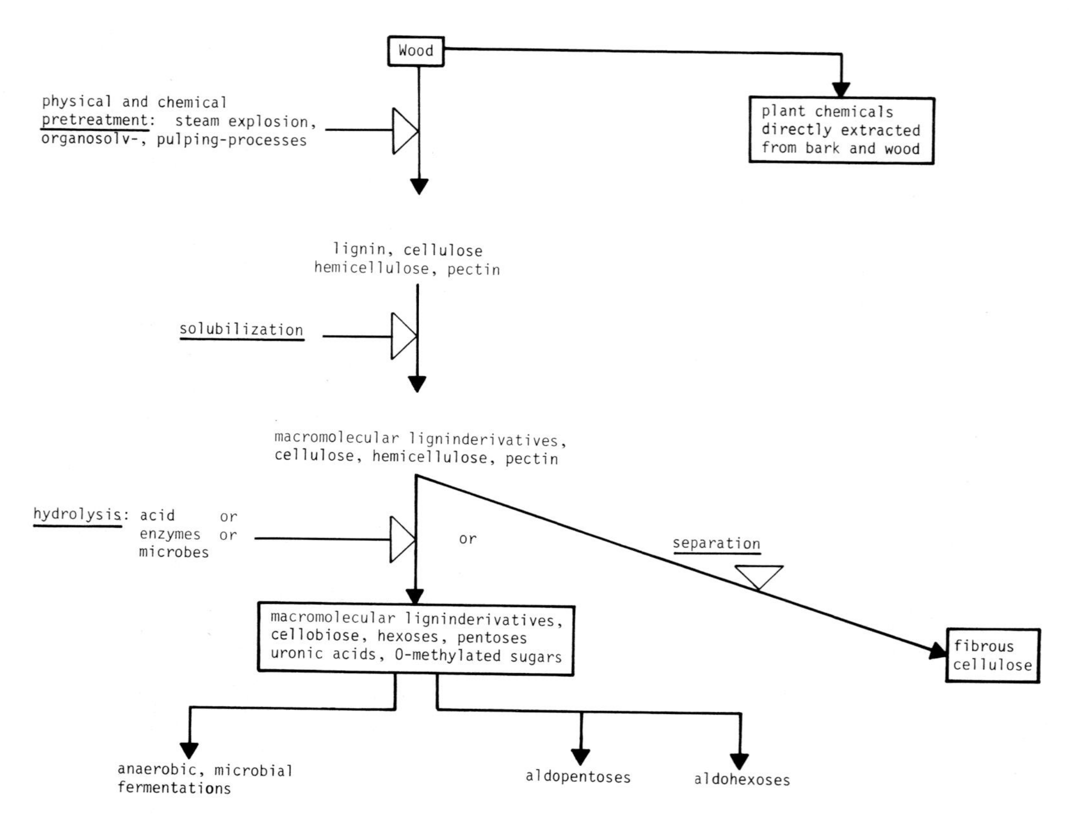

Figure 5. *a* Extraction, partial solubilization and separation of some wood components.

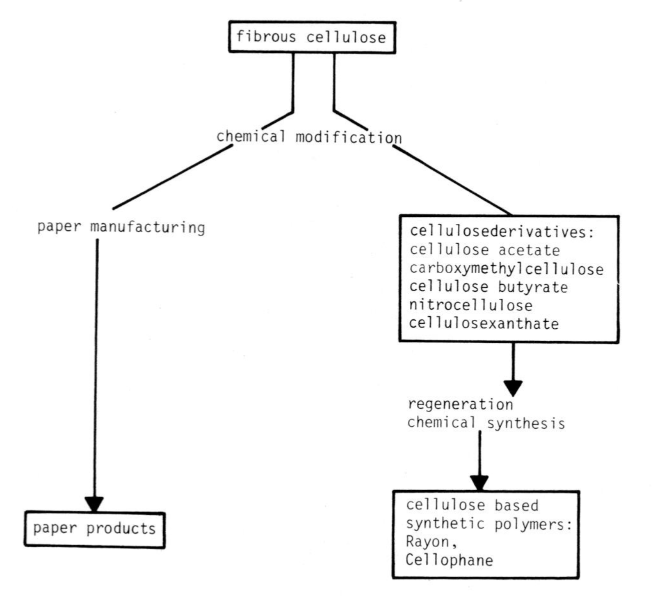

Figure 5. *b* Synthetic polymers based on fibrous cellulose.

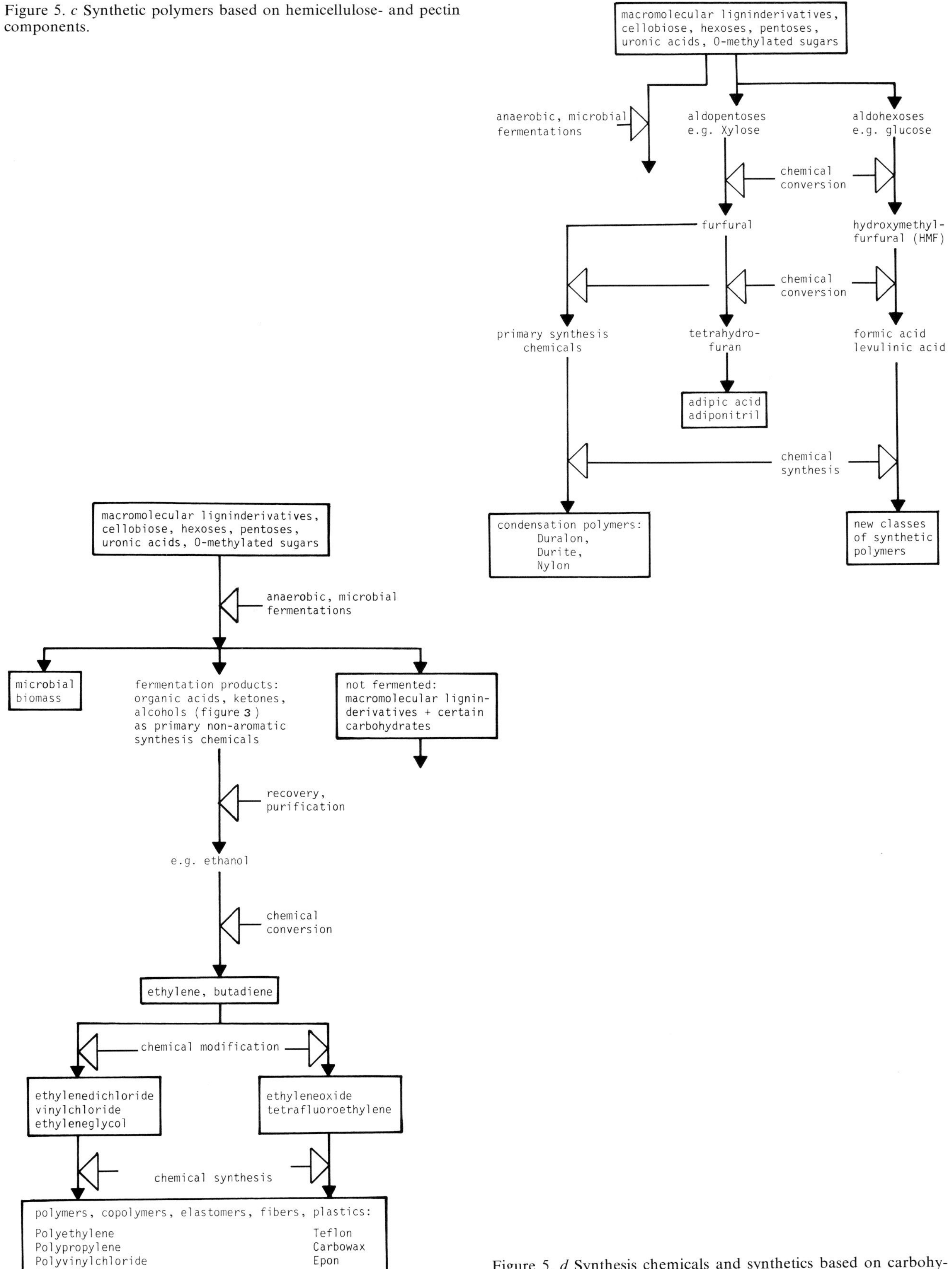

Figure 5. *c* Synthetic polymers based on hemicellulose- and pectin components.

Figure 5. *d* Synthesis chemicals and synthetics based on carbohydrate to ethanol fermentation.

Haraguchi and Hatakeyama[33] have synthesized polystyrenes with building blocks that are structurally related to lignin monomers: Poly(p-hydroxystyrene), poly(3-methoxy-4-hydroxystyrene) and poly(3,5-dimethoxy-4-hydroxystyrene). Under aerobic conditions these substances were biodegradable by mixed microbial cultures enriched from soil. Thus, introduction of functional groups which are characteristically present in lignin monomers change polystyrene into a biodegradable polymer. These few encouraging results on production and application of lignin-derived aromatics support the technological feasibility of using wood as a resource for aromatic synthesis chemicals. Some ingenuity in chemical synthesis may be needed to find ways to use other lignin-derived aromatics.

Economic aspects

Lignochemicals become economically attractive if the costs for the raw material and the production of desired chemicals are competitive with the costs for comparable chemicals obtained from fossil oil. Today the technology and the plants for the refining of oil and gas are available while corresponding facilities for lignochemistry would still have to be developed. This means large investments into an enterprise that will presumably not become commercially interesting within the next few decades[3]. A proper cost-benefit analysis cannot be made for most of the proposed biological processes since many have not even reached the pilot-plant stage.

Although there are countries with large reserves of wood, this alone would not justify the buildup of a lignochemical industry. We will not solve our materials problems for the future, and probably will cause great ecological damage by using the biomass accumulated over several centuries at a faster rate than at which it can be reproduced. For a balanced lignochemical industry, the source for raw materials has to be secured through silviculture. Selection and planting of the species best adapted to a particular climatic region and their growth until harvest, however, takes many years (see Schwarzenbach and Hegetschweiler, this issue).

The schedule to be followed for the next decades, therefore, will demand restricted use of oil and natural gas, reserving them for the production of chemical feedstocks. Meanwhile, the developing

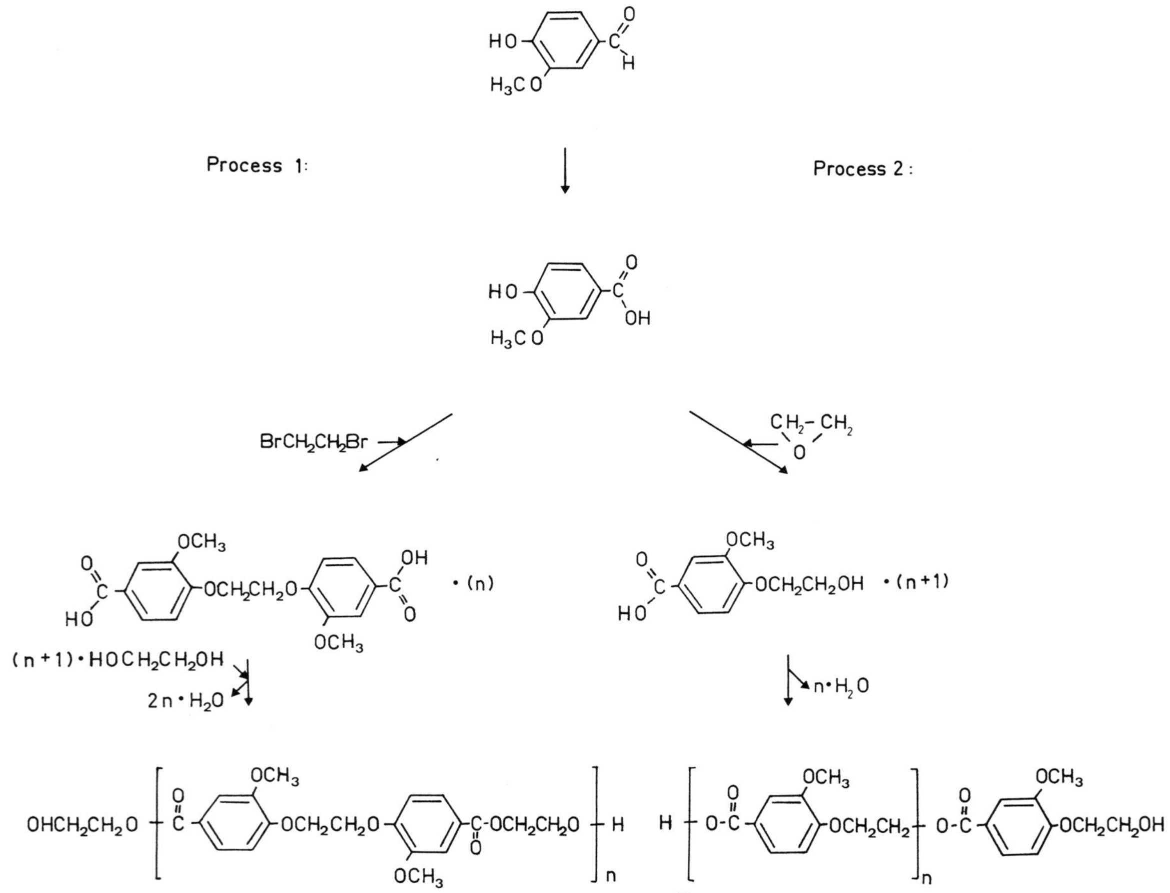
Process 1:
Process 2:
HO
H₃CO
OH
BrCH₂CH₂Br
CH₂-CH₂
O
OCH₃
OCH₂CH₂O
• (n)
OCH₂CH₂OH
• (n+1)
(n+1)•HOCH₂CH₂OH
2n•H₂O
n•H₂O
OHCH₂CH₂O
C-OCH₂CH₂O
H
n
O-C
OCH₂CH₂
OCH₂CH₂OH

Figure 6. Polymer synthesis with vanillic acid monomer derived from lignin[32].

wood-based chemicals industry should concentrate on specialty products that can be made from lignocellulose. A more efficient use of wood could increase revenues and make existing wood processing plants economically more attractive. An example shall serve as illustration. The production of 1 m^3 of ethanol by the Inventa process requires 5.2–7.7 tons of wood[30]. Assuming an average lignin to cellulose ratio of 2:3 (table 3) one is left with 2–3 tons of lignin per m^3 of ethanol. In the Inventa process the lignin serves as fuel for process energies. If alternative energy sources can be applied, lignin would be available for hydrolysis and microbial conversion to valuable aromatic synthesis chemicals. In this way one could achieve the production of olefins (from ethanol) and aromatic hydrocarbons in the same plant, an option that would appear profitable. Lignin residues produced by paper mills could similarly increase revenue for this group of wood users.
Goldstein[31] showed that about 95% of the polymers produced today theoretically could be manufactured from monomers derived from lignocellulose: 47% with ethylene, 12% with butadiene and 36% based on phenolic compounds as building blocks. An average of 2.62 tons of lignocellulose would be required to produce 1 ton of lignin-derived synthetics (table 5). This ratio could be improved if the yields of aromatics obtained from lignin in physico-chemical procedures could be increased in combination with microbial conversion.
It will be difficult to find substitutes for the large quantities of fossil fuels that are used today for transportation and heating[12]. To replace petrochemicals by lignochemicals, today, is not so much a question of quantity as a problem of processes and economics.
Petrochemical products account for approximately one third of the sales turnover and half of the total capital investments of the chemical industries[2]. Worldwide production of petrochemicals is divided presently among Western Europe (30%), North America (25%), Japan (10%), countries with central economic planning (25%) and other countries (10%). Oil producing countries are now developing their own petrochemical industry rapidly in order to decrease their dependence on imported chemical products and to increase their share in the world market for petrochemicals. Oil and natural gas resources are plentiful there and the gas, for example, is presently not at all being used economically. With regard to raw materials, energy and production costs, the OPEC countries are in an enviable situation. The financial

Table 5. Estimate of lignocellulose raw material required for the production of plastics, synthetic fibers and synthetic rubber*

Product	Tons of lignocellulose required per ton of product synthesized	Principal biomass source of monomers
A) Plastics		
Epoxies	2.84	Lignin
Polyesters	2.68	Lignin
Phenolic resins	2.86	Lignin
Polyamides	2.86	Lignin
Styrene and copolymers	2.97	Lignin
Polyethylene	4	Cellulose
Polypropylene and copolymers	4	Cellulose
Polyvinylchloride	1.74	Cellulose
Other vinyl resins	2.51	Cellulose
Average lignin-derived plastics	2.84	
Average cellulose-derived plastics	3.06	
B) Synthetic fibers		
Polyamide fibers (Nylon)	2.86	Lignin
Polyester fibers (Dacron)	2.68	Lignin
Acrylic fibers	2	Cellulose
Olefinic fibers	4	Cellulose
C) Synthetic elastomers (rubbers)		
Butadiene-styrene	3.53 + 1.19	Cellulose + lignin
Isoprene-isobutylene (butylrubber)	5.89	Cellulose
Nitrile	2	Cellulose
Polybutadiene	5.89	Cellulose
Ethylene-propylene	5.89	Cellulose
Average cellulose-derived synthetic elastomers	4.64	
Average lignin-derived synthetics	2.62	
Average cellulose-derived synthetics	3.77	

* Calculated from data given by Goldstein[31], based on 'optimistic approximate yields of the monomers obtainable from wood'.

assets which they have accumulated since the drastic increase of the oil prices allow those countries to finance the transfer of technological know-how and the chemical plants from industrialized nations. The European petrochemical industry will therefore be pressured from two sides: by the dwindling availability and the rising prices of petroleum on one hand and by the market competition for refined petrochemicals and synthetic products on the other. Two of the options the chemical industry might consider are:

1. A partial substitution of fossil hydrocarbons by renewable biomass as the alternative basis of chemical resources and,
2. a reorientiation towards development of advanced chemical and biochemical technologies for the production of more refined basic chemicals for medicinal, agricultural and chemical uses.

Intelligently applied biochemical capabilities of microorganisms could play a decisive role in this transition.

Conclusions and outlook

Would there be enough wood to supply the raw material for chemical feedstock production? Fulfilling the demands of the plastics industry in Switzerland, for example, would require 10 times the amount of wood which is presently processed by the paper industry. Today's silviculture, mostly directed towards paper manufacturing, fire wood production and the lumber industry could not supply the additional quantities needed for lignochemical production. Large forest areas would have to be carefully cultivated, without upsetting their ecological functions, to yield the amount of wood needed by a lignochemical industry. Silviculture with new species, well adapted to local conditions could become a reservoir for renewable lignochemical feedstocks. It is assumed that low grade wood and woody shrubs could be used for the production of chemicals while wood of high quality should be reserved for lumber, plywood and pulp. Also the lignin residues that remain after cellulose extraction in paper mills constitute an unused source for the production of synthesis chemicals.

In the future we will be forced to make more intensive use (and, it is to be hoped, more intelligent use) of the ability of green plants to synthesize a multitude of chemical structures from CO_2 and H_2O. Three routes of development might be persued concurrently:

1. Selection of plants that produce high yields of a desired biomass;
2. search for microbiological processes to convert plant raw materials into 'useful' chemicals; and
3. development of chemical techniques to process new classes of chemicals in synthesis.

The transition from fossil hydrocarbons to renewable biomass requires farsighted decisions and long term investments.

1 Acknowledgment. I thank Ms H. Müller, Mr J.P. Kaiser and Dr G. Hanselmann-Mason for their help in preparing the illustrations and the manuscript.
2 Shell Briefing Service, Die Zukunft der Petrochemie in Europa. Shell Switzerland, Zürich 1981, 11 pp.
3 R.S. Wishart, Industrial energy in transition: a petrochemical perspective. Science *199*, 614–618 (1978).
4 P.H. Abelson, Energy and chemicals from biomass. Science *213*, 605 (1981).
5 L.C. Bratt, Wood-derived chemicals: Trends in production in the U.S. Pulp and Paper *53*/6, 102–108 (1979).
6 A.L. Compere and W.L. Griffith, Industrial chemicals and chemical feedstocks from wood pulping wastewaters. Tappi *63*/2, 101–104 (1980).
7 R.L. Crawford, Lignin, Biodegradation and Transformation. J. Wiley, New York 1981; 154 pp.
8 I.S. Goldstein, ed., Organic chemicals from biomass. To be published by CRC Press, Boca Raton, FL.
9 F.W. Herrik and H.L. Hergert, Utilization of chemicals from wood: retrospect and prospect. Rec. Adv. Phytochem. *11*, 443–515 (1977).
10 A. Hollaender, R. Rabson, P. Rogers, A. San Pietro, R. Valentine and R. Wolfe, eds., Trends in the biology of fermentation for fuels and chemicals. Plenum Press, New York 1981.
11 E.S. Lipinsky, Chemicals from biomass: Petrochemical substitution options. Science *212*, 1465–1471 (1981).
12 K.V. Sarkanen, Renewable resources for the production of fuels and chemicals, in: Materials: Renewable and non-renewable resources. Ed. P.H. Abelson and A.L. Hammond. Special Science Compendium No.4, 184–188. AAAS, Washington, DC 1976.
13 R.L. Pruett, Synthesis gas: a raw material for industrial chemicals. Science *211*, 11–16 (1981).
14 Shell Briefing Service, Oil and gas in 1980, No.3, 9 pp. Shell International Petroleum Co. Ltd, London 1981.
15 D.E. Eveleigh, The microbial production of industrial chemicals. Scient. Am. *245*, 120–130 (1981).
16 R.A. Buchanan, F.H. Otey and M.O. Bagby, Botanochemicals. Rec. Adv. Phytochem. *14*, 1–22 (1980).
17 D.W. Goheen, Low molecular weight chemicals, in: Lignins, Occurrence, Formation, Structure and Reactions, p.797–832. Ed. K.V. Sarkanen and C.H. Ludwig. Wiley Intersci., New York 1971.
18 C.H. Hoyt and D.W. Goheen, Polymeric products, in: Lignins, Occurrence, Formation, Structure and Reactions, p.833–865. Ed. K.V. Sarkanen and C.H. Ludwig. Wiley Intersci., New York 1971.
19 National Academy of Sciences, Firewood crops: shrub and tree species for energy production. NAS, Washington, DC 1980; 237 pp.
20 E. Sjöström, Chemicals from wood and by-products after pulping, in: Wood Chemistry: Fundamentals and Applications, chapter 10, p.190–208. Academic Press, New York 1981.
21 E.S. Lipinsky, Fuels from biomass: Integration with food and materials systems. Science *199*, 644–651 (1978).
22 K. Hanselmann, Wie Pflanzen Reservestoffe speichern. Biologie in unserer Zeit *9*/*4*, 103–111 (1979).
23 P.L. McCarthy, L.Y. Young, J.M. Gossett, D.C. Stuckey and J.B. Healy, Heat treatment for increasing methane yields from organic materials, in: Microbial energy conversion, p.179–199. Ed. H.G. Schlegel and J. Barnes. Pergamon Press, Oxford 1977.
24 J. Nakano, Trends of utilization of pulping spent liquor in Japan. Appl. Polymer Symp. *28*, 85–91 (1975).
25 I.S. Goldstein, Perspectives on production of phenols and phenolic acids from lignin and bark. Appl. Polymer Symp. *28*, 259–267 (1975).
26 I.A. Pearl, The chemistry of lignin. Marcel Dekker, New York 1967.

27 P. Ander and K.-E. Eriksson, Lignin degradation and utilization by microorganisms. Prog. ind. Microbiol. *14*, 1–58 (1978).
28 I.S. Goldstein, New technology for new uses of wood. Tappi *63/2*, 105–108 (1980).
29 Inventa, Ethanol process by wood saccharification. Process Description 81-Dl. Inventa, Domat-Ems 1981.
30 P. Wettstein and B. Domeisen, Production of ethanol from wood. 1st int. energy agency (IEA) conference on new energy conservation technologies and their commercialization, 1981.
31 I.S. Goldstein, Potential for converting wood into plastics, in: Materials: Renewable and nonrenewable resources, p. 179–184. Ed. P.H. Abelson and A.L. Hammond. Special Science Compendium No. 4. AAAS, Washington, DC 1976.
32 J.J. Lindberg, V.A. Erä and T.P. Jauhiainen, Lignin as a raw material for synthetic polymers. Appl. Polymer Symp. *28*, 269–275 (1975).
33 T. Haraguchi and H. Hatakeyama, Biodegradation of lignin-related polystyrenes, in: Lignin Biodegradation: Microbiology, Chemistry and Potential Applications, vol. 2, p. 147–159. Ed. T.K. Kirk, T. Higuchi and H. Chang. CRC Press, Boca Raton, FL 1980.
34 B.O. Palsson, S. Fathi-Afshar, D.F. Rudd and E.N. Lightfood, Biomass as a source of chemical feedstocks: an economic evaluation. Science *213*, 513–517 (1981).
35 J.G. Zeikus, Fate of lignin and related aromatic substrates in anaerobic environments, in: Lignin Biodegradation: Microbiology, Chemistry and Potential Applications, vol. 1, p. 101–109. Ed. T.K. Kirk, T. Higuchi and H. Chang. CRC Press, Boca Raton, FL 1980.

The formation of methane from biomass – ecology, biochemistry and applications

R.E. Hungate explains why methane as a product and cellulose as a substrate command so much attention today; he then examines the ecology of one natural methane producing system, the rumen. Other natural ecosystems are discussed by K. Wuhrmann.
The known biochemical pathways of methanogenesis are reviewed by R.S. Wolfe. It is astonishing how some of the structural and chemical properties of the cells are uniquely and specifically restricted to the group of methanogens. The practical engineering, operational and economic aspects of the methane production are reviewed by J.T. Pfeffer who discusses substrate properties, process characteristics, residue disposal and costs.
The special case of biogas production from agricultural wastes, especially those from animals, is presented by P.N. Hobson and practicable high- and low-technology systems are compared, particularly in regard to their application in underdeveloped countries.
The article by M. Gandolla and co-workers describes a small experimental landfill which, in producing methane, is a solid state fermentation system. This paper stresses many of the practical problems and considerations associated with such a system (composition of gas, leakage at the landfill, purification, storage, utilization and safety).

Methane formation and cellulose digestion – biochemical ecology and microbiology of the rumen ecosystem

by R.E. Hungate
Department of Bacteriology, University of California, Davis (CA 95 616, USA)

In postulates[1] on the Earth's origin, gaseous chemical elements combined with each other during cooling; compounds with the highest boiling points condensed first, followed by those containing lighter elements. Living material, formed gradually by chemical reactions of the lighter elements and traces of the heavy ones, was peculiar in its tendency to revert to non-living material unless chemical work maintained its living state. Abundant non-living compounds of C, H, and O also formed in various proportions. The relatively large energy changes involved in the oxido-reduction of these elements equipped them to be agents for the chemical work.
Because the Earth was initially anaerobic, with insufficient oxygen to combine completely with the available carbon and hydrogen, much of the carbon must have been in an intermediate state of oxidation. In the absence of O_2, energy was not available through the oxidation of carbon to CO_2 and H_2O, but it could be derived by converting C atoms at an intermediate state of oxidation to CO_2 and CH_4. These molecules are in a low energy state anaerobically, incapable of further redox reactions except that CO_2 can be reduced with H_2 to CH_4. Thus methane was, presumably, an important waste product of early metabolism, and methanogenesis was a primitive phenomenon, accomplished by possibly many diverse forms.

The primitive carbon compounds at intermediate states of oxidation do not have an equal potential for chemical work. This is evident from the following comparisons of single-C compounds containing both H and O[2].

$$4\ HCOOH\ (\text{formic acid}) \rightarrow 3\ CO_2 + CH_4 + 2\ H_2O \quad (I)$$
$$\Delta G_0' = -120\ kJ$$
184 daltons -0.65 kJ/dalton

$$4\,CH_3OH\ (\text{methanol}) \rightarrow 3\,CH_4 + CO_2 + 2\,H_2O \qquad (II)$$
$$\Delta G_0' = -311\ kJ$$
128 daltons $\quad -2.43$ kJ/d

$$2\,CH_2O\ (\text{formaldehyde}) \rightarrow CO_2 + CH_4 \qquad (III)$$
$$\Delta G_0' = -176\ kJ$$
60 daltons $\quad -2.93$ kJ/d

Simultaneous oxidation and reduction of the intermediate state, CH_2O, gives 20% more work potential per unit weight than does that of methanol. This may have been sufficient to select carbohydrate as a preferred substrate for metabolic reactions supporting primitive anaerobic life, with CO_2 and CH_4 as low energy waste products[3]. This low energy state anaerobically can be inferred also from the fact that they are the end products of the anaerobic fermentation of most kinds of organic matter, not just carbohydrates.

$$C_6H_{12}O_6 \rightarrow 3\,CO_2 + 3\,CH_4 \qquad (IV)$$
$$\Delta G_0' = -393\ kJ$$

This is the maximum energy derivable anaerobically from glucose, as compared to the 2650 kJ released aerobically by oxidation to CO_2 and H_2O, or the 2255 kJ available from the oxidation of 3 CH_4.

Phosphorus and nucleic acids were drawn early on into metabolism, and participated in oxido-reductions in such a way that adenosine triphosphate (ATP) became a convenient high energy (44 kJ) molecule[2] for accomplishing biochemical work.

Carbohydrates such as glucose, less reactive than formaldehyde, have the advantages that they are less toxic and that the energy of redox reactions can be channeled through ATP to transform individual carbon atoms into the carbon skeleton precursors of amino acids and other cell components. The sugars can also be polymerized into large insoluble molecules such as starch and glycogen, well suited for storage, and common in anaerobes.

Photosynthesis with O_2 evolution made carbohydrate extremely cheap, manufacturable in great quantity from light, water and CO_2. In the form of cellulose it was used to strengthen the walls of aquatic plant cells, a non-energy-requiring mechanism preventing osmotic plasmoptysis. At the time land plants appeared, cellulose was used to support aerial organs, and has become an abundant and almost universal component of higher plants. Although digestible to sugar by many organisms, it is hydrolyzed less readily than starch or glycogen.

Lignin. Do we then have huge supplies of cellulose? The answer is yes. Annual world power consumption has been estimated at 10^{17} kJ. The maximum annual agricultural production (possibly 50% cellulose) has been estimated to be equivalent to 2×10^{17} kJ, if oxidized. This is a large amount, but unfortunately much of it is not readily exploitable through microbial fermentation. The lignins, less digestible than cellulose, are linked with it in a way that diminishes cellulose digestibility.

It is commonly believed that lignin cannot be digested and fermented anaerobically, but the aromatic ring structure may actually be susceptible, as is benzoate[4], to microbial oxido-reduction to CO_2 and CH_4. Perhaps the insolubility of lignin is the impasse blocking lignin fermentation. The achievement of successful fermentation of lignin would greatly increase the amount of fermentable cellulose. Alkali treatment of lignified plant materials increases their digestibility, but the process is still incomplete.

Cellulose conversion to methane. Methanogenesis is commonly encountered in fermenting systems containing cellulose. Early investigators assumed that methane was a waste product of the cellulolytic bacteria, but lacked knowledge of anaerobiosis and of the nutrients needed to grow axenic cultures of the cellulolytic and the methanogenic bacteria. In their agnotobiotic enrichment cultures containing pure cellulose, aerobic and euryoxic bacteria mopped up the traces of O_2 initially present. Various members of the microbial consortium produced small amounts of needed nutrilites, and the metabolic products of the total biota helped to create a highly reduced environment.

Because of the slow digestibility of the cellulose and the limited synthesis of nutrilites, the acid fermentation products of the cellulolytic bacteria did not exceed the buffering capacity of the medium. Methanogens could develop fast enough to keep pace with acid production, and the cellulose was gradually converted to CO_2 and CH_4.

Improved culture methods have yielded pure cultures of both methanogenic[5,6] and cellulolytic[7,8] bacteria, and revealed the existence of the microbial consortium concerned. Modern methanogens are stenotrophic. All except one[12] isolated pure cultures can grow at the expense of the reaction[2],

$$4\,H_2 + HCO_3^- + H^+ + \tfrac{1}{2}\,P_i + \tfrac{1}{2}\,ADP \rightarrow CH_4 + 3\tfrac{1}{2}\,H_2O + \tfrac{1}{2}\,ATP \qquad (V)$$
$$\Delta G_0' = -96\ kJ$$

Some strains can use also formate, methanol and acetate[5,6]. Also, most of the actively cellulolytic anaerobic bacteria are stenotrophic, fermenting cellulose and its derivatives and only a few other substrates. Various organic nutrilites are either required by most strains of both these groups or are stimulatory.

The rumen ecosystem

The rumen is an essentially anaerobic ecosystem[7] containing hundreds of kinds of microbes in total concentrations as high as 2×10^{10} bacteria and 10^6 protozoa/ml. 40–60% of the cellulose and hemicellulose and most of the starch and pectin in the plant material are digested, and the sugars formed, together with those in the feed, are fermented to volatile fatty acids (vfa), CO_2, CH_4 and microbial cells.

The ruminant is an open system fermentor, not requiring sterilization or expensive procedures for growing pure cultures. It supports continuous fermentation by taking in water, gathering plant bodies, comminuting and retaining them for microbial action in a large fermentation chamber; it maintains a constant favorable temperature and anaerobiosis; regulates the pH with bicarbonate and by absorption and utilization of fermentation acids; transports and voids undigested residues; and digests the microbial cells and converts them, with the acids, into meat, milk, wool and hides[8].

Each fermentation unit is small and mobile enough for adequate substrate collection, yet the total vast capacity on earth, realized through a multiplicity of units, produces enough methane to meet the average requirements of one-fourth of the world's human population. Unneeded units have a high sale value, and pairs of units can periodically produce new fermentors, all this with no human intervention except for exploitation.

These attributes are not easy to simulate. In times when energy and materials (also representing much energy) are to be conserved, industrial conversions of plant materials can profitably be compared with the fiber economy of the ruminant. In turn, industry might conceive of ingenious means to collect ruminant methane. If ruminant husbandry moves toward the massive battery stage already characteristic of pig and fowl production, methane collection may not be entirely fanciful.

Most plant material is fibrous and fiber cannot be fermented until it has been digested to soluble sugars. Digestion is slow because enzymes act only at the surface, and even there they do not act until the surface of ingested fibers is covered by fibrolytic enzymes and bacteria. Electron microscopy has disclosed that some plant tissues digest completely without bacterial attachment but bacteria attach to other tissues, eat out 'Frassbetten' in the cell walls, multiply and ultimately dissolve out the digestible fibrous components. Attachment may increase the chance of absorbing the soluble products. Enzyme action is hampered also by the chemico-physical bonding of lignin in the fiber.

In spite of these limitations microbes can ultimately digest and metabolize most plant material. The holes digested through cell walls of lignified wood by higher fungi suggest the presence of a formidable array of enzymes. These fungi grow slowly and use fixed nitrogen with great economy. Could this be associated with high energy costs for enzyme elaboration?

The present analytical categories for plant fiber (cellulose, hemicellulose and lignin) do not correspond to biological digestibility. An improvement in the knowledge of plant fiber composition might be obtained by applying various permutations and combinations of pure enzymes to the fiber itself as substrate. The false concept of a C_1 enzyme (not cellulolytic but necessary to prepare 'native' cellulose for digestion) has arisen largely from use of artificial soluble derivatives of fiber instead of the more slowly digested natural material.

Microbes can use the sugars from fiber digestion faster than they can be formed. The usual low sugar concentrations in the rumen limit the speed of microbial growth, with biochemical mechanisms increasing the growth rate through maximal microbial synthesis per unit of substrate fermented.

Apparently a single species cannot perform at maximal rate all possible biochemical work reactions. A community of specialized microbes, linked by various interactions, is more effective. Possible interactions include: a) excretion by some of wastes used by others, b) leakage of useful metabolites into the milieu, c) commensalism, as in the escape of fiber digestion products before they are captured by the cell elaborating the responsible enzymes, d) predation, e) parasitism, and f) synergistic colony formation.

Changes in the ruminant's feed can induce changes in its microbiota, but as long as the substrate remains limiting, the biochemistry remains stable, in that ATP production is maximal for a wide variety of feeds. Acetic, propionic and butyric acid are the chief volatile fatty acid products of the rumen fermentation, in the proportions 65, 20 and 15%, respectively.

The biochemical work possible when acetate is formed can be summarized as follows[2]:

$$C_6H_{12}O_6 + 4\,P_i + 4\,ADP \rightarrow 2\,CH_3COO^- + 2\,HCO_3^- + 4\,H^+ + 4\,H_2 + 4\,ATP \quad \Delta G_0' = -51\text{ kJ} \qquad \text{(VI)}$$

Pyruvate (2 molecules) is an intermediate. 2 ATP and 4 H (2 NADH $\times$ 2 H^+) arise during its formation, and 2 ATP and 2 H_2 are formed in the conversion of the pyruvate to acetate. In theory the 8 H can reduce 1 CO_2 to 1 CH_4, a 3rd of that in equation IV, but the actual rumen methane production is always less because some H is used for cell synthesis (cells are more reduced than carbohydrate) and some for other reductions, the chief one being propionate formation. When a molecule of hexose is fermented to propionate, the 2 H atoms needed can be generated endogenously along with acetate, giving

$$C_6H_{12}O_6 + 4\,P_i + 4\,ADP \rightarrow \tfrac{4}{3}\,CH_3CH_2COO^- + \tfrac{2}{3}\,CH_3COO^- + 2\,H^+ + \tfrac{2}{3}\,CO_2 + 4\tfrac{2}{3}\,H_2O + 4\,ATP \quad \Delta G_0' = -131\text{ kJ} \qquad \text{(VII)}$$

(or exogenously) from the H_2 in rumen liquid,

$$C_6H_{12}O_6 + 2\,H_2 + 4\,P_i + 4\,ADP \rightarrow 2\,CH_3CH_2COO^- + 2\,H^+ + 6\,H_2O + 4\,ATP \quad \Delta G_0' = -173\text{ kJ} \qquad \text{(VIII)}$$

The ATP in propionate production is formed by electron transport in fumarate reduction to succinate, a precursor of propionate. The $\Delta G'$ of -147 kJ in the combined equations V and VI,

$$C_6H_{12}O_6 + 4\tfrac{1}{2}\,P_i + 4\tfrac{1}{2}\,ADP \rightarrow 2\,CH_3COO^- + HCO_3^- + 3\,H^+ + CH_4 + 3\tfrac{1}{2}\,H_2O + 4\tfrac{1}{2}\,ATP \quad \Delta G_0' = -147\ kJ \qquad (IX)$$

is about the same as in equations VII and VIII for propionate formation, but the theoretical yield of ATP is slightly more. This may explain the prevalance of acetate in the rumen, where the acetate/propionate ratio is 3, as compared to $\tfrac{1}{2}$ in equation VII.

This advantage in acetate production is a good example of increased growth through species interaction. The rumen methane bacteria have a high affinity for H_2[9], oxidizing it with CO_2 at 10^{-3} atm of H_2, at which concentration a conversion of $NADH + H^+$ to $NAD + H_2$ can occur.

It is difficult to understand biochemically how the work accompanying butyric acid production can compete with that when acetate and propionate are formed. The recent report[10] that CO_2 increases the growth yield of *Butyrivibrio*, the chief rumen producer of butyrate, may be pertinent.

In addition to the reactions summarized in equation IX, complete conversion of hexose to CO_2 and H_2O as in equation IV, must include the following:

$$2\,CH_3COO^- + 2\,H_2O \rightarrow 2\,HCO_3^- + 2\,CH_4 \quad \Delta G_0' = -62\ kJ. \qquad (X)$$

The work of ATP synthesis which should be integrated into this reaction has not been measured, but it is even less than that in equation V. Perhaps for this reason methanogenic growth on acetate is extremely slow, slower than the average rumen dilution rate, *k*.

The rate of dilution of rumen contents is also important in fiber digestion. Slow dilution (passage) rates allow the fiber to remain in the rumen longer, and its digestion is more complete. But at the concomitant slower growth rate, with longer intervals between cell division, more work is expended for microbial maintenance, and the microbial cell yield per molecule of fermented substrate diminishes. This lower cell yield leaves more H for methanogenesis, and the proportion of methane formed increases.

As *k* increases, the concentration of soluble substrate increases slightly, and the microbes grow faster and more efficiently because of reduced maintenance costs. Young forages are rapidly digested, and transported (*k* is relatively large) and promote ruminant growth more efficiently than do the older, dried and more lignified plant materials composing most of earth's photosynthate.

Pure cultures of rumen bacteria often produce lactate, ethanol, formate, succinate and H_2, as well as the final rumen products, the vfa, CO_2, and CH_4. Formate appearing in the rumen is rapidly decarboxylated to CO_2 and H_2; succinate is decarboxylated to propionate, and H_2 reduces CO_2 to methane, all as part of efficient ATP-producing systems. But lactate and ethanol are not important intermediates in the rumen; their production is associated with formation of only 2 ATP per hexose.

When a forage-fed rumen is suddenly supplied an excess of sugar (or easily digested starch) the soluble carbohydrate concentration is no longer limiting, and lactacidigenic bacteria, developing almost explosively[11], metabolize hexose to produce ATP faster than can the predominant microbiota. The rate of lactic acid production soon exceeds the rate of its conversion by other microbes and by the ruminant, and the ecosystem is destroyed. But with gradual adaptation of a ruminant to starch and sugar a very dense microbial population slowly develops, capable of a fermentation sufficiently rapid to keep soluble sugar at limiting concentrations, and a highly productive though somewhat unstable rumen ecosystem with few lactic acid bacteria results.

The ability of the lactic acid bacteria to outgrow the other microbes indicates that anaerobic survival success is not due simply to the ability to obtain more ATP/hexose, but to obtain more ATP per unit time. At limiting rumen sugar concentrations, more ATP/hexose gives more ATP/unit time, but not when the sugar supply is ample. 'Too easy living' wrecks the biochemical stability of the rumen microbial ecosystem.

1 A.L. Oparin, Life, Its Nature, Origin and Development. Oliver and Boyd, Edinburgh and London 1961.
2 R.K. Thauer, K. Jungermann and K. Decker, Bact. Rev. *41*, 100 (1977).
3 R.E. Hungate, in: Biochemistry and Physiology of Protozoa, vol. II, p. 195. Ed. Hutner and Lwoff. Academic Press, New York 1955.
4 J.G. Ferry and R.S. Wolfe, Archs Microbiol. *107*, 33 (1976).
5 R.A. Mah, D.M. Ward, L. Baresi and T.L. Glass, A. Rev. Microbiol. *31*, 309 (1977).
6 R.S. Wolfe, Adv. Microbiol. Physiol. *6*, 107 (1971).
7 R.E. Hungate, The Rumen and Its Microbes. Academic Press, New York 1966.
8 R.E. Hungate, Bact. Rev. *14*, 1 (1950).
9 R.E. Hungate, W. Smith, T. Bauchop, I. Yu and J.C. Rabinowitz, J. Bact. *102*, 339 (1970).
10 B.D.W. Jarvis, C. Henderson and R.V. Asmundson, J. gen. Microbiol. *105*, 287 (1978).
11 R.E. Hungate, R.W. Doughterty, M.P. Bryant and R.M. Cello, Cornell Vet. *42*, 423 (1952).
12 S.H. Zinder and R.A. Mah, Appl. environ. Microbiol. *38*, 996 (1979).

Ecology of methanogenic systems in nature

by K. Wuhrmann

Swiss Federal Institute of Technology, Zürich, and Federal Institute for Water Resources and Water Pollution Control, CH-8600 Dübendorf (Switzerland)

Ecological significance of methanogenesis

Decomposition in terrestrial and aquatic environments of biogenic or synthetic organic material to inorganic products (also referred to as 'mineralization') is predominantly accomplished by microbial oxidations. Under anaerobic conditions, protons, sulfur or carbon atoms are the exclusive electron sinks, products being H_2, H_2S or CH_4 respectively. In this last case, the hydrogenation of CO_2 to CH_4 or the protonation of the methyl group in acetate (or methylamins) to CH_4 are the final reactions:

[1] $CO_2 + 4\,H_2 \rightarrow CH_4 + 2\,H_2O$ $\qquad \Delta G^{0'} = -139.1$ kJ/reaction

[2] $CH_3COO^- + H_2O \rightarrow CH_4 + HCO_3^-$ $\qquad \Delta G^{0'} = -31$ kJ/reaction

Methanogenic bacteria are those unique microbes which can utilize anoxic ecological niches where only H_2 or such ultimate fermentation products as acetate are left over as major sources for energy gain, because other organisms have already exploited all more readily available organic compounds in the medium for ATP formation[1]. Hydrogenotrophic methanogens are also, by virtue of their specific physiological capacities, the pioneer organisms which settle those anoxic loci on earth, where only geochemical H_2 and CO_2 are available as primary substrates. – The present rate of global methanogenesis by microbial activity in terrestrial and aquatic ecosystems as well as in animal intestines has been estimated to be 5.5–11 · 10^{14} g/year[2].

Development of the knowledge on methanogenesis in nature

The present biochemical concept for methane formation (equations [1] and [2]) is the result of two centuries of studies on the educts of methanogenesis, or more generally, on the fundamental question of the recycling of dead plants and animals under anoxic conditions. In his second letter on the origin of the 'aria infiammabile nativa delle paludi' Alessandro Volta, the discoverer of methane, wrote in November 21, 1776: '... Egli è adunque non poco verosimile che dà vegetali macerati e corrotti nell'aqua, e fors' anque dagli animali ... e non dalla pura terra o da altra fossile sostanza ...'[3]. When Béchamp in 1867[4] postulated that methane is a product of microbial fermentation, he also took it for granted that the organic substances used in his experiments (mutton meat and/or ethanol), were the immediate educts of the mixture of methane and hydrogen which he detected as fermentation products. Hoppe-Seyler[5] was the first to demonstrate that acetate decarboxylation is a mechanism of microbial CH_4-formation. This was later confirmed by Söhngen, who further suggested CO_2-reduction by hydrogen as a second reaction leading to methane[6,7]. Both authors, working with fairly crude enrichment cultures, could not exclude, however, other organic compounds as immediate substrates for methane bacteria. Barker, who first gained highly enriched (if not pure) cultures of several methane producing strains, demonstrated clearly hydrogen as an educt for microbial methane, CO_2 being the oxidant[8,9]. He further considered methane bacteria in general as the last links in anaerobic degradation chains, utilizing – besides H_2 – a very restricted and species specific number of low molecular weight end products of fermentation (e.g. formate, ethanol, acetate, propionate, butyrate)[10]. This also confirmed the earlier conclusions drawn by Buswell[11,12] from his extensive studies on sewage sludge digesters. Unequivocal confirmation of the chemolithotrophic nature of part of the methane formation in nature had still to wait, however, for the first uncontestably pure strain of such a methane producer. This was *Methanobrevibacter ruminantium,* isolated by Smith and Hungate[13] in 1958 from the rumen fluid of cattle. Their technique represented a breakthrough in anaerobic methodology and enabled other workers within a relatively short period to isolate in pure culture numerous other methane bacteria from a variety of sources. The present status of biochemical knowledge on methane formation, mostly elaborated by Wolfe within the last 12 years, is outlined in Wolfe's chapter of this review and in earlier summaries of his work (e.g. reference 14). – Interestingly enough, all these new strains were chemolithotrophic hydrogen oxidizers, with some exceptions such as *M. formicicum* which used formate, and strains of *Methanosarcina barkeri,* found to metabolize acetate or methanol besides H_2. None of the strains enriched and named by Barker which were assumed to oxidize fatty acids could be refound. Even the chief witness for organotrophic methane formation from ethanol, *M. omelianskii,* proved to be an association of a heterotrophic ethanol oxidizer (called organism S) and a strictly lithotrophic methane bacterium, *Methanobacterium bryantii*[15]. It was obvious from these findings that, on the one hand, the traditional concept of methane formation in sediments, muds, sewage sludge digesters and similar environments from alcohols or fatty acids was no longer tenable in this generalized form. On the other hand, however, there remained unexplained such clear cut observations from material balances or experiments with isotope labelling which demonstrated methane formation from acetate up to 60% in systems with cellulose as educts, or up to 70–75% with substrates rich in fats[12,16]. There also lacked the link between the oxidation of the terminal fermentation products propionate and butyrate with methane formation in the natural ecosystems.

Hydrogenotrophic and acetotrophic methane bacteria

Hydrogen is the substrate for methane formation by the families Methanobacteriaceae, Methanococcaceae and Methanomicrobiaceae[1]. About half of the presently known species can also use formate. The metabolism of these hydrogenotrophs will be described by Wolfe in the following article. Those properties, pertinant in an ecological context, may be summarized as follows (unfortunately very few kinetic data are available): hydrogenotrophic species are found in environments with a pH around neutrality and at temperatures from about 0 °C up to 65 or 70 °C (e.g. *M. thermoautotrophicum*[14,17]). The mesophilic strain isolated in our laboratory (*M. arboriphilus* strain AZ[18,19]) absorbs hydrogen at rates up to 115 mmoles/g · h. In

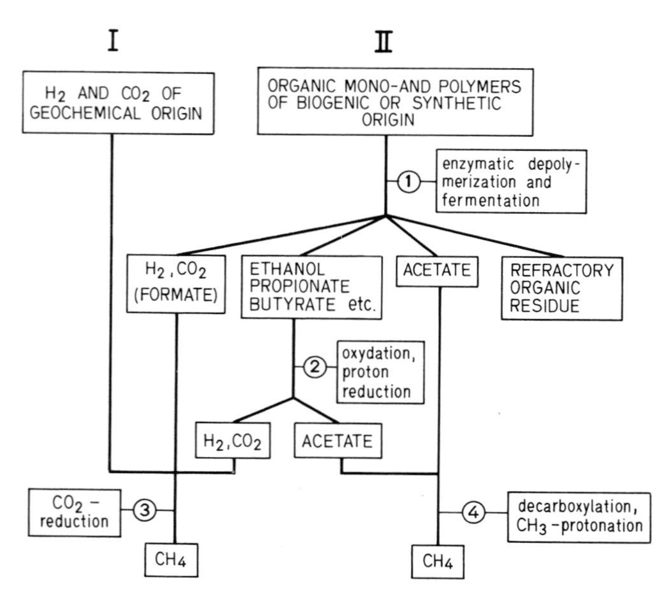

Flow of primary substrates to CH_4 in lithotrophic (I) and organotrophic (II) methanogenic ecosystems. Educt I with reaction 3: thermal springs and lakes with intrusion of volcanic gases. Educts II (a) with reactions 1 and 3: rumen of ruminating wild or domestic animals. Educts II (b) with reactions 1 to 4: sediments, bogs, muds, sewage sludge digesters, garbage dumps, 'biogas' plants. Organism groups in reaction 1: ubiquiceous facultative and obligate anaerobic bacteria; in reaction 2: proton reducing, strict anaerobes not yet in pure culture; in reaction 3: obligate or facultative chemolithotrophic (hydrogenotrophic) methane bacteria; in reaction 4: organotrophic (acetotrophic) methane bacteria.

digested sewage sludge, containing a mixed hydrogenotrophic flora, a maximum uptake rate of ca 15 mmoles/l of sludge/h at 33 °C and at a hydrogen pressure of 0.02 atm was observed. The half rate hydrogen concentration K_s was 0.078 mmoles H_2/l[19]. These figures indicate that the actual H_2-consumption rates in such mixed methanogenic systems are about two orders of magnitude lower than the potential uptake capacity of the hydrogenotrophic community. There exists obviously a high 'buffer capacity' for maintaining hydrogen pressures at very low levels. It is shown below that this is of utmost significance ecologically.

Acetate as a precursor of methane was already considered in the early studies of methane fermentation by Hoppe-Seyler in 1876[5] and others. Söhngen indicated two, morphologically different types of organisms as possible methane formers from acetate: 1. a coccoid, sarcina forming species and 2. a large rod, aggregating to long twisting filaments[6,7]. The sarcina type organism was enriched by Barker in 1936[8] and later gained in pure culture by several authors. The various strains are assembled in the genus *Methanosarcina* with the type species *M. barkeri*[1]. Their common property is a more or less developed capability to produce methane from acetate although lithotrophic methanogenesis from H_2 is the principle energy source for most strains. Their affinity for acetate seems to be rather small. For strain 227 for instance, a $K_{s(Ac)}$= 5 mmoles/l was reported[21] which exceeds acetate concentrations in natural environments (sediments etc.) by many times. Thus the ecological significance of the *Methanosarcina* as acetate remover is doubtful. The thermophilic strain TMl[22] might be an exception in so far as no other acetotrophic methane bacterium has been found which would eliminate acetate from high temperature environments (e.g. thermophilic sludge digesters). All experience indicates that the filamentous bacterium, already described by Söhngen and observed in mass occurrence in experimental and natural methanogenic systems by numerous authors, is mainly responsible for the fermentation of acetate to methane in psychrophilic and mesophilic environments. Barker[8] enriched such an organism and named it *Methanobacterium söhngenii;* neither he nor later authors succeeded, however, in obtaining pure cultures. It was only recently that isolation and growth in pure culture was achieved in our laboratory[23,24]. The organism agrees perfectly with the earlier descriptions by Söhngen and Barker. The bacterium transfers 98–99% of the methyl group of acetate absorbed to methane. Hydrogen is not consumed, the temperature optimum is around 37–40 °C and a pH in the range of pH 7.4–7.8 is required. In contrast to the *Methanosarcina* it has a high affinity to acetate ($K_{s(Ac)}$=0.5–0.7 mmoles/l) and a remarkable consumption rate of 1.6–2 mmoles/g biomass · h. Methane is formed from the methyl group of acetate according to the stoichiometry of equation [2]. The organism grows very slowly (generation time at 37 °C about 8–10 days!). No other methane bacterium, consuming acetate as exclusive substrate has been reported and we assume that this organism, first found by Söhngen, is mainly responsible for the terminal removal of acetate in methanogenic ecosystems. The name *Methanothrix söhngenii* was proposed.

All methane bacteria described up to now require in pure culture a pH for maximum growth in the range of about pH 7–8. This contradicts the many observations on methane formation in water logged, acidic peat bogs[25]. Russian investigators described a *Methanobacterium kuzneceovii* from such environments which reportedly grew at pH 4[26,27].

Types of methanogenic ecosystems

The methane generating ecosystems can be classified into two groups when their starting conditions are considered (fig.).

The chemolithotrophic ecosystems (pathway I in the figure) are the functionally most simple ones. Hydrogen and CO_2 from geochemical sources in terrestrial, sublacustrian or submarine thermal springs are exploited by chemolithotrophic methane bacteria according to reaction [1] for ATP synthesis. The energy conserved is used for CO_2-assimilation. Large populations of methane bacteria such as in Lake Kiwu (an African rift lake)[28] or in thermal springs in Yellowstone Park[17] characterize these ecosystems as typical sites of autotrophic production independant from photosynthesis. The methanogens function as pioneers in environments unexploitable by any other organism.

Two ecotypes must be discussed in group II of the figure, namely: 1. the rumen of cloven-footed wild or domestic animals and 2. natural or technical systems for anoxically degrading organic matter (organic muds, sediments, water saturated soils, sewage sludge or liquid waste digesters, 'biogas' installations, garbage dumps etc.). The educts for methanogenesis in these chemoorganotrophic, methanogenic systems are – in contrast to group I – fermentation products of organic matter of (mostly) biogenic origin. The primary substrates in rumen systems as well as in littoral swamps and muds are terrestrial or aquatic macrophytes containing roughly 35–45% cellulose and other polymeric carbohydrates, 20–30% lignin, 12–20% protein and 2–4% lipids. The sewage sludge fed to sludge digesters in municipal waste treatment plants is composed (on the average) of 10–15% cellulose and other structural carbohydrates, 6–7% lignin, 20–25% protein and 15–30% lipids. So-called 'biogas' plants at farms receive mostly animal excreta with about 14–25% cellulose, 8–15% lignin, 5–10% protein and 1.5–2.5% lipids (all figures on a dry weight basis). Low molecular weight dissolved compounds are present in negligible quantities in these natural substrates. In treatment systems for concentrated liquid wastes (mostly spent liquors from fermentation industries) sugars, short-chain fatty acids, amino acids and peptides represent the bulk of primary substrates.

Since the substrates in group II systems are in general insoluble polymers, the fermentation flora consists largely of species with exoenzymatic properties. Except for the rumen of domestic animals where extensive investigations on its species composition exist[29], the biocenosis responsible for the 'liquefaction' and fermentation in nature or in technical methane fermentations is still terra incognita. The common biochemical knowledge on pathways and products of fermentations gathered from pure strains of anaerobes, isolated from soils and muds, is insufficient for a mechanistic understanding of the dynamics of these complicated microbial communities. Although it may be assumed that many analogies exist to the rumen fermentation, a thorough microbiological investigation of sludge digesters and other methanogenic systems in nature would be gratifying.

Published data on the number and species of organisms active in the initial 'liquefaction/fermentation' of the primary substrates are to be considered with caution due to obvious methodological difficulties in 'counting' organisms (usually with MPN-methods) of highly different metabolic properties. Overall figures (excluding methane formers) in the order of magnitude of $2\text{–}3 \cdot 10^9$/ml were reported for the rumen[29,30]. In sludge digesters similar figures of $2\text{–}6 \cdot 10^9$/ml are indicated[31]. Siebert and Toerien[33] isolated more than 50 strains of proteolytic species of *Clostridium, Peptococcus, Eubacterium, Bacterioides* and others from digesting sludge. Their total number was estimated to be at least $6.5 \cdot 10^7$/ml, mostly strict anaerobes. The proportion of facultative anaerobic bacteria in digesters does not seem to exceed some 1% of the total count[31,32]. Anaerobic counts in the uppermost strata of the sediment in a hypertrophic lake amounted to $2\text{–}6 \cdot 10^6$ cells/g dry sediment (e.g. about $5\text{–}8 \cdot 10^5$ cells/ml) with 72% *Clostridia (C. bifermentans, C. sporogenes, C. butyricum)* and 5% *Eubacterium*[33]. Lignin decomposers have never been found although indications exist for the fermentation of some of the monomers to be expected from lignin break down[29,35]. In general, the relative concentration of lignin in the remains of a fermentation mixture increases (e.g. in sewage sludge from 9–10% up to about 17%[12]), showing that this important biogenic material is not noticeably attacked in methanogenic systems.

The fermentations of carbohydrates, proteins and fats yield the C_1 to C_4 monocarboxylic acids as the most oxidized terminal products plus H_2 and CO_2. Other compounds such as ethanol, lactate, succinate or valerate have been found occasionally in insignificant amounts in fermentations well equilibrated around neutral pH. The main products are hydrogen and acetate. – The principle source of hydrogen is the phosphoroclastic reaction of pyruvate to acetylphosphate via acetyl-SCoA. The oxidant and hydrogen carrier in anaerobiosis is ferredoxin[36,37]. Further H_2 is released by formatelyase from formate produced by enterobacteria in the splitting of pyruvate to acetate and formate. The population density of these bacteria in sludge digesters and other anaerobic systems and, hence, the importance of this reaction, is not known. – A considerable part of the total of acetate formed is due to the splitting of pyruvate already mentioned above. Minor amounts of acetate may also be produced in this reaction sequence when reduced ferredoxin is reoxidized by CO_2 (homoacetate fermentation with e.g. *Clostridium formicoaceticum* and *Acetobacter woodii*). The oxydation of H_2 with CO_2 to acetate by *C. aceticum* may further contribute to the acetate pool, and the β-oxidation of higher fatty acids in substrates rich in fats (sewage sludge) will add appreciable amounts of acetate in such specific environments. The same pathway to acetylphosphates and ATP from other acyl-CoA esters (propionyl-SCoA, butyryl-SCoA etc.) yields probably most of the propionate and butyrate respectively. Insufficient knowledge of the species composition in sludges and mud makes it difficult to identify the specific metabolic reactions leading to the various products observed when, for some reason, their immediate further oxidation is inhibited. A last and pertinant source of both acetate and hydrogen is the final oxidation of ethanol, propionate and butyrate in non-rumen systems[39] which will be discussed below.

The analogy of the rumen and other methanogenic systems in group II of the figure ends at this point. In the rumen the organic acids and alcohols are absorbed by the rumen wall and transferred to the bloodstream of the animals, thus maintaining a favorable and well equilibrated environment for continuous fermentation, buffered at about neutrality by the rumen fluid. The hydrogen is respired to methane by the lithotrophic *M. ruminantium* at the same rate as it is formed.

In sediments, muds, sludge digesters etc. the terminal fermentation products cannot be eliminated as in the rumen and hence, would accumulate and acidify the environment if they were not further oxidized. Long experience shows indeed that large amounts of acetate, propionate, butyrate and other fatty acids accumulate at times in sludge digesters. This 'failing' of the fermentation process by acidification is accompanied by a notorious decrease in methane production and an increasing percentage of hydrogen in the digester gas. It can easily be demonstrated that acid accumulation and hydrogen pressure in the system are closely related[20,40,41]. This is obvious when the stoichiometry and the thermodynamics of the oxidation of propionate and butyrate is considered:

$$[3]\ CH_3CH_2COO^- + 2\,H_2O \leftrightarrows CH_3COO^- + 3H_2 + CO_2 \qquad \Delta G^{0'}\ (pH\ 7) = +81.6\ kJ/mole$$

$$[4]\ CH_3CH_2CH_2COO^- + 2H_2O \rightleftarrows 2CH_3COO^- + 2H_2 + H^+ \qquad \Delta G^{0'}\ (pH\ 7) = +41.6\ kJ/mole$$

The reactions become thermodynamically feasible ($\Delta G^0 \pm 0$ kJ/reaction) at reactant concentrations of 10^{-3} moles/l, pH 7 and 25 °C when the hydrogen pressure in the medium is held below 10^{-2} atm with butyrate or 10^{-4} atm with propionate respectively. For good growth of organisms using these acids as substrates, hydrogen pressures in the order of less than 10^{-5} atm must be maintained in the external medium when the oxidations should yield useful energy. The biochemistry of the oxidations [3] and [4] is not known. Protons are assumed to act directly as electron acceptors (hence the term 'obligate proton-reducing bacteria'[37]). However, neither the initial reaction with the acids nor the mechanism of electron disposal have been formulated as yet. Speculations on these reactions[36,42] cannot be substantiated unless pure cultures of the organisms involved are available. The above thermodynamic considerations point to the fundamental difference between the rumen and the other methanogenic systems in nature: in the rumen the elimination of hydrogen by methanogenesis is only a facultative reaction. When the lithotrophic methanogens are absent after fasting of the animal or when they are suppressed by inhibitors, hydrogen instead of methane is eructated, apparently with no noticeable detriment to the feed utilization by the animal[29]. In the other methanogenic systems, however, oxidation of hydrogen by CO_2 to methane represents a reaction which decides on the existence or non-existence of anaerobic mineralization in nature, at least under conditions of low sulfate content of the environment. In regard to the problem of culturing the acid oxidizers, the analogy to the earlier experience with *M. omelianskii* which was found to be a 'syntrophic' association of a lithotrophic methane bacterium and an ethanol oxidizing, organotrophic microbe[15], is obvious. The anaerobic oxidation of ethanol is endergonic, although much less so than the oxidation of propionate or butyrate. The organism using ethanol (organism S) obviously found its ecological niche in close commensalism with the methane bacterium which provided for the low H_2 pressure, required for exergonic utilization of ethanol (so-called interspecies hydrogen transfer). It was but a logical consequence when Bryant's group tried to enrich acid oxidizers in co-culture with hydrogen consuming organisms. Reports of successful two-strain co-cultures of a propionate[43] or a butyrate oxidizer[44-46] with a hydrogenotrophic methane bacterium or a sulfate reducer were recently published (*Syntrophobacter wolinii* and *Syntrophomonas wolfei* respectively). No culture technique has been found up to now for the separation of such 'syntrophic' pairs and getting the acid oxidizers in pure culture, because physico-chemical methods to absorb hydrogen from the medium at a rate, exceeding the rate of its production, have not been detected yet.

To sum up, it is now well established with these last findings that natural and technical methanogenic systems (except the rumen) require three groups of microbes (not two as previously thought) for accomplishing the transfer of organic matter to methane and CO_2, i.e. a) organisms for the fermentative oxidation of organics to the terminal products organic acids, H_2 and CO_2, b) organisms oxidizing the acids to acetate, H_2 and CO_2 and c) bacteria oxidizing H_2 with CO_2 and decarboxylating acetate to CH_4 and CO_2. It is further indispensable that the three groups form a biocenosis within the same ecological niche, to assert an environment equilibrated at a neutral pH. It is not amazing, therefore, that systems with large inputs of primary substrates (sludge digesters!) work best under continuous mixing (continuously stirred fermenters). For the same reasons, the many attempts to split sludge digestion into separated steps (e.g. 'liquefaction'/ fermentation within one compartment, methanogenesis within a second one) have never been successful.

Reactions competing with methane output in anaerobic ecosystems

Up to this point methanogenic systems low in sulfate, nitrite or nitrate have been considered. In marine environments and in many technical fermentations, however, these anions are present in the medium in appreciable concentrations and might affect methane output through 1. consumption of the educts for methanogenesis i.e. H_2 and/or acetate by organisms reducing sulfate or nitrogen oxides and 2. oxidation of methane within the system. The problem arose with observations in anaerobic marine sediments, indicating that dissolved methane in the interstitial water only appeared at depths where sulfate was nearly depleated[47,48]. Martens and Berner[49] concluded 'that sulfate reduction and methane production are mutually exclusive processes'. Since potentials of key redox reactions in desulfurication and methane formation do not sustain such an exclusion from a biochemical point of view, substrate competition must be considered as a mechanism. New isolations of desulfuricating bacteria from marine environments[50-52] which have considerably enlarged our knowledge on substrate requirements of this group, demonstrate the existence of chemoorganotrophic and chemolithotrophic species respiring H_2, propionate or butyrate and acetate with inorganic sulfur compounds as terminal electron acceptors and excreating H_2S. Others oxidize numerous organic, low molecular weight compounds to acetate, with again sulfur oxides as electron sink. Taking together the metabolic capacities of all presently known genera, the desulfuricating bacteria represent a group which is able to completely replace the biocenosis of methanogens and acid oxidizers for full 'mineralization' in an organotrophic, anaerobic ecosystem, when the supply of sulfate enables acceptance of all the electrons from the oxidation of the organic compounds introduced into the system. The higher metabolic versatility of this organism group as a whole, and the slightly more favorable thermodynamic situation in comparison to the restrictive biological conditions for methanogenesis (e.g. the obligate commensalism with H_2-oxidizers!) accounts for an obvious advantage of desulfuricating biocenoses over methanogenic associations. This does not exclude, however, simultaneous methane production and sulfate reduction in marine sediments as reported[53].

There remains the question of methane oxidation concurrent with methane production. In addition to studies on the possible impact of sulfate on methanogenesis, some evidence of the oxidation of methane in marine and fresh waters or sediments has been documented[54-57]. Since it seems well established that no organisms exist which oxidize methane anaerobically with NO_3^-, NO_2^- or SO_4^{2-} as terminal electron acceptors, some other mechanisms must be involved. Zehnder and Brock[58,59] observed that after injection of $^{14}CH_4$ into growing and methane producing pure cultures of a number of hydrogenotrophic methane bacteria, ^{14}C could be found in the culture vessels as $^{14}CO_2$ as well as in labelled biomass. With cultures of acetotrophic *M. barkeri* and *Methanothrix söhngenii* growing on acetate, the injected $^{14}CH_4$ produced labelled acetate, CO_2 and biomass. At low partial pressures of CH_4 and low rates of methanogenesis, the percentage of CH_4 'reconversion' was small ($< 1\%$). Increasing the pressure of $^{14}CH_4$ up to 20 atm in very actively digesting sewage sludge resulted, however, in losses of up to 90% of produced methane. The classical inhibitor of methanogenesis, 2 Br-ethanesulfonic acid in concentrations not to affect methanogenesis, inhibited methane oxidation to a great extent. This surprising reconversion of CH_4 to its educts is, of course, not a simple back reaction of the product formation. At present neither a satisfactory mechanism of this oxidation is known, nor is its ecological significance understood in regard to the functioning of a methanogenic system.

1 W.E. Balch, G.E. Fox, L.J. Magrum, C.R. Woese and R.S. Wolfe, Microbiol. Rev. *43*, 260 (1979).
2 D.H. Ehhalt, in: Microbial production and utilization of gases, p. 13. E. Goltze, Göttingen 1976.

3 A. Volta, Seconda lettera 'Sull aria infiammabile nativa delle palludi' Como, 21 November 1776.
4 A. Béchamp, Annls Chim. Phys. *13*, 103 (1868).
5 F. Hoppe-Seyler, Z. physiol. Chem. *2*, 561 (1887).
6 N.L. Söhngen, Thesis. Techn. Hochschule, Delft 1906.
7 N.L. Söhngen, Recl Trav. chim. *29*, 238 (1910).
8 H.A. Barker, Archs Microbiol. *7*, 420 (1936).
9 H.A. Barker, Proc. natl Acad. Sci. *29*, 184 (1943).
10 H.A. Barker, Bacterial Fermentations, Ciba lectures in microbial biochemistry. J. Wiley, New York 1956.
11 A.M. Buswell and S.L. Neave, Illinois St. Wat. Surv. Bull. 30 (1930).
12 A.M. Buswell and W.D. Hatfield, Illinois St. Wat. Surv. Bull. 32 (1939).
13 P.H. Smith and R.E. Hungate, J. Bact. *75*, 713 (1958).
14 R.S. Wolfe and I.J. Higgins, Microb. Biochem. *21*, 268 (1979).
15 M.P. Bryant, E.A. Wolin, M.J. Wolin and R.S. Wolfe, Archs Microbiol. *59*, 20 (1967).
16 J.S. Jeris and P.L. McCarty, J. WPCF *37*, 178 (1965).
17 J.G. Zeikus and R.S. Wolfe, J. Bact. *109*, 707 (1972).
18 A.J.B. Zehnder, Thesis. Eidg. Techn. Hochschule Zürich No.5716 (1976).
19 A.J.B. Zehnder and K. Wuhrmann, Archs Microbiol. *111*, 199 (1977).
20 H. Kaspar, Thesis Eidg. Techn. Hochschule Zürich No.5984 (1977).
21 R.A. Mah, M.R. Smith and L. Baresi, Appl. envir. Microbiol. *35*, 11 (1978).
22 S.H. Zinder and R.A. Mah, Appl. envir. Microbiol. *38*, 996 (1979).
23 A.J.B. Zehnder, B.A. Huser, T.D. Brock and K. Wuhrmann, Archs Microbiol. *124*, (1980).
24 B.A. Huser, Thesis Eidg. Techn. Hochschule Zürich, No.6750 (1981).
25 B.H. Svensson, in: Microbial production and utilization of gases, p.135. Ed. H.G. Schlegel, G. Gottschalk and N. Pfennig. E. Goltze, Göttingen 1976.
26 E.S. Pantskhava and V.V. Pchelkina, Dokl. biol. Sci. *182*, 552 (1968).
27 E.S. Pantskhava, Dokl. biol. Sci. *188*, 699 (1969).
28 W.G. Deuser, E.T. Degens and G.R. Harwey, Science *181*, 51 (1973).
29 R.E. Hungate, The rumen and its microbes. Academic Press, New York–London 1966.
30 P.A. Henning and A.E. van der Walt, Appl. envir. Microbiol. *35*, 1008 (1978).
31 R.A. Mah and C. Sussmann, Appl. envir. Microbiol. *16*, 358 (1967).
32 J.P. Kotzé, P.G. Thiel, D.F. Toerien, W.H.J. Hattingh and L. Siebert, Water Res. *2*, 195 (1968).
33 M.L. Siebert and D.F. Toerien, Water Res. *3*, 241 (1969).
34 J.J. Molongoski and M.J. Klug, Appl. envir. Microbiol. *31*, 83 (1976).
35 J.B. Healy and L.Y. Young, Appl. envir. Microbiol. *38*, 84 (1979).
36 K. Jungermann, M. Kern. V. Riebeling and R. Thauer, Microbial production and utilization of gases. p.85. E. Goltze, Göttingen 1976.
37 R.K. Thauer, K. Jungermann and K. Decker. Bact. Rev. *41*, 100 (1977).
38 H. Heukelekian and P. Mueller, Sewage ind. Wastes *30*, 1108 (1958).
39 R.I. Mackie and M.P. Bryant, Appl. envir. Microbiol. *41*, 1363 (1981).
40 H. Kaspar and K. Wuhrmann, Microb. Ecol. *4*, 241 (1978).
41 H. Kaspar and K. Wuhrmann, Appl. envir. Microbiol. *36*, 1 (1978).
42 M.J. Wolin, in: Microbial production and utilization of gases, p.141. E. Goltze, Göttingen 1976.
43 D.R. Boone and M.P. Bryant, Appl. envir. Microbiol. *40*, 626 (1980).
44 M.J. McInerney, M.P. Bryant and N. Pfennig, Archs Microbiol. *122*, 129 (1979).
45 M.J. McInerney, R.I. Mackie and M.P. Bryant, Appl. envir. Microbiol. *41*, 826 (1981).
46 M.J. McInerney, M.P. Bryant, R.B. Hespell and J.W. Costerton, Appl. envir. Microbiol. *41*, 1029 (1981).
47 W.S. Reeburgh and D.T. Heggie, in: Natural gases in marine sediments, p.27. Ed. I.R. Kaplan. Plenum Press, London 1974.
48 W.S. Reeburgh, Earth Planet. Sci. lett. *15*, 334 (1976).
49 C.S. Martens and R.A. Berner, Science *185*, 1167 (1974).
50 N. Pfennig and H. Biebl, Archs Microbiol. *110*, 3 (1976).
51 F. Widdel and N. Pfennig, Archs Microbiol. *112*, 119 (1977).
52 F. Widdel, Thesis. Univ. Göttingen (1980).
53 W.S. Reeburgh, Earth Plant. Sci. Lett. *47*, 345 (1980).
54 T.E. Cappenberg, Hydrobiology *40*, 471 (1972).
55 T.E. Cappenberg, Ant. v. Leeuwenhoek *40*, 285 (1974).
56 W.S. Reeburgh and D.T. Heggie, Limnol. Oceanogr. *22*, 1 (1977).
57 C.S. Martens and R.A. Berner, Limnol. Oceanogr. *22*, (1977).
58 A.J.B. Zehnder and T.D. Brock, J. Bact. *137*, 420 (1979).
59 A.J.B. Zehnder and T.D. Brock, Appl. envir. Microbiol. *39*, 194 (1980).

Biochemistry of methanogenesis

by R.S. Wolfe

Department of Microbiology, University of Illinois, Urbana (Illinois 61801, USA)

The recent and unexpected finding that methanogenic bacteria occupy an isolated biochemical island in the sea of procaryotes has added a touch of excitement to the study of these organisms[1]. This island is defined by such diverse biochemical qualities as: a very restricted range of oxidizable substrates coupled to the biosynthesis of methane; synthesis of an unusual range of cell-wall components; synthesis of biphytanyl glycerol ethers as well as high amounts of squalene; synthesis of unusual coenzymes and growth factors; synthesis of rRNA that is distantly related to that of typical bacteria; possession of a genome size (DNA) approaching ⅓ that of *E. coli*.

Our purpose here is to focus on those aspects of the biochemistry of methanogens that are related directly to the biosynthesis of methane. Barker[2] and his students made fundamental contributions to knowledge of the mechanism of methane formation. They showed that for certain methanogenic bacteria carbon dioxide is the precursor of methane. That is, carbon dioxide serves as the final electron acceptor and is reduced to methane. A stepwise scheme was postulated for this process in which 8 electrons were consumed in the reduction of 1 molecule of carbon dioxide. In another contribution they documented that in certain other methanogenic bacteria the methyl group of acetate or methanol was converted to methane. By use of deuterated acetate or methanol they proved that the hydrogen or deuterium atoms on the methyl carbon remained attached to the carbon atom. So the methyl group was transferred intact and was reduced, accepting 1 proton from the medium.

These 2 mechanisms represent the 2 major routes for methane formation in nature. In sediment and sludge digesters 60–70% of methane is formed from the methyl group of acetate, whereas 30–40% of the methane arises from reduction of carbon dioxide. These 2 routes of methane formation also reflect the 2 major routes of substrate oxidation by methanogenic bacteria. Oxidation of hydrogen or formate is coupled to the reduction of carbon dioxide. All of the 13 species of methanogens now in pure culture are able to oxidize hydrogen and reduce carbon dioxide to methane[1]. The acetophilic methanogens are represented in pure culture at present only by *Methanosarcina*, an organism that is the most metabolically diverse methanogen, converting hydrogen and carbon dioxide as well as methanol, acetate, and methylamines to methane. Other acetophilic methanogens are known, and some of them are in a highly purified stage of culture[3].

Of the 2 major substrate systems of methanogens, a) the hydrogen-carbon dioxide (and formate) system and, b) the acetate system, the former has yielded to fractionation and biochemical studies of subcellular components, whereas the latter is poorly studied. The reason for this is very simple; hydrogenophilic methanogens have yielded to mass culture; acetophilic methanogens are difficult to mass cultivate, having a long generation time and poor cell yield. Of the minor substrate systems methanol produces a better growth response for *Methanosarcina* than does acetate; growth on methylamines is a relatively recent finding[4].

When we began to study the reduction of carbon dioxide to methane by use of $^{14}CO_2$, counts were found in the reaction mixture associated with a small, acidic molecule. McBride fractionated this compound and showed that it was converted to methane. He named this compound coenzyme M, since it was involved in methyl transfer[5]. Its structure was determined by Taylor[6] to be 2-mercaptoethanesulfonic acid. This molecule accepts a methyl group to become 2-(methylthio)ethanesulfonic acid which is the substrate for the methylreductase of methanogens. This molecule is unique in that it is the smallest of the coenzymes, having the most oxidized sulfur atom on one end and the most reduced sulfur atom on the other end separated by a CH_2CH_2 moiety. *Methanobrevibacter ruminantium*[1] (formerly *Methanobacterium ruminantium*, Ml) requires coenzyme M as a growth factor[7], a vitamin. So this compound has a classical vitamin-coenzyme relationship. To test the specificity of this compound Romesser and Gunsalus[8] synthesized a wide variety of analogues and derivatives that were tested in cell extracts as well as in the *M. ruminantium* vitamin assay system.[9]. If substitutions were made for either sulfur atom, the derivatives were neither active as vitamin nor coenzyme. For example, taurine or isethionic acid were completely inactive. Bromoethanesulfonic acid and chloroethanesulfonic acid were powerful inhibitors of methanogenesis at 10^{-6}M. Addition of an extra C_1moiety between the sulfur atoms destroyed activity. Methyl, ethyl, or hydroxymethyl-coenzyme M could be metabolized. For example, extracts produced ethane from ethyl-coenzyme M at 20% of the rate that methane was produced from methyl-coenzyme M[8].

To study the biosynthesis of coenzyme M we decided to use *Escherichia coli*, but to our surprise the coenzyme was not to be found in this organism. This seemed odd, since water-soluble vitamins and coenzymes were known to have a universal distribution in the biological world. Balch[9] then performed a long series of careful experiments documenting the distribution of coenzyme M. A wide variety of organisms and tissues were extracted under a variety of conditions. The sensitivity of the vitamin assay was 10 pmoles. The results were clear-cut; coenzyme M was not found elsewhere but was present in all methanogens tested. We were forced to conclude that methanogens were different. This was the first indication (besides the fact that methanogens produce methane) that these organisms had unique properties.

With the discovery of methyl-coenzyme M it became possible to study the methylreductase system, and it is in this area that most progress has been made. Taylor[6] developed a small reaction vial that was sealed with a rubber septum and in which the volume of the reaction mixture was 0.25 ml. Components were added by syringe, and the gas atmosphere was made anaerobic by use of gassing needles. As methane was formed it escaped into the atmosphere of the reaction vial from which samples could be transferred to a gas chromatograph by syringe. For the methyl group of methyl-coenzyme M to be converted to methane, hydrogen and ATP were required.

To fractionate oxygen-sensitive components of methanogens we found it necessary to take exceptional care to exclude oxygen. Gunsalus developed a system that employed an anaerobic Freter-type chamber[10] that contained an atmosphere of 97% nitrogen and 3% hydrogen. In this flexible plastic chamber oxygen that diffused through plastic walls was scrubbed out by circulating the gas atmosphere over palladium catalyst. Deoxygenated solutions were transferred into the chamber through the air lock. A chromatographic column with o-ring seals was poured aerobically. The inlet and outlet of the column were connected by thick polyethylene tubing to connector ports in the wall of the chamber. After pumping 3–4 column-volumes of anaerobic buffer through the column, the sample was pumped onto the column. Elution buffer was pumped through the column, with the eluate being returned to a fraction collector inside the chamber. By use of this procedure Gunsalus[11] separated the methylcoenzyme M methylreductase into 3 components. They were labeled A, B and C in order of their elution by a salt gradient from a DEAE-cellulose column. Component A was a large protein complex of about 500,000 daltons that possessed hydrogenase activity; component C was a protein of about 130,000 daltons; component B was a heat-stable cofactor. In a reconstituted system all three components were required for methane formation from methyl-coenzyme M. Components A and B were found to be oxygen-labile. At present, purification of component B is in progress but is extremely difficult; once the cofactor is exposed to oxygen it is inactivated, and no reducing conditions that we have tried regenerate any activity. The factor has no visible or UV-absorption spectrum; so the methylreductase is the only assay presently available. One of the most interesting findings in our study of the methyl-coenzyme M methylreductase was made by Gunsalus[12]. When methyl-coenzyme M was added as substrate to cell extract in the presence of excess hydrogen and ATP, methane was formed in stoichiometric amounts, 1 mole of methane being formed from 1 mole of methyl-coenzyme M added. Sequential addition of more substrate yielded the same result. However, if the same experiment was carried out in the presence of hydrogen and carbon dioxide, the rate of methane formation increased 30-fold with a 12-fold increase in the amount of methane formed. At each new addition of substrate the same effect was seen again and again. This effect we have named the RPG effect after R.P. Gunsalus who discovered it; each mole of methyl-coenzyme M generated an active complex through which 11 moles of carbon dioxide was activated and reduced to methane. So in some manner the terminal reaction in methane formation is coupled with the first, the activation of carbon dioxide, suggesting a definite cycle. Results of additional studies showed that the role of ATP in the methylreductase reaction was that of an activator, about 15 moles of methane being formed per mole of ATP added.

To explore the possibility that coenzyme M might be a carrier of C_1 moieties more oxidized than the methyl level Romesser[13] synthesized formylcoenzyme M, but it was not converted to methane by cell extracts. Hydroxymethylcoenzyme M also was synthesized and the C_1 moiety was converted to methane. However, hydroxymethylcoenzyme M was found to hydrolyze to formaldehyde and coenzyme M. Formaldehyde was converted to methane by cell extracts, but this conversion required coenzyme M. So the picture is a bit fuzzy; hydroxymethyl-coenzyme could be an intermediate, perhaps in a hydrophobic area of an enzyme, but definitive experiments have not yet been done.

In studying the RPG effect Romesser was able to resolve cell extracts for a factor that was required for carbon dioxide reduction to methane. Resolved extracts were not able to exhibit the RPG effect; 1 mole of methane was formed from 1 mole of methyl-coenzyme M. When the factor (CDR factor) was added back, carbon dioxide was reduced to methane. This unknown factor is under study at present. The C_1 carrier at the formyl level of reduction remains unkown at the present time.

Hydrogenase was found to be a component of the methylreductase system, and no soluble electron acceptor has been implicated in this reaction. Although no direct evidence has been obtained for specific electron donors for the reduction of carbon dioxide to the formyl, formaldehyde, and methyl levels, another interesting factor, coenzyme F_{420} may be involved[14]. Coenzyme F_{420} has a strong maximal absorption at 420 nm and is a characteristic of all methanogens now in pure culture; so far it has not been found elsewhere. It is a 2-electron carrier that handles electrons at a low potential between hydrogenase and NADP, or formate and NADP. The F_{420}-NADP oxidoreductases of methanogens are rather specific for coenzyme F_{420} whereas the hydrogenases show typical activity with a wide range of natural and aritificial electron acceptors. The structure of coenzyme F_{420} was determined by Eirich to be an 8-hydroxy, 7-demethyl. 5-deaza derivative of FMN with lactyl-diglutamyl moieties attached to the phosphate of the side chain. The 5-deaza chromaphore cannot act as a semiquinone; so the coenzyme serves as a 2-electron donor.

For hydrogen-grown methanogens ATP appears to be formed by generation of a proton motive force[15]. However, ATP generation during the conversion of acetate to methane appears to be more complicated. This area has not moved in the last 20 years due to the difficulty of growing *Methanosarcina* on acetate.

To emphasize the biochemical properties of the methanogenic bacteria the following summary may be of value. These organisms are strict anaerobes that produce methane at a potential near the hydrogen electrode from mainly acetate or hydrogen and carbon dioxide; formate, methanol, and methylamines also serve as substrates for certain species. A new group of cofactors, so far found only in methanogens, includes coenzyme M[6], coenzyme F_{420}[14], undescribed factors F_{430} and F_{342}[16], component B of the methylreductase[11″], the CDR factor[13], and an unknown vitamin required for growth of *Methanomicrobium mobile* (formerly *Methanobacterium mobile).* Nature's biochemical strategy for the metabolism of small molecules has been to invent coenzymes to participate in enzymic catalyses. When a pathway involves a novel sequence, such as the reduction of carbon dioxide to methane, then nature appears to have evolved a series of special coenzymes. The biochemical chapter on coenzymes was supposed to have been closed; we have been forced to reopen it. Methanogens appear unusual in that they apparently carry out electron transport phosphorylation in the absence of quinones, since they lack these compounds[17]. The work of Woese and colleagues has shown that the 16S rRNA of methanogens is only distantly related to typical procaryotes[18]. Kandler's laboratory has documented the wide diversity of cell-wall types among the methanogens[19]. No D-amino acids have been found, and muramic acid is absent; in one species N-acetyltalosaminuronic acid replaces muramic acid. Tornabene and Langworthy[20] have shown that the polar lipids of methanogens are non-saponifiable diphytanyl and dibiphytanyl glycerol ether-linked lipids. Squalene is found as a major component of the neutral lipids. Klotz's laboratory[21] has shown that the DNA complexity of a methanogen approaches ⅓ that of *Escherichia coli.* The mechanism of carbon dioxide activation for fixation into cell carbon is unknown. It would appear that we have only scratched the biochemical surface of these interesting organisms. For example, at the present time not a single mutant or phage has been isolated. The technology for handling these organisms is now at hand, and more of nature's biochemical secrets should be revealed in the near future. Perhaps we shall eventually understand nature's strategy for maintaining such a unique group of organisms. Why have the methanogens remained as an isolated biochemical island apparently not in genetic equilibrium with the microbial world?

1 W.E. Balch, G.E. Fox, L.J. Magrum, C.R. Woese and R.S. Wolfe, Microbiol. Rev. *43,* 260 (1979).
2 H.A. Barker, Bacterial Fermentations, John Wiley, New York 1956.
3 A. Zehnder, B. Huser, T. Brock and K. Wuhrmann, Archs Microbiol. *124* (1980).
4 H. Hippe, D. Caspari, K. Fiebig and G. Gottschalk, Proc. natl Acad. Sci. USA *76,* 494 (1979).
5 B.C. McBride and R.S. Wolfe, Biochemistry *10,* 2137 (1971).
6 C.D. Taylor and R.S. Wolfe, J. biol. Chem. *249,* 4879 (1974).
7 C.D. Taylor, B.C. McBride, R.S. Wolfe and M.P. Bryant, J. Bact. *120,* 974 (1974).
8 R.P. Gunsalus, J.A. Romesser and R.S. Wolfe, Biochemistry *17,* 2374 (1978).
9 W.E. Balch and R.S. Wolfe, J. Bact. *137,* 1329 (1972).
10 A. Aranki and R. Freter, Am. J. clin. Nutr. *25,* 1329 (1972).
11 R.P. Gunsalus, Ph.D. Thesis, University of Illinois, Urbana, Il. 1977.
12 R.P. Gunsalus and R.S. Wolfe, Biochem. biophys. Res. Commun. *76,* 790 (1977).
13 J.A. Romesser, Ph.D. Thesis, University of Illinois, Urbana, Il. 1978.
14 L.D. Eirich, G.D. Vogels and R.S. Wolfe, Biochemistry *17,* 4583 (1978).
15 R.K. Thauer, K. Jungermann and K. Decker, Bact. Rev. *41,* 100 (1977).
16 R.P. Gunsalus and R.S. Wolfe, FEMS Microbiol. Lett. *3,* 191 (1978).
17 R.S. Wolfe and I.J. Higgins, in: Microbial Biochemistry, p. 267. Ed. J.R. Qualyle MTP Press Ltd, Lancaster, England, 1979.
18 G. Fox, L.J. Magrum, W.E. Balch, R.S. Wolfe and C.R. Woese, Proc. natl Acad. Sci. USA *74,* 4537 (1977).
19 O. Kandler and H. König, Archs Microbiol. *118,* 141 (1978).
20 T.G. Tornabene and T.A. Langworthy, Sciene *203,* 51 (1978).
21 R.M. Mitchell, L.A. Loeblich, L.C. Klotz and A.R. Loeblich III, Science *204,* 1982 (1979).

Engineering, operation and economics of methane gas production

by John T. Pfeffer

Department of Civil Engineering, University of Illinois, Urbana (Illinois 61801, USA)

Processing of biomass for the production of a fuel gas containing methane requires a complex system. The degree of complexity is, in part, a function of the biomass utilized. In general, this system consists of 3 main subsystems;

- Raw material preparation
- Methane fermentation
- Residue processing, utilization and/or disposal

Gas scrubbing for carbon dioxide removal to produce a gas that is essentially 100% methane is not considered in this discussion.

Certain biomass materials such as animal manure from a confined and enclosed beef feeding operation can be added directly to the fermentation subsystem without any preparation. Conversely, urban solid waste requires extensive preparation including size reduction and various separation processes for removal of those materials that have the potential for creating operational difficulties with the physical processes employed in the fermentation and residue processing subsystems.

The essence of this processing system is the methane fermentation subsystem. The ability to convert a major portion of the organic material to methane is paramount to the success of this system. This conversion efficiency has an impact on 3 separate costs. First is the raw material cost. If the biomass cost is $20 per t, the methane cost at a 75% conversion efficiency will be about $6.5 per 100 m^3. At a 50% conversion efficiency, the raw material cost alone is $10 per 100 m^3.

A 2nd cost factor is associated with the reactor

volume. When processing a relatively dry material, the required reactor volume for a given retention time is fixed by the feed slurry solids concentration. The feed slurry solids in turn are a function of the conversion efficiency and the concentration of solids in the reactor slurry that will permit good mixing.
The 3rd cost factor relates to the cost associated with the residue processing. The fermentor slurry must be dewatered and either processed for material recovery or disposed in an acceptable manner. The processing costs are directly related to the mass and/or volume of slurry in the effluent from the fermentor.

Substrate characteristics

The chemical and physical characteristics of the biomass will have a significant effect on the process conversion efficiency and the economics of the system. Several substrate characteristics should be considered and they are listed as follows: - biodegradability, - chemical composition and structure, - moisture content.
Each of these characteristics affects one or more of the costs associated with the processing system. The biodegradability is the most important factor. In addition to the lower product yield per unit of substrate processed, this factor will have a significant impact on the reactor volume, size of dewatering system and residue disposal costs.
The impact on reactor volume can be illustrated as follows. There is a limit to the level of solids in the reactor slurry that can be efficiently mixed. This level will vary with the type of substrate, but in general, will be between 5 and 10%. Preliminary studies on mixing of reactors receiving urban refuse for methane production suggest that a slurry containing 8% solids is the near optimum for mixing. This solids level and the biodegradability control the solids concentration in the feed slurry. This in turn sets the feed volume for a given quantity of substrate and the required reactor volume for a given reactor retention time. Table 1 shows the reactor volume required per t of dry substrate containing 15% ash. The reactor has a 10-day retention time resulting in 80% conversion of the biodegradable solids to methane and carbon dioxide. The limit on the reactor slurry solids concentration was set at 8%.
In addition to a significant reduction in the reactor cost, other savings result from the smaller reactor volume and feed slurry volume.
Heat losses are lower with a smaller reactor volume. The thermal energy required to elevate the temperature of the feed slurry is reduced since less water is needed to maintain the solids level in the reactor slurry. Lower feed slurry volumes result in lower pumping costs because less water is circulating in the process. Also, less power is required to mix the smaller reactor volume associated with a substrate having a high biodegradability. These cost reductions are not of the magnitude of those associated with the reduction in reactor volume, but they will result in lower costs.
The substrate biodegradability is a major factor in determining the costs of fermented residue processing and disposal. As shown in table 2 the mass of dry solids as well as the volume of slurry remaining after the fermentation is complete is much less when the organic solids in the raw material are more biodegradable. Slurry dewatering processes are sized either on a volumetric flow rate or a solids mass flow rate. In either case, a much larger dewatering system is required for the less biodegradable material. Direct application of the fermented slurry to land will also be more expensive because of the larger volume of slurry to be handled from the less biodegradable substrate.
The chemical nature of the raw material to be used as a substrate for the production of fuel gas will affect the costs in 2 areas. The chemical structure of the material can significantly alter the rate of conversion to gas. Carbohydrates in the form of simple sugar have a much higher rate of conversion than cellulose. Even the crystalline nature of cellulose will affect the kinetics of cellulose fermentation. The fermentation process is a biological process that requires a balanced substrate. If the raw material does not contain adequate nitrogen, phosphorus and micronutrients, it will be necessary to add these nutrients for fermentation.
Control of the fermentation pH may also require chemical addition. The gas produced by the fermentation contains 30-50% carbon dioxide. Maintaining a neutral pH will require a significant level of alkalinity. With some substrates, sufficient natural alkalinity is formed to maintain the pH. In other cases, lime, soda ash or caustic must be added to obtain the desired pH.
The moisture content of the substrate can alter the economics in much the same way as biodegradability. When the moisture content is very high, as with sewage sludges, the reactor volume per t of dry solids

Table 1. Effect of substrate biodegradability on reactor volume

Biodegradability (% organic solids)	Feed slurry Solids (%)	Volume (m^3/t)	Reactor volume (m^3/t)
40	10.7	9.37	93.7
60	12.8	7.81	78.1
80	16.0	6.24	62.4
100	21.4	4.68	46.8

Table 2. Effect of substrate biodegradability on quantity of residue

Biodegradability (% organic solids)	Residue Dry solids (t)	Volume (m^3/t substrate)
40	0.73	9.1
60	0.59	7.4
80	0.46	5.7
100	0.32	4.0

processed will be between 200 and 300 m^3. This imposes a severe cost penalty on gas production. In general, a substrate with a moisture content of 90% or greater will not yield cost competitive gas. Credits such as those applied to waste disposal systems are necessary to cover the costs of gas production from these sludges. Because of the cost in energy for drying these wet substrates, this is not a viable option. The substrate will have to be used as received. Conversely, the moisture content of the dry substrate can be increased by the addition of water. Frequently, this water originates from an internal recycle and, as such, does not impose an added cost.

Process characteristics

In an attempt to improve the kinetics of conversion of organic material to gas, much effort has been invested in trying to determine the optimum reactor type and geometry. The following reactors have received the most attention:
CSTR – no cell recycle
CSTR – cell recycle
Fixed film
Plug flow – multi stage
(CSTR refers to a completely-stirred-tank-reactor).
Since it has been recognized that methane fermentation is a multiphase process, researchers have attempted to separate the acetogenic from the methanogenic phase. With this approach, it should be possible to operate each stage under conditions that optimize the growth of the specific cultures. However, cellulose hydrolysis rather than acetogenesis or methanogenesis has been found to be the rate limiting step when fermenting complex natural substrates[1]. McBee[2] found thermophilic cellulolytic bacterial growth occurs in a pH range of 6.4 to 7.4, which is generally the optimum range for acetogens and methanogens. Stranks[3] reported extremely high rates of cellulose hydrolysis when using a mixed culture of thermophilic microorganisms. Pure culture cellulose hydrolysis rates were much lower[2].
Cellulose hydrolysis as well as acetogenesis is an enzymatic reaction. The simplest enzyme reaction (equation 1) can be described by the Michaelis-Menten relationship.

$$S+E \underset{k_2}{\overset{k_1}{\rightleftarrows}} SE \overset{k_3}{\rightarrow} P+E \tag{1}$$

The substrate (S) and enzyme (E) are in equilibrium with substrate-enzyme complex (E–S). However, an irreversible reaction resulting in the product (P) and enzyme is assumed. This reaction is the rate limiting step having a constant, k_3. The Michaelis-Menten expression (equation 2) is developed from this equilibrium.

$$-\frac{dS}{dt}=\frac{k_3(S)(E_0)}{K_a+(S)} \tag{2}$$

Many enzyme reactions do not satisfy the restriction that the enzyme-substrate complex breaks down irreversibly. The complex may also form from the product side as shown in equation 3.

$$S+E \underset{k_2}{\overset{k_1}{\rightleftarrows}} SE \underset{k_4}{\overset{k_3}{\rightleftarrows}} P+E \tag{3}$$

As the concentration of the product increases, most enzymatic reactions slow down. This is due to the phenomenon of product inhibition. The rate equation shown in equation 4 is developed from equation 3. This equation shows that the substrate utilization rate is a function of not only the substrate and enzyme concentration, but also the product concentration.

$$-\frac{dS}{dt}=\frac{dP}{dt}=\frac{[k_1 k_3(S)-k_2 k_4(P)]E_0}{[k_4+k_3]+k_1(S)+k_4(P)} \tag{4}$$

In any basic biochemistry text such as Mahler and Cordes[4], one will find information that an enzymatic reaction slows down as equilibrium is approached, not only by virtue of the thermodynamic back reaction, but also because, as the product concentration increases, an increasing proportion of the enzyme is immobilized as an EP complex. This kinetic effect of product inhibition is thus an intrinsic property of any realistic, i.e. reversible, mechanism of enzyme catalysis.
Consequently, process configurations that approach a plug flow reactor will be much less efficient than completely mixed reactors. With a multistaged biochemical process such as methane fermentation, the best reactor design is one that allows these reactions to occur concurrently with the final product being methane and carbon dioxide. These gases have limited solubility and are lost from the reacting medium thereby reducing the effect of product inhibition on kinetics.
A completely stirred reaction tank is generally the most efficient reactor design. One limitation of this reactor type is the inability to operate with a mean cell residence time greater than the hydraulic retention time of the tank. This deficiency has been overcome by employing sludge (cell) recycle, either internal or external, or by adding a packing material to the reactor vessel to form a fixed-film reactor. These reactor types have been successfully applied to substrates that are either soluble or essentially 100% biodegradable.
However, if the substrate contains a quantity of suspended biologically inert material, the fixed film or CSTR with cell recycle may not be able to efficiently process the material. The packing material in the fixed film reactor provides a multitude of small quiescent settling chambers where the suspended material can accumulate. If these solids are nonbiodegradable, the reactor will fill with these solids.
A similar problem exists with the CSTR-sludge recycle system. A recirculation factor (R), defined as the

ratio of mean cell residence time (θ_c) to the hydraulic residence time (θ), is used to calculate the accumulation of inert material in a reactor employing sludge recycle. In a system operating with θ of 1 day and θ_c of 10 days, the value of R is 10. If the feed stream to the reactor contained 10 g/l of inert solids, the equilibrium concentration of these solids in the reactor would be 100 g/l or 10%. The volume of the reactor is simply occupied by these inert solids. Most of the substrates available for the production of a fuel gas will contain a substantial quantity of inert material and the only reactor type that can be expected to function efficiently would be the CSTR.
The fermentation temperature has also been found to significantly effect the conversion efficiency. When urban solid wastes were used as a substrate, thermophilic fermentation (60 °C) yielded much higher gas production[5]. These data were evaluated using a first-order kinetic expression for substrate utilization. A mass balance on a CSTR yields equation 5.

$$\frac{S_0}{S} = 1 + K\theta \qquad (5)$$

In this equation, K is the first-order rate constant, S_0 is the initial substrate level, S is the final substrate level and θ is the hydraulic retention time.
Measurement of S_0 and S when the substrate is a complex material such as plant fiber is difficult. S_0 must be the initial biodegradable substrate. Volatile solids are frequently used as a measurement of S and S_0. In order to determine the portion of the volatile solids that are biodegradable, the following technique was employed. At an infinite retention time, θ, the value of S will be zero and S_0-$S = S_0$. A semi-log plot of volatile solids destroyed, S_0-S, versus θ^{-1} will provide the value for S_0 when θ^{-1} equals zero.
Once the value of S_0 is obtained for each fermentation temperature, the value of K can be determined by plotting S_0/S versus θ. The slope of the line is K. Table 3 lists the values for S_0 and K obtained from this analysis. The apparent increase in the biodegradability of the volatile solids is substantial, increasing from 44% at 40 °C to 55% at 60 °C. This is a 25% increase in the portion of the volatile solids that are biodegradable. When this is combined with an increased rate constant, the thermophilic fermentation temperature yields a more efficient and economical processing system.

Residue disposal

The discharge from the fermentor must be processed in order to eliminate adverse environmental effects. The unfermented solids must be removed and the liquid stream can be recycled back through the system, disposed by land application or treated for discharge to receiving water bodies. Recycling of this water offers significant cost savings as well as conservation of heat and of chemicals that may be necessary for pH control and for microorganism nutrition. The solids may have some value. For example, the organic solids still have energy that can be recovered by incineration. The calorific value of the residue from the fermentation of city refuse was found to be 18.5 MJ/kg total solids (24 MJ/kg volatile solids). Recovery of these solids with a centrifuge will produce a cake having a solids content between 30 and 40%. This cake will provide self-sustaining combustion in a properly designed incinerator. Recovery of a significant quantity of energy in the form of steam is possible[6].
Other material recovery may be possible. Based on studies reported by Turk and Coe[7], a number of researchers are investigating the recovery of protein from the residue of fermentors processing animal manures. The results have not been encouraging because of the poor efficiency of protein recovery with centrifuge systems. Hashimoto et al.[8] have only been able to recover about 20% of the protein (as measured by organic nitrogen) in the fermentor residue when using a commercial centrifuge operating at a centrifugal force of 2300 × g. Other uses may be found for the fermentor residue. This will in part depend upon the type of raw material fed into the process.

Economics of methane fermentation

The economics associated with methane recovery from biomass is highly dependent upon the raw material. There are 2 cost factors. First, the cost of the raw material is significant. When this material must be purchased, a significant cost is added to the methane production costs. However, if this is a waste material, the producer may pay a fee for disposal of this material. The 2nd factor is the degree of preparation required before the material can be added to the fermentation reactor. Urban solid waste must undergo size reduction, separation of undesirable constituents such as plastics, metals, glass, etc. and preparing a slurry with water. Conversely, animal manure from certain confined animal feeding units can be added directly to the fermentor.
An earlier paper[6] presented an economic analysis for a system designed to convert 900 t per day of urban refuse into methane and steam. The 1980 capital costs

Table 3. Biodegradability and rate constant as a function of temperature

Temperature (°C)	Biodegradability (% of volatile solids)	Rate constant - K (day^{-1})
35	36	0.53
40	44	0.58
45	40	0.47
50	49	0.63
55	51	0.78
60	55	0.95

for this plant are given in table 4. Costs should be reduced significantly by not processing the gas for removal of carbon dioxide. As produced, this gas is a good fuel gas that can be used in any system that is designed to burn gaseous fuels.

The allocation of the costs to the various unit processes is shown in table 5. Major capital expenditures are required for residue processing and disposal. The centrifuge and incineration account for 50% of the capital costs. The size of these units is directly related to the mass of the solids passing through the system. Consequently, the higher the efficiency of converting the raw material to methane, the lower the cost associated with these processes.

Gas processing also requires capital investment. The product gas from the fermentor has a calorific value of about 22.4 MJ/m^3 (600 Btu/ft^3). Removal of carbon dioxide increases the heating value of this gas to that commonly employed in the U.S. natural gas system. However, this is an unnecessary requirement as the gas as produced is an excellent fuel.

Energy recovery with this system is good if a use for the steam produced by the residue incineration can be found. A portion of this steam can be used for process heat. However, a considerable excess remains. The energy balance for the system is shown in table 6. Input energy includes all of the energy content of the organic material, the thermal energy and the electrical energy.

A cost analysis for fuel gas production is site- and raw material-specific. Such an analysis is shown in table 7 for the urban solid waste system. Capital costs are amortized at 10% interest over a 20-year period. Public financing is assumed so no tax or profit is included in this analysis. The fuel gas is priced at $3.00/GJ and steam at $1.00/GJ. The net processing cost is the dipping fee required for the process. This fee is substantially lower than most fees for refuse disposal, so there is a margin for profit in this analysis. Careful economic analysis must be conducted for any installation. In general, one can expect that methane production from any waste biomass that is relatively biodegradable will be economically attractive. However, if the biomass has an acquisition cost, it is probable that the economics will not be attractive unless the energy costs escalate to near $10 per GJ.

The U.S. Department of Energy has funded 2 demonstration plants to determine the true economics associated with the methane recovery process. One plant located in Pompano Beach, Florida, processes city refuse for methane recovery. Operation of this plant was initiated in July 1978. A 2nd plant located in Bartow, Florida, is processing animal manure from a confined beef cattle feeding operation. This plant became operational in late 1978.

Table 4. 1980 capital cost for methane fermentation

Raw material processing system	$5,248,000
Fermentation system	7,700,000
Incineration system	9,105,000
Gas purification system	3,655,000
Total	$25,708,000
Additional capital required	4,890,000
Total capital	$30,598,000

Table 5. Allocation of costs to unit processes used in methane recovery

Unit process	% Capital costs	% Operation and maintenance
Shredder	9.6	22.4
Separation	4.1	4.1
Storage	6.6	2.0
Fermentation	15.6	44.9
Centrifuge	14.3	10.0
Incineration	35.4	5.0
Gas processing	14.2	11.6

Table 6. Energy balance for refuse fermentation

Energy in	11.82 GJ/Mg
Methane produced	3.87 GJ/Mg
Steam produced	3.64 GJ/Mg

Table 7. Fuel gas production cost analysis – 1980 cost index

Annual capital cost ($/yr)	3,121,000
Labor ($/yr)	1,101,000
Operation and maintenance ($/yr)	2,081,000
Total annual costs ($/yr)	6,303,000
Tonnes/yr processed	331,000
Processing costs ($/t)	19.04
Gas production (J/yr)	1.3×10^{15}
Gas value ($/yr)	3,900,000
Steam production (J/yr)	1.2×10^{15}
Steam value ($/yr)	1,200,000
Total revenue ($/yr)	5,100,000
Net processing costs ($/t)	3.65

1 J.T. Pfeffer, Reclamation of Energy from Organic Refuse, EPA-670/2-74-016, U.S. Environmental Protection Agency, National Environ. Research, Cincinnati, Ohio, 1947.

2 R.H. McBee, The Culture and Physiology of Thermophilic Cellulose-Fermenting Bacterium, J. Bact. 56, 653 (1948).

3 D.W. Stranks, Microbiological Utilization of Cellulose and Wood. I. Laboratory Fermentation of Cellulose by Rumen Organisms. Can. J. Microbiol. *2*, 56 (1956).

4 H.R. Mahler and E.H. Cordes, in: Basic Biological Chemistry, p. 158. Harper and Row, New York 1966.

5 J.T. Pfeffer, Temperature Effects on Anaerobic Fermentation of Domestic Refuse, Biotech. Bioengng *16*, 771 (1947).

6 J.T. Pfeffer and J.C. Liebman, Energy from Refuse by Bioconversion, Fermentation and Residue Disposal Processes. Resource Recovery Conserv. *1*, 295 (1976).

7 M. Turk and W.B. Coe, Production of Power Fuel by Anaerobic Digestion of Feedlot Waste. Phase II Final Report, Contract No. 15-14-100-109-98(71), U.S. Dept. of Agric. Northern Regional Research Center, Peoria, IL, 1974.

8 A.G. Hashimoto, R.L. Prior and Y.R. Chen, Methane and Biomass Production Systems for Beed Cattle Manure. Great Plains Extension Seminar on Methane Production from Animal Manure, Liberal Kansas, 1979.

Biogas production from agricultural wastes

by P. N. Hobson

Rowett Research Institute, Microbiology Department, Greenburn Road, Bucksburn, Aberdeen AB2 9SB (Great Britain)

'Biogas' is the word used to denote the mixtures of methane and carbon dioxide produced by bacterial action, in vitro, on various organic substrates. 'In vitro', because similar mixtures of methane and carbon dioxide are formed by essentially the same bacterial actions in the gut of animals (in particular the rumen and similar organs of the herbivore gut; see previous papers) and in decaying vegetation in marshes and river beds.

In most of the natural systems formation of methane is undesirable, and if the reactions can be controlled, say in ruminant feeding, control is in the direction of reduction of methane production. On the in vitro man-made systems, production of methane is the objective and control is, for 2 reasons, in the direction of increasing this. Production of biogas from wastes in a structurally-defined system (the 'anaerobic digester') is not a new technology: a controlled anaerobic digestion has been used for 60 years or so in the treatment of municipal sewage sludges for the reduction of pollution and nuisance caused by these sludges. An index of the extent of pollutant reduction is the amount of biogas produced, so, obviously, maximizing gas production maximizes pollution control. However, in many sewage works the biogas is used as a fuel for powering the aerobic side of the sewage treatment and for lighting, heating, etc., in the works, as a substitute for mains electricity, gas or oil fuels. So maximum gas production is again desirable.

Since biogas production on a large scale has been going on for so many years, one might reasonably ask why an energy system based on biogas is not now widespread? There are 2 or 3 main reasons. While gas production in a big sewage works is large, it is only obtained by collecting and processing the excreta from many thousands, or millions, of people. This involves miles of sewers with large tanks and other plant at the sewage works. All this is expensive and the construction and costing was not based on producing fuel gas at a price competitive with low-cost oil and coal-based fuels, but on doing the pollution control needed for the healthy living of a city population; the biogas fuel helped to cut running costs. Secondly coal and coal-gas, and particularly oil, were in plentiful supply. While it was known that other waste organic matter could be used to produce biogas, the topic was not, generally, pursued. Biogas production from agricultural wastes was attempted in times of crisis, for instance war time, to obtain a substitute for scarce petrol for vehicles, but with the coming of peace the plants were largely forgotten about. They were forgotten because the plants had running problems, they were not very efficient in gas production and overall energy balances, and with apparently never-ending and cheap supplies of oil and other fuels there was no incentive to undertake systematic and long-term research and development to produce a proper plant.

Two things have lately changed the picture in agriculture and its related industries. In the older (but still practised) mixed farming systems animal production was related to the land available for disposal of the animal excreta as fertilizer. Farms were relatively small and so were quantities of wastes. Pollution from the wastes was little, if any, and localized. Biogas production from the small individual quantities of wastes, even if wanted, would have been small and it was not worth collecting waste from different farms to a central gas-production unit.

The introduction in the last 20 years or so of the 'intensive', specialist farm, or intensive unit to existing farms, and particularly the 'slurry systems' of waste collection used in intensive pig- and cattle-units, has changed the picture. These farms and units have hundreds, thousands, or even millions, of animals or birds kept in small land areas and the collection, storage and disposal of the excreta from these animals can cause pollution problems, especially in a world increasingly conscious of pollution and health hazards. And it was from the point of view of pollution control on such farms that the present author and colleagues started investigations on anaerobic digestion of agricultural wastes some 15 years ago. The large amounts of wastes involved in these farming operations made comparatively large-scale production of biogas possible, but the previously mentioned considerations applied and the biogas was thought of mainly as a fuel for running the plant and making the pollution control energetically and economically cheap.

The more recent realization that supplies of fossil fuels, particularly oil, are finite and that whatever the outcome is in terms of reserves the prices of conventional fuels will continue to rise, has changed the picture again. While pollution control remains a big consideration (although in many cases economically unquantifiable), present fuel prices have made it possible by using construction methods different from the conventional sewage plant to produce biogas at a price competitive with conventional fuels and which will pay for the digester plant in terms of a fuel source. The intensive farm is a big energy-user and has fuel needs and costs commensurate with this, but the numbers of animals in some cases are such that production of gas to supply not only the farm but towns and villages is possible. Development in countries like Britain is based on the biogas supplying energy for the farm and perhaps some houses or a factory in its locality; in countries such as the USA the much larger feedlot units could supply gas for more widespread distribution. But the picture is also changed in other countries where small, 'peasant', farming operates. Supplies of the conventional fuels such as wood, are decreasing and cannot be replaced by expensive oil. Biogas production on a small scale could provide a substitute fuel and a good fertilizer, and in passing one might mention that a conventional substitute for wood, the burning of dried animal-excreta, gives not only a poorer fuel than biogas but destroys the fertilizer value of the excreta.

Agricultural wastes of all kinds are a renewable feedstock for energy production, and because of demands for food the intensive farming units will continue in the future, so in the last few years research on biogas production has increased considerably.

The amount of information on biogas production has increased in the same way, and in a short paper such as this it is impossible to give a detailed description with bibliographical annotations of the microbiology of the process and of results of all the tests on various substrates, and of the engineering research which have been carried out. The bibliography to this paper suggests a few papers and books which can lead the reader more deeply into the subject and all that is attempted here is a short review of the overall state of research and development.

The reactions of anaerobic digestion

The reactions of anaerobic digestion are, as the name suggests, carried out in a highly reduced liquid under an atmosphere devoid of oxygen. The reactions are complex and a large and very heterogeneous population of bacteria is involved. But one of the properties of this mixed population is that, provided it is contained in a vessel to which the access of air is limited, some members of the population and their reactions can provide the highly-reduced conditions needed by other members. Thus no chemical reducing agents, or exacting precautions to keep air from the feedstock or digester, are needed. Although the individual bacteria when grown in laboratory pure culture are among the most exacting known, in the mixed population of the digester interactions and symbiotic growth lead to a stable population which can withstand many adversities. Since, as indicated before, the bacteria occur in many habitats, inoculation of a feedstock, to start the growth of the bacterial population, is not generally necessary. Fecal wastes contain the bacteria, although they are not in the correct proportions of the final digester population, so a stable digestion can be obtained by feeding the wastes slowly into a digester and allowing time, and the right conditions, for the correct population to develop. Purely vegetable-waste digestions can be started by addition of fecal wastes. On the other hand, as the bacteria develop in slurries of waste vegetable-matter, or the effluents from factory processing of vegetable matter of different kinds, digesters for these wastes have been started by using feedstock which has been allowed to stand in open ponds for some months and which has begun to decompose. Of course, a digestion will start off more quickly if the digester is inoculated from the contents of a digester already working on that feedstock, but, as it is not often possible to obtain such an inoculum for agricultural waste digesters, the previously-mentioned development of a digester flora ist the usual process used.

The microbial reactions in digestion can be divided into 2, or more generally 3, stages although, of course, in the usual digester these stages all occur at the same time and intermediate products are immediately further metabolized.

The digester feedstock, fecal or vegetable waste, contains polymeric carbohydrates, proteins and fats from the vegetation or residues of feed undegraded in the animal gut, or from the intestinal bacteria which make up a large proportion of feces, or from intestinal secretions. The carbohydrates are hydrolyzed to monomer or dimer units such as glucose, xylose, maltose, and proteins are split to amino acids some of which are then deaminated to give ammonia. Fats are split to give long-chain fatty acids and these are degraded to give acetic acid and hydrogen.

In the 2nd stage the sugars are fermented to produce mainly acetic acid, hydrogen and carbon dioxide. The fermentation reactions and possibly other reactions produce energy utilized by the bacteria to form new cells incorporating amino acid or ammonia nitrogen.

In the 3rd stage the acetic acid, hydrogen and carbon dioxide are converted into methane and carbon dioxide: the biogas.

This is only a very brief summary; many other actions and interactions involving salts and other components (non-protein nitrogen compounds, etc.) of the feedstocks take place, and the actual fermentation stage is not as simple as presented here. But these reactions lead to the feedstock being converted to gas, with the bacterial cells, organic materials unattacked or not completely degraded, grit, stones and other debris, forming in the water of the feedstock the digested sludge. In the process, polluting materials in the feedstock are converted to gas or relatively nonpolluting residues.

If the feedstock were a factory waste containing sugars and not polymeric carbohydrates then the initial hydrolysis step would not be needed and the feedstock would be directly fermented.

While the reactions occurring during digestion are

known, the bacteria involved have not been completely identified, and details remain to be investigated. Microbiological investigation of digestion has developed from rumen microbiology and significant advances have been made only comparatively recently. Nevertheless, although digesters have been, and are being run and the practical details of breakdown of a particular feedstock can be experimentally determined without detailed knowledge of digester microbiology, the microbiological details are needed for, among other things, mathematical modelling and prediction of digester behaviour. Future application of microbiological data may bring about more efficient digester operation.

Types of digesters, and gas production

Digestion like any bacterial culture, can be done as a batch or as a continuous process. The role of batch digesters is very limited. As in most cases the feedstock is produced continuously (e.g. farm animal effluents) and energy demand is continuous, most digesters are run on a continuous basis. High-volume, low-solid-content, waste waters from food processing or other factories may require a feed-back ('contact') digester system and large-scale plants of this type are being tested in Britain and elsewhere. However, agricultural wastes are mainly animal excreta and/or vegetable matter and these are slurries of relatively high solids content. Just as for the sewage sludges the most generally applicable type of digester for these feedstocks is the single-stage, stirred tank. Other configurations such as the tubular digester are being tested, but the tank digester is being applied to all sizes of plants from the simple 2 or 3 m^3 digester of India or China to the hundreds or even few thousand m^3 automated digesters of the large farms or feedlots.

The large-scale, single-stage, stirred tank system consists, first, of a holding tank for the feedstock. In the case of animal excreta this tank may be under the animal house and be the sump at the end of floor, or under-floor, channels collecting the excreta from animal stalls or pens or, because of the arrangement of the farm buildings, be a separate tank near the digester with excreta pumped or flowing from the animal-house tanks. From the tank, the feedstock is pumped by an appropriately programmed pump to the digester tank. The digester is sized on the basis of the feedstock flow and the necessary detention time for digestion of the particular feedstock, and it is heated and stirred. Stirring is required only to ensure mixing in of the input and reasonable homogeneity of the tank contents, and is usually done intermittently for a short time every few hours. Mechanical stirring is used, but most large digesters now use some form of gas bubbling, the biogas being taken from the digester head-space and reinjected at the bottom of the digester. Heating is by warm water, usually passing through internal heat-exchangers of various designs, and coming either from a biogas-fired boiler or the cooling water of a biogas engine. The digester contents are removed at the rate of input feeding, usually by gravity flow over some weir or standpipe system which retains a gas pressure of a few inches water-gauge in the digester. The overflow then runs to storage tanks from where it can be distributed to fields as fertilizer, or treated in some other way. A small part of the daily gas production is stored either in a separate gas-holder or in a floating gas-holder forming the digester top.

There are many variations in detail on this basic design. The tank may be above or below ground, some digesters have pumped outflow, a few a gravity inflow, some are run at more than the few inches water-gauge pressure, and so on. The simple 'peasant-farm' digester is, of course, entirely hand- or gravity-loaded and unloaded, and hand-stirred, if stirred at all.

Digesters can be run in the 'mesophilic' range at temperatures usually about 30–42 °C, or in the 'thermophilic' range at about 55–60 °C or 65 °C. But the bacterial populations must be developed and stabilized for the operating range and within that range the temperature must be controlled as accurately as possible. While some laboratory-scale work has been done on thermophilic digestion and there are some advantages in rates of reactions in using this temperature range, there are energetic and other reasons for preferring the mesophilic range for most practical applications, so the greatest amount of research has been done on mesophilic digestion and most, if not all, practical digesters are running in this range.

The absolute amount of gas produced, and the methane content, varies with substrate, and the actual amount per digester volume/time unit depends on temperature, detention time, solids content and composition of feedstock, etc. In a short paper all the results for all the feedstocks which have been tested under different conditions cannot be given. However, the figure for piggery waste of 0.3 m^3/kg total solids (dry wt) fed to the digester, with gas of 68–70% methane content is, roughly, the kind of gas production expected from many agricultural wastes. As 1 fattening pig produces about 0.5 kg of fecal solids then 1000 pigs will produce approximately 150 m^3 of gas per day. This will have an energy value of the order of 1000 kWh.

While in a warm climate little, if any, heat will be needed to keep a digester at a mesophilic temperature, in a temperate climate artificial heating of a digester will be needed for most of the year. This heating is a debit to be set against the useful energy production of the digester. The heat may be taken from a biogas boiler which may also be linked to some building or other heating system. However, many agricultural-waste digesters are designed to run gas-engine generators and here waste engine heat can be used for producing hot water for digester, and possibly other, heating purposes. Some 25% of the fuel energy can be obtained as electrical energy and up to 65% more of the fuel energy can be recovered from engine and exhaust heat-exchangers. A very approximate figure of 30% of the biogas energy is usually used to suggest the heating requirements of a digester in a temperate climate, but this obviously varies.

The situation all over the world at the moment with regard to the bigger agricultural-waste digesters is one of development and testing. There is no difficulty about the microbiological side of the process; digestion once started is stable. What problems are being encountered concern the handling of the feedstocks, and sometimes on farms provision of feedstocks of suitable consistency and uniformity from day to day. Other aspects being tested are the life of the gas engines being used as these differ froœ the large dual-fuel engines which have been used for many years in most of the sewage works using biogas energy. The life of pumps and other components of the actual digester construction has still to be assessed. While some digesters are having more difficulties than others, many are now running successfully and a problem in some cases is to find uses for the energy being produced.

It is difficult to find out how many of the small-scale, 'developing country', digesters are continuing to run successfully, but there is no doubt that many are running. However, one of the problems here is to find a design cheap enough for the very poor farmer, as well as to improve the efficiency of the digesters.

While there remains research work to be done on optimum conditions for digestion of some feedstocks and, particularly, mixed waste feedstocks, testing of different digester designs, use of digested sludge for other than fertilizer purposes, and so on, digestion has now got into the large-scale testing and development stage and the next few years should see the number of working digesters increasing. Digestion is the alternative fuel source of most widespread possible application, using, as it does, almost any organic waste, and its ability to use fibrous vegetable matter is putting the energy-farm nearer practical application in countries where crop production and energy requirements of the population are suitable. Digestion can also be used in conjuction with other forms of bio-energy production to utilize vegetable residues from sugar or starch production - there are many possibilities now being or to be exploited.

Some books and review papers on anaerobic digestion:

P.N. Hobson, S. Bousfield and R. Summers, The anaerobic digestion of organic matter. Crit. Rev. environ. Control *4*, 131 (1974).

M.P. Bryant, Microbial methane production - theoretical aspects, J. Anim. Sci. *48*, 193 (1979).

P.N. Hobson, Biogas production - agricultural wastes, in: Energy from the biomass, p.37. Watt Committee on Energy Report 5, London 1979.

P.N. Hobson, S. Bousfield and R. Summers, Methane production from agricultural and domestic wastes. Applied Science Publishers, Barking, England, 1980.

Proceedings 1st Int. Symposium on Anaerobic Digestion, Cardiff 1979. Applied Science Publishers, Barking, England, 1980.

P.N. Hobson and A. Robertson, Waste treatment in agriculture. Applied Science Publishers, Barking, England, 1977. (Amounts of wastes produced, anaerobic, aerobic, physical and chemical methods for pollution control).

Possibilities of gas utilization with special emphasis on small sanitary landfills

by Mauro Gandolla*, Ernst Grabner* and Romano Leoni*

Swiss Federal Institute of Technology, Zürich, and Federal Institute for Water Resources and Water Pollution Control, CH-8600 Dübendorf (Switzerland)

Summary. Based on general observations on gas production in sanitary landfills, properties of some important landfill gases are discussed, especially with regard to the potential heat recovery. It has been shown, based on practice oriented considerations, that the utilization of landfill gases can be worthwhile even in small landfills. A scheme has been given which shows all the possibilities for collection, pretreatment, storage and combustion of the gases. The question of energy storage and energy utilization has also been addressed. The scheme has been discussed, as well as some of the processes, using the example of the hot water generation plant in the Croglio sanitary landfill in Tessin (Switzerland).

Calculation of the running costs shows that this plant, which is designed for a 335 MJ/h production, is working economically.

The term 'refuse management' implies the re-utilization of materials and/or energy. It is very seldom however, that one encounters good examples for such re-use, especially when dealing with small refuse disposal areas.

Since even in small landfills (landfill volume up to

10^6 m^3, refuse quantity up to 100 t/d), a considerable gas quantity can be produced, a pilot plant has been constructed[1] to study the possibility of utilizing the generated gas. The data thus obtained on gas utilization in a sanitary landfill serve as the basis of the following report.

Methane production

Knowledge of the production of methane, including that created in landfills, has been available for a long time. However, only recently, now that it has become clear that such gases can create safety hazards as well as problems in recultivation of the land[2,3], has one started paying much attention to its generation in landfills. In answer to the question why these problems have arisen only recently, there are any number of hypotheses. The following, however, seems to be most probable: In the past, refuse was buried by the communities in small, uncontrolled dumpings, which allowed aerobic decompositon. Today, we have large sanitary landfills, where the organic materials are mostly decomposed anaerobically.
Aerobic fermentation is caused by microorganisms which are also responsible for sludge digestion[4]. These organisms are able to convert part of the organic materials present in refuse into methane (table 1).
For a detailed summary of organic substances in refuse, refer to Stegman[3].

Table 1. Waste composition[5]

Waste classification	Switzerland 1973 (%)	Federal Republic of Germany 1971 (%)
Paper	36	22–25
Synthetics	4	2–3
Textiles, leather	8	2–4
Rubber, wood Kitchen scrap	20	10–20
Glass, gravel Ceramics	12	10–16
Metals	5–8	4–9
Miscellaneous	4–7	–

Table 2. Constituents so far found in landfill gases (from Winter[7])

Constituents		
Methane	CH_4	0–85% by vol.
Carbon dioxide	CO_2	0–88% by vol.
Carbon monoxide	CO	2.8% by vol.
Ammonia	NH_3	0–0.35 ppm
Hydrogen	H_2	0–3.6% by vol.
Oxygen	O_2	0–31.6% by vol.
Nitrogen	N_2	0–82.5% by vol.
Hydrogen sulphide	H_2S	0–70 ppm
Ethylmercaptan	C_2H_5SH	0–120 ppm
Acetaldehyde	CH_3CHO	150 ppm
Acetone	C_2H_6CO	100 ppm
Benzene	C_6H_6	0.08% by vol.
Argon	Ar	0.01% by vol.
Heptane	C_7H_{16}	0.45% by vol.
Nonane	$C_6H_5CH_3$	0.09% by vol.

Besides methane, a large amount of carbon dioxide is produced in landfills. Other gases are present only in small quantities (table 2).
With respect to the total attainable gas quantity, the information given in the literature varies from 60 to 290 m^3 gas/t refuse[3]. From our own experience, a gas quantity of 100 m^3 methane/t refuse can be expected to accumulate during 20 years.
Immediately after the deposit of refuse, a high production of CO_2 begins which reduces gradually in favor of methane production. The gas production varies greatly in quantity and composition during the first months. The primary unstable phase is followed by a second stable phase, where CH_4: CO_2 can reach a constant value of 3:2 (fig. 1).

Properties of some landfill gases and gas mixtures

It has already been mentioned that the gas mixture of landfill refuse consists mainly of methane and carbon dioxide. Relatively large amounts of nitrogen and oxygen are often present as well, especially in cases of forced suction, or in areas directly influenced by the atmosphere. Carbon monoxide and hydrogen, if they appear at all, are present at a low concentration (see also table 2).
Even in the stable phase the concentrations can vary, thus changing the ignition and combustion parameter. Table 3 shows the physical and chemical properties of the individual gases. The properties of the various gas mixtures are not yet well known.

Due to the high CH_4 content, the combustion parameter of methane itself is often used for the entire mixture. This yields on the one hand, an overestimation of the energy properties of the concerned mixtures, and on the other hand, an under-estimation of the danger caused by the presence of a small quantity of CO and/or H_2. From the experiences with

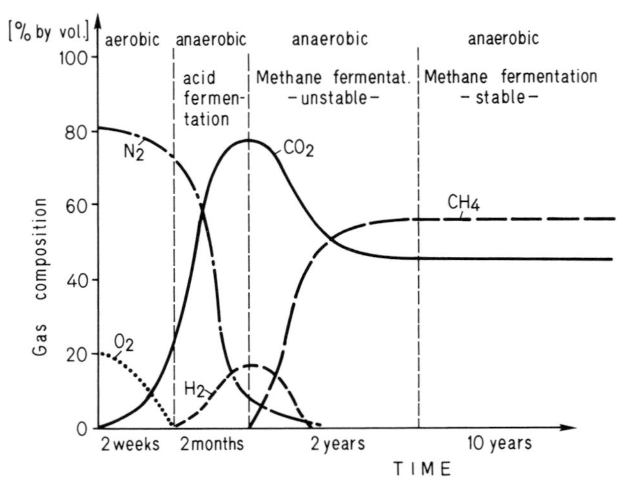

Figure 1. Gas composition during decomposition of municipal refuse (from Farquar et al.[6]).

Table 3. Properties of some gases[7-11]

Gas	Formula	Density (kg/m³)	Net calorific value (kJ/m³)	Critical temperature (°C)	Ignition range in air lower/upper (% by vol.)	Flame velocity (m/sec)	Minimum ignition energy (mJ)	Ignition temperature (°C)	Water solubility (g/l)	Common properties
Methane	CH_4	0.717	35,600	− 82.5	5/15	0.4	0.6–0.7	600	0.0645	Odorless, colorless, non poisonous
Carbon dioxide	CO_2	1.977	31.1	31.1	–	–	–	–	1.688	Odorless, colorless, non poisonous at low concentrations[7]
Oxygen	O_2	1.429		− 118.8	–	–	–	–	0.043	Odorless, colorless, non poisonous
Nitrogen	N_2	1.250		− 147.1	–	–	–	–	0.019	Odorless, colorless, non poisonous, non inflammable
Carbon monoxide	CO	1.250	12,640	− 139	12.5/74	0.5	–	600	0.028	Odorless, colorless, poisonous, inflammable
Hydrogen	H_2	0.090	10,760	− 239.9	4/74	2.8	0.05	560	0.001	Odorless, colorless, non poisonous, inflammable
Hydrogen sulphide	H_2S	1.539		100.4	4.3/45.5				3.846	Colorless, poisonous
Air		1.29	–	–	–	–	–	–	–	Odorless, colorless, non poisonous, non inflammable

the Croglio sanitary landfill, where no considerable release of CO and H_2 was detected, the following statements can be made:
1. Under normal conditions, the flame of methane mixtures and air can reach a velocity of 0.4 m/sec. The combustion of the landfill gases released directly into the air yields a lower flame velocity (< 0.2 m/sec). This usually causes the flame to be unstable, leading to lifting and extinction.
2. While an ignition energy of 0.7 mJ is required for the methane and air mixture, a higher energy is necessary for the directly released gases from landfills. The gas ignition in the Croglio sanitary landfill has an electric arc of 15,000 V and 10 mA by an electrode gap of about 10 mm.

Problems with release of landfill gases

The gases produced through anaerobic refuse decomposition are easily ignited, and their combustion in absence of air can quickly set off an explosion. Control and preventive measures are therefore necessary in order to avoid this risk. Since the gases migrate underground, one should not under-estimate the danger for people and objects in the immediate vicinity. A strict management of the landfill is required to prevent such dangers. Through technical means, the leakage and underground movements of the gases can be reduced, controlled or eliminated.
Specific safety measures are required for the landfill personnel in order to ensure safety in the working areas. The gas mixtures of landfill contain a small amount of malodorous materials, which can be a nuisance to the personnel, as well as the residents in the neighborhood, if a leakage occurs. This unpleasant odor can be counteracted by the application of suitable chemical substances, or better yet, by careful coverage of the refuse with soil, and by collecting, processing or combusting of the gases.

Advantages of methane production from landfills

It is quite understandable that methane represents an important energy source. As we have seen from the above, 1 kg CH_4 is equivalent to 1.18 kg fuel oil, and 1 m^3 CH_4 to 1 l fuel oil.
Assuming: a) that 1 kg refuse can generate about 100 l CH_4, b) that 40% of the produced methane can be recovered, whereas the rest is lost in the soil and atmosphere, and c) that each person produces on the average 0.75 kg refuse per day (this value has been calculated for the area of 'Consorzio eliminatione rifiuti del Luganese'), the potential methane production can be calculated to 75 l · $person^{-1}$ d^{-1}, of which about 30 l would be usable. The following figures illustrate the amount of methane, i.e., the fuel oil equivalence of the population of a small Swiss region (Lugano).

Population, approximately	100,000
Theoretical methane production	7500 m^3/d
Recovery (40%)	3000 m^3/d
Fuel oil, approximately	3000 l/d

The above shows, that even in such areas, it is worthwhile to consider the possibility of energy recovery from refuse landfill.

In the following discussion, some various means of gas recovery and utilization of communal refuse in organized landfills by anaerobic decomposition are examined.

The blockscheme in figure 2 shows the most important elements of a full-scale plant for gas recovery and utilization. It should be noticed that some elements are not absolutely necessary. The different elements separated by dashed lines, numbered 0–5, indicate the elements within the boundaries of the landfill.

The quality of landfill gases can be compared with that of pure methane. A landfill produces gas of optimal quality if the composition of the mixture has more than 60% of vol. CH_4, and less than 50% of vol. CO_2 and 1% of vol. for the remaining gases. It is undesirable to have a high CO_2 content, and for N_2 or O_2 to be present; especially if the ratio $O_2:N_2$ is different from that of the atmosphere. In such cases, the atmospheric oxygen can diffuse into the landfill, causing the CH_4 to oxidize while the atmospheric nitrogen remains unchanged.

Recovery of gases

Water drains are often the means by which the gases

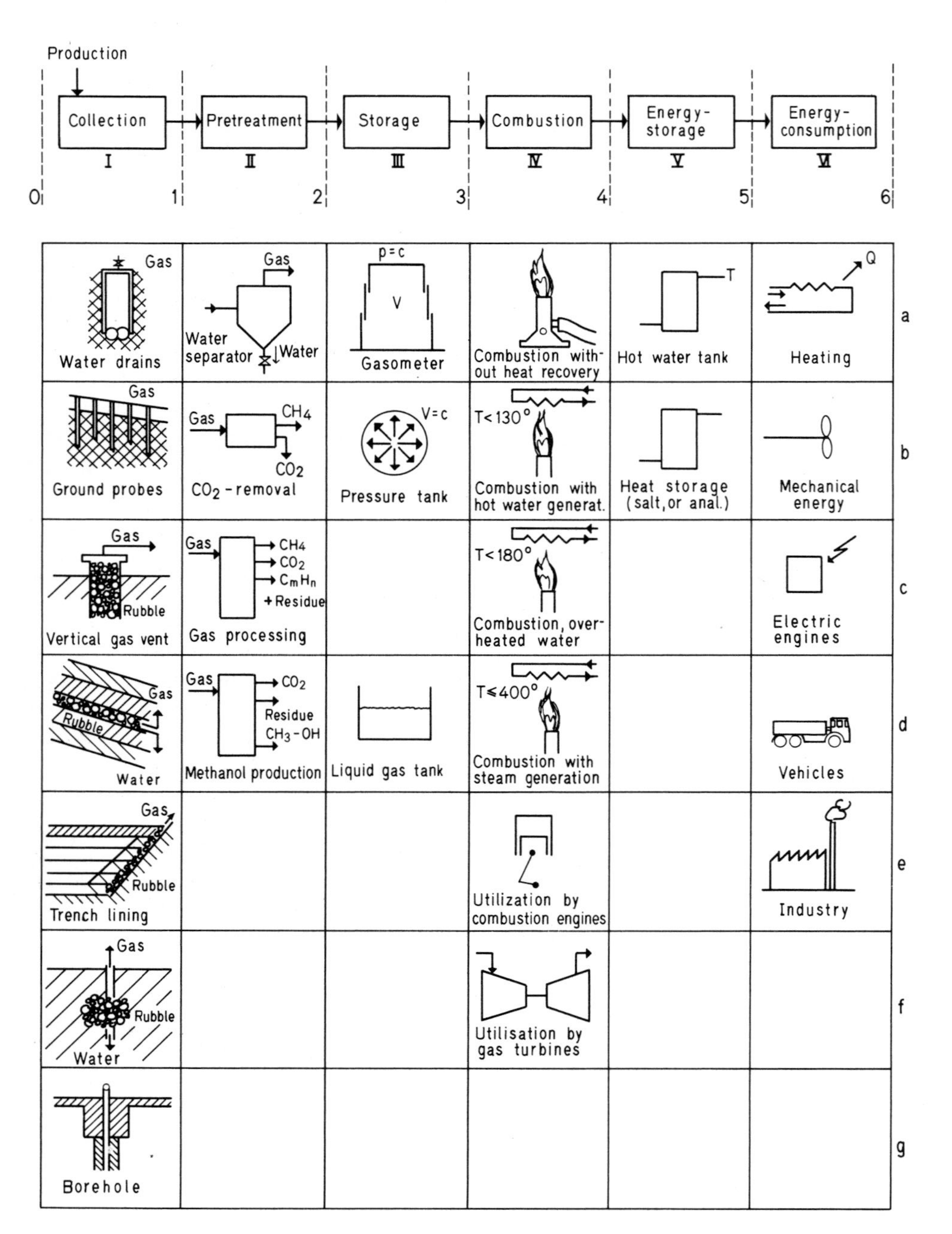

Figure 2. Possible elements of a landfill gas utilization plant.

tend to travel (fig. 2-I/a). This causes a release of unpleasant and easily flammable materials at the outlets. It is therefore advisable to use a siphon system to enable drawing of the gases without their being mixed too much with air. Experience from the Croglio sanitary landfill shows that atmospheric conditions can easily influence the composition of gases drawn. Since it is generally impossible to avoid a gas exchange between atmosphere and drainage system, the gases obtained have a large amount of nitrogen and oxygen. This means that the energy content of the gas mixture is very low. If a large amount of water flows through the drains, more CO_2 than CH_4 are dissolved (under normal conditions, about 1 l CO_2/l H_2O and 0.01 l CH_4/l H_2O). This causes the ratio $CH_4:CO_2$ in mixed gases to change in favor of CH_4. The drainage pits can eventually become potential explosion chambers and strict safety measures must therefore be taken.

The gases can also be sucked through ground probes (fig. 2-I/b) which are driven into the landfill. This allows a relatively inexpensive collection system. At the Croglio sanitary landfill, steel-tipped, 5-cm diameter ground probes were used, driven by a pneumatic hammer and reaching 5–10 m in depth.

From an area of about 1000 m^2, about 40–60 m^3 CH_4/d were extracted in 2 months time by using 10 ground probes. At the time of experiment, the landfill had been closed for 3 years.

The quality of the gases drawn off depends strongly on the air diffused into the landfill. This effect can be influenced by covering the landfill. The parameters for the gas suction as well as the distances between the individual probes should be determined separately for each landfill.

This system has the advantage that it can be installed immediately after completing the landfill, and without special preparations. Also, even if the collected gases seldom show outstanding qualities, gas mixtures of relatively constant composition can be easily combusted. This system is especially suitable where there is limited vegetation caused by high gas content in the upper layers of soil.

Gas recovery by way of a vertical stone-filled gas vent (fig. 2-I/c) is only effective if the upper part is well isolated from the atmosphere. If not, the gas quality is expected to be bad or the gas mixture can even be uncombustible. If there is not enough suction, the undesirable, malodorous gases can be an additional problem. To prevent possible water infiltration, this drainage should be connected to a water drain located above the soil layer. Already during the landfill operation, the gas vents should be filled with stones, which can disturb the truck traffic within the landfill. Since the gas vents must remain opened during this phase, they present a danger and are often a source of malodorous fumes.

Gas recovery from slanted combined drains (fig. 2-I/d) is possible when, during the filling phase of landfill, canals are built to let the water and gases travel through. These canals are easily built, are constructed in the form of a dry stone wall, and are slanted towards the surface of the landfill. These drains have the same properties as the gas vents. However, they can be more easily isolated from the atmosphere and disturb the landfill less during the filling phase.

Due to the fact that the landfill body is sinking while the surrounding ground stays stable, canals are often formed where gases travel easily to the surface. Such gas emissions can be accelerated by building a wall of stones or gravel (fig. 2-I/e). The extraction technology should proceed similar to the stone gas vent. The underground gas migration to the open should be controlled or reduced.

The gas recovery out of a landfill via a specially incorporated stone lens (fig. 2-I/f), can yield good results if this is located deep enough (beyond atmospheric influence), and if it can function as a collecting place for natural and man-made preferred flow channels (see also fig. 2-I/d). Seepage accumulation, however, should be avoided by, for example, connecting the base of the lens to a water drain or to a safe percolation zone. Experiences in the Croglio and other sanitary landfills confirm that the stone lenses will retain seepage if not drained. The recovered gases are generally of good quality as long as no air is sucked in through the drains.

Recovery with boring probes (fig. 2-I/g) is a relatively expensive system but gives the best results. Such borings are usually possible even after the completing of the landfill. The sucked gas is of a very good quality containing 95–99% of CH_4 and CO_2. In order to prevent the gas quality from deteoriating by infiltrated air, the upper part of the bores must be isolated from the atmosphere with 1–3-m-thick layer of concrete surfacing or other similar materials. Even with such boreholes there is a danger that they will eventually fill with water so that they must be taken out of service. According to our experience, the diameter of the borehole is determined by the gas quantity to be drawn off. However, the costs of the borehole increase disproportionally to the diameter. It is therefore not economical for small and medium landfills to have a borehole more than 20 cm in diameter.

Pretreatment of gases

The recovered landfill gases, depending on their utilization must be pretreated to a greater or lesser extent.

The emanating landfill gases usually have a temperature of about 30 °C and a relative humidity of 100%. The cooling down of gases in the pipes causes water condensation such that the deepest places fill with water. It is therefore necessary to separate water from the gases. How much needs to be taken out depends on how the gases will be utilized. A certain amount must be removed to prevent retention of water in measuring apparatus, pipes, ventilators, etc. The simplest water separation technique is to place a sufficiently large tank in the coolest area. These tanks should also have an emptying device, and are to be located at the lowest part of the pipes. When such tanks are filled additionally with fine gravel, they can function as a safety device against backfiring (fig. 2-II/a).

The CO_2 content must often be reduced in order to optimize combustion requirements or because the customers request it. One possible way of removing CO_2 is through the absorption by liquids, for an example, trietylamine (fig. 2-II/b). The CO_2 can be then released by heating up the liquid. The evaporated CO_2 can be reutilized in the event that this is economical. Unfortunately, such plants are only economical if an amount over 1000 m^3 CH_4/h is produced. They are thus unsuitable for small and medium landfills.

In some cases the separation of CO_2 is not sufficient, and other components (for example, C_nH_m) must be removed as well (fig. 2-II/c). In these cases, advance processes are needed (for example, by using a molecular sieve). Such a plant with an advance processing stage exists in Palos Verdes landfill near Los Angeles[13]. Also these plants are not economical for small and medium landfills.

The conversion of gaseous methane into liquid methanol (fig. 2-II/d) has significant advantages with regard to transportation and storage. Processes for chemical conversions have been known for a long time, but are not economical. Processes for biological conversions are presently being tested, and depending on the results produced, they could provide a good alternative in the future[14].

Storage of gases
Since the gas production in a landfill is relatively constant and cannot be influenced within a short time, and since the energy utilization is nevertheless subject to changes, the following points must be considered: a) If gas production is higher than peak demand, intermediate storage is not required. During the remaining time, a considerable loss of gas is to be expected. b) If gas production is lower than peak demand, a possibility to store the gas must be foreseen, so that the gas supply can best be utilized. The different possibilities of gas storage are shown in the following paragraphs. In principle, only gas mixtures of good quality should be stored, whereby the oxygen content should be kept minimal for safety reasons.

Gasometers (fig. 2-III/a) are systems which have for a long time been known as digesters in sewage treatment plants. They consist of a container with a variable volume, in which the gas is stored almost without pressure. Such containers seem to be uneconomical for small landfills.

Compression containers (fig. 2-III/b) are unsuitable for storing landfill gases. This is due to the fact that the critical temperature of these gases is too low to be liquidized at room temperature ($T_k = -82.5$ °C for methane). This is also true for propane and butane. If these gases are to be utilized as fuel, they should be stored in compression containers which can endure a pressure up to 200 bar. Compression containers can also be installed in small landfills if the gas quality is good, and if safety measures are taken.

The conversion of gaseous methane into liquid methanol would enable easier storage, since a normal liquid container could be used. Unfortunately, such conversion procedures are still in the development stage (fig. 2-III/d).

Combustion
The gas obtained from refuse landfill is usually burned in order to eliminate foul odors or to utilize it as an energy source.
In most cases, the heat produced by combustion of landfill gases is not utilized (fig. 2-IV/a). This type of energy waste is only reasonable if there are no consumers for the heat produced, and if the gas quantity is small or the gas quality is poor (a minimum heat equivalent of 125 MJ/h is required).

The easiest way to obtain energy from landfill refuse gases is by producing hot water of relatively low temperature (< 130 °C) (fig. 2-IV/b). The water must be pressurized so that a temperature above 100 °C can be reached. If the hot water is not utilized in the area of production, this system is limited because large, well isolated pipes are required for transportation. Such an installation was built in the Croglio sanitary landfill and is working successfully (see fig. 4).

Combustion with production of super-heated water (fig. 2-IV/c) differs from the above in that the temperature of the water reaches 180 °C. The advantage of this system lies in the fact that the produced heat can be transported more economically.

Combustion with steam generation (fig. 2-IV/d) is suitable for combustion of a large quantity of gases, and in cases where the heat has to be transported over a long distance (several km). Such a system works up to a pressure of 50 bar and a temperature up to 400 °C. Among the 4 systems mentioned, this system is the most expensive and is not economical for small and medium landfills.

The combustion of digester gases has been in practice

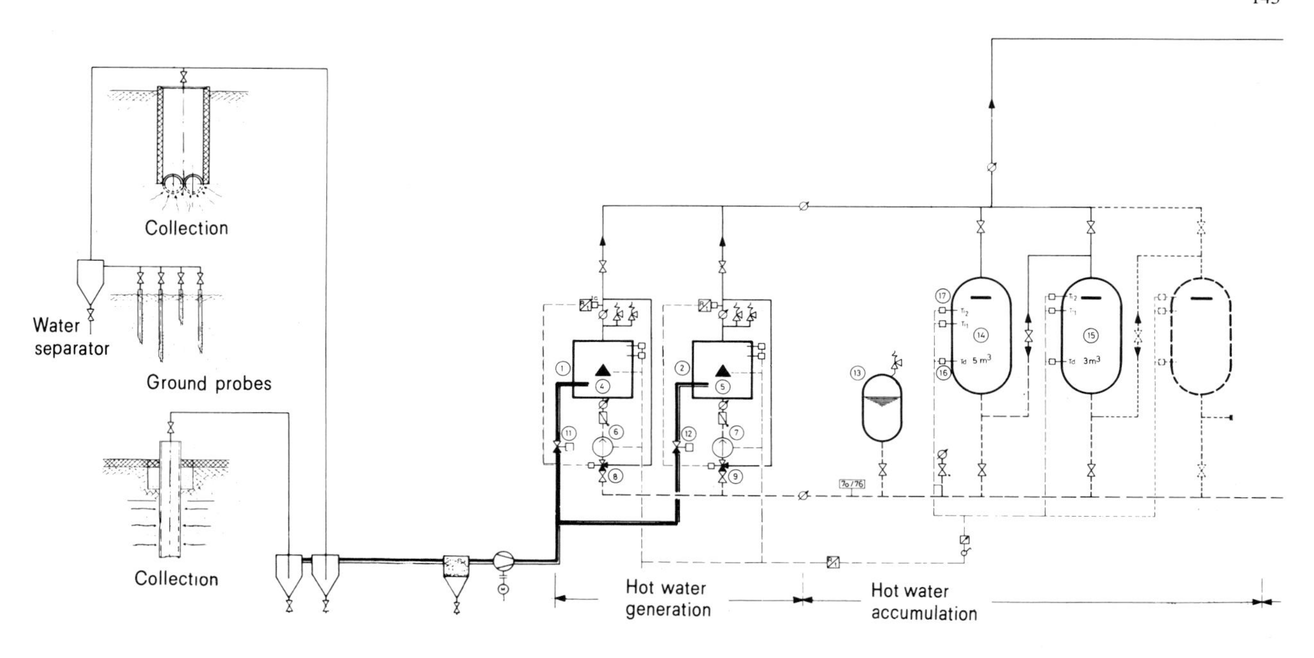

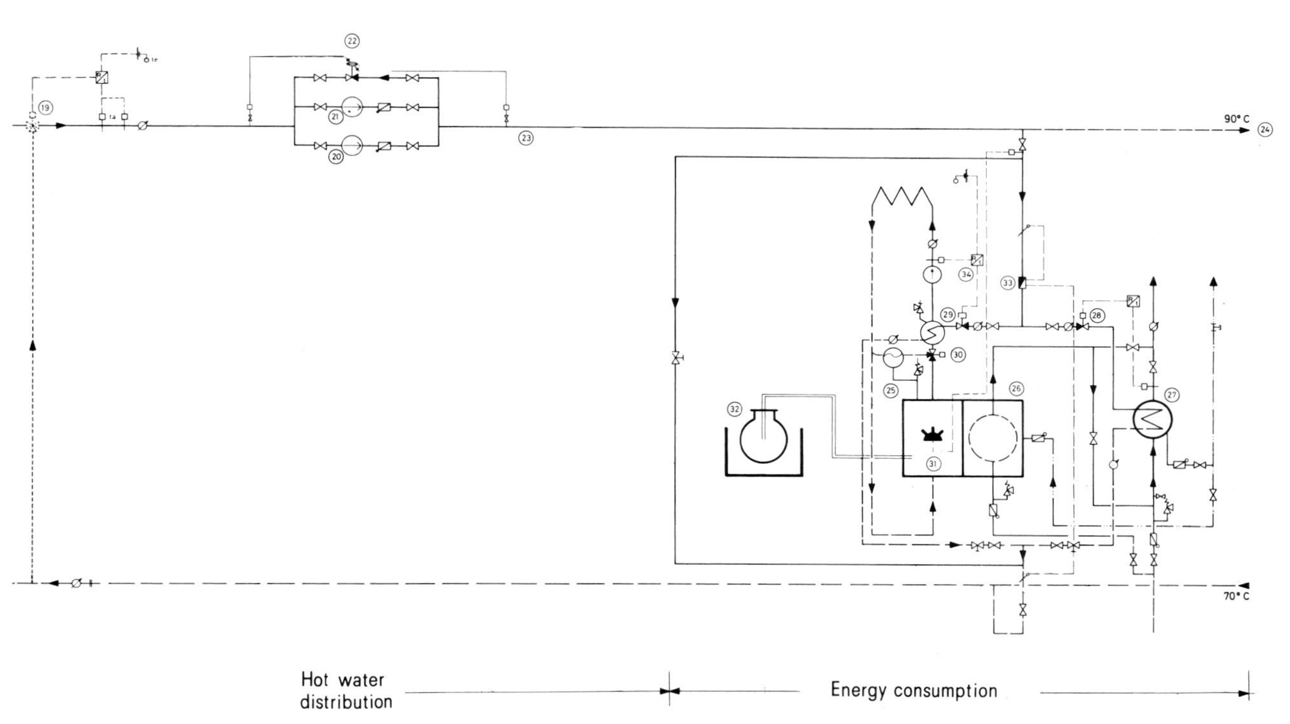

Figure 4. Hot water supply plant system of Croglio's sanitary landfill.

for a long time. Such systems are often found in sewage treatment plants, but little is known about the application in landfills. This combustion of digester gases usually produces mechanical energy (fig. 2-IV/e), which can of course be transformed into electrical energy. The heat waste of the engines is also utilized in some cases. Examples are the 'Fiat Totem'-engine for smaller plants and the 'Caterpillar'-engine for larger plants.

Such a system is now realized in Palos Verdes, where a gas compressor is driven by a gas engine. In the landfill near Turin, a Fiat Totem-engine is utilized and in Croglio, a modified 'Renault'-engine has been recently experimented with. Since the sulfur content in a landfill's biogas is generally very low, in contrast to sewage treatment plants, desulferization is not needed (table 2). The disadvantages of such a system, however, are the high initial and operational costs. It should also be taken into consideration that constant adjustments of vaporizer and ignition are required, due to the changing concentration of the gases. This system is also applicable in small landfills.

Although the initial costs of gas turbines (fig. 2-IV/f) are high, the low operational costs and their durability enables a more economical system than the alternative engines. The disadvantage of gas turbines is the necessity of processing the gases in order to reach a minimum heating value of approximately 22,000 KJ/m^3 [15]. In addition, the efficiency of gas turbines is very low if the waste heat is not utilized, and the system can only be operated by specially trained personnel. Therefore, the use of turbines is only worth-while and economical for large landfills.

Heat storage systems
The hot water produced could be stored in order to optimize the utilization of the gases, or to minimize the boiler's dimension.
The simplest of heat storage systems makes use of good insulated water tanks (fig. 2-V/a). The water temperature depends on the recovery system (see 'Combustion section'). The large storage volume required is the main drawback. In Croglio, for example, 8000 l of water are stored with Δt of 70 °C. This corresponds to a heat amount of about 2500 MJ (equivalent to 60 kg fuel oil). Because of these drawbacks, this system is only economical for small and medium landfills.

To increase the storage capacity of the water tanks, special materials, such as naphtalin or mixtures of diphenyls, could be added (fig. 2-V/b). In this way, the heat capacity could be increased by a factor of 20 using the same volume. The costs, however, limit the usage of such heat storage systems to small and medium landfills only.

Utilization of energy
To utilize the energy, the following possibilities are conceivable: direct utilization of hot water or steam, as practised in the Croglio sanitary landfill (fig. 2-VI/a); usage of mechanical energy, as for example in Palos Verdes (fig. 2-VI/b); conversion to electrical energy, as it is often practised in sewage treatment plants (fig. 2-VI/c); operation of vehicles with compressed gas. (This possibility is limited practically to vehicles belonging to the plant facilities (see 'compression containers'). Processes for methanol production would greatly increase the advantages of such a system (fig. 2-VI/d)); and utilization as raw material in industries (fig. 2-VI/e).

Standard of individual components of System I–VI
The blockscheme in figure 2 is divided into 6 sections representing various elements either inside or outside of the landfill. The dashed lines 0–6 indicate possible boundaries between each element. It is of course not possible to formulate rules which are always applicable, however, the following guidelines can be given:
- It is not recommended to combust landfill gases directly by using normal commercial means. This is because they are often unsuitable to be operated with such gases. In addition, most of the suppliers are not able to provide specially trained service personnel.
- If normal commercial means are nevertheless desired, the gases should be processed so that they have the same properties as conventional gases (for example, natural gas). These processes are only economical for landfills from which more than 1000 m^3/h methane can be recovered.
- Since the concentration of individual components of gases are subject to wide variations, combustion problems can occur if alternative engines are operated.
- If landfill gases are to be transported, a careful removal of water is necessary since the condensation of the water in the pipes creates problems.
- There is no problem with direct transportation of heat, at least for small amounts (up to 0.2 Mcal/h) over short distances. For amounts larger than 0.2 Mcal/h) and for longer distances, super-heated water or vapor should be used.
- Because of hygienic reasons, one should refrain from supplying gases directly to consumers.

In summary, it can be said that ideally the processes I–V (shown in blockscheme of paragraph 2) should be located directly in or near the landfill.

The hot water generation in the Croglio sanitary landfill

Using the above mentioned guidelines in Croglio, the utilization of energy is realized in the form of a hot water generation plant according to the following scheme.
The plant consists of 4 units, hot water generation, hot water storage, water distributor, and energy consumer; these are designed to enable further expansion.
Two boilers with a total production of 335 MJ/h are installed for the *hot water generation.* They are

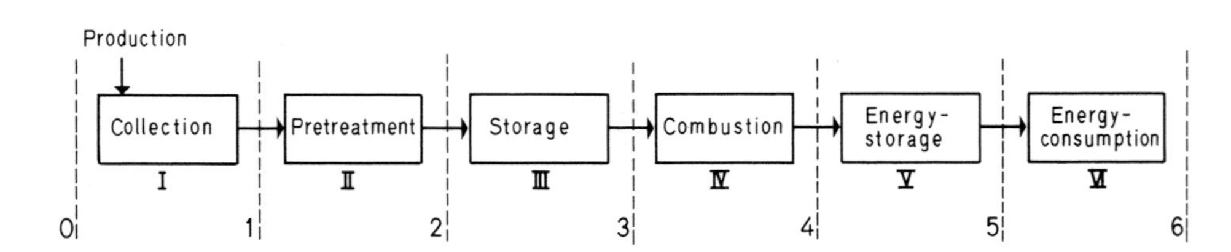

Figure 3. Blockscheme of the Croglio sanitary landfill.

especially equipped with atmospheric burners for combustion of landfill gases. Each boiler has its own individual water cycle and feeds into the hot water tank.

The *hot water storage* consists of 2 isolated cylindrical tanks, which function according to thermal stratification, and are placed in series. The storage volume is 8 m^3, and the temperature between 60 and 130 °C. The tanks are connected to a pressure equalizing tank.

Hot water distribution consists of fixed and flexible isolated pipes (heat transfer coefficient of polyurethanfoam isolation at 20 °C: 0.084 kJ/mh °C)[16], of hot water pumps and additional pressure regulators. The flexible pipes to the consumers are put under ground. A part of the hot water produced is used directly for the operation of the landfill heating and sanitary installations, while the rest is transported to private consumers. Heat exchangers are installed between the hot water cycle and the consumers for operational safety reasons. In order to ensure independence of consumers during disturbances, the heat exchangers are installed parallel to the existing heating plant. If the hot water supply cuts out, the heating plant of the consumer starts functioning automatically.

Costs

Based on the data obtained from the Croglio sanitary landfill, several cost aspects are given below to show the amount of investment required for a plant which can function economically if hot water consumers are in the immediate vicinity.

Investments (including salaries)	
Burners and boilers (pipes 335 MJ/h)	SFr. 15,000.–
Hot water storage (2 tanks with a 3-m^3 content)	SFr. 11,000.–
Installations for heat transport (pipe lines, circulating pumps, valves, etc.)	SFr. 6,000.–
Underground pipes to the consumers (flexwell-pipes 30 m each way)	SFr. 10,000.–
Additional installations at the consumer (heat exchanger, boiler, electrical controls)	SFr. 15,500.–
Miscellaneous (costs for planning, etc.)	SFr. 7,500.–
Total investments	SFr. 65,000.–
Revenues	
Savings at the landfill operation itself by the conversion from electricity to landfill gas combustion for the production of hot water	SFr./y 8,000.–
Estimated energy supply to the consumers (introduction price)	SFr./y 1,570.–
Total revenues and savings respectively	SFr./y 9,570.–
Amortization	
Yearly amortization of 6% interest in 10 years	SFr./y 8,450.–

Experiences with the plant

The mentioned plant has been operating automatically since November 1979 without failures. The daily methane consumption is presently 150 m^3, and the estimated efficiency is around 0.6.

* Authors' addresses: M. Gandolla, Dipl. Ing. ETH, CH-6934 Bioggio. E. Grabner, EAWAG, CH-8600 Dübendorf. R. Leoni, Dipl. Ing. ETH, CH-6900 Lugano.

1 M. Gandolla, Gasverwertung bei der verdichteten Deponie Croglio, ISWA Journal *25*, 16 (1978).
2 W. Ryser, Überlegungen zur Gasentsorgung und praktische Hinweise zur Zwangsentgasung, ISWA Journal *25/26*, 25 (1978/79).
3 R. Stegmann, Gase aus geordneten Deponien – Allgemeine Problemstellung, Entstehung und Anfall, ISWA Journal *26/27*, 11 (1978/79).
4 A.J.B. Zehnder, B.A. Huser, Th.D. Brock and K. Wuhrmann, Characterization of an acetate-decarboxylating, non-hydrogen oxidizing methane bacterium, Archs Microbiol. *124*, 1 (1980).
5 Stand der Abfallbeseitung in der Schweiz. Stand Januar 1978. Bundesamt für Umweltschutz, Bern, Switzerland, 1978.
6 G.J. Farquhar and F.A. Rovers, Gasproduction during refuse decomposition, Water, Air and Soil Poll. 2 (1973).
7 K. Winter, Kriterien bei baulichen Einrichtungen auf Deponien hinsichtlich Gefährdung durch Gase. Forschungsbericht 10302102. Erich Schmidt Verlag, Berlin 1979.
8 Sicherheitstechnische Kennzahlen von Flüssigkeiten und Gasen, Schweiz. Unfallversicherungsgesellschaft, eds. Luzern 1976.
9 F. Schuster. G. Leggenwie und I. Skunca, Gas-Verbrennung-Wärme. Essen 1964.
10 P. Hostalier, Die industriellen Gasbrenner. München/Wien 1974.
11 Handbook of Chemistry and Physics, Chemical Rubber Company Publishing.
12 W. Foerst, ed., Ullmanns Enzyklopädie der technischen Chemie, Bd 1, S. 333. München 1951.
13 R.H. Collins, Upgrading landfill gas to pipeline specifications, ISWA Journal *28/29*, 1 (1979).
14 J. Nuesch, personal communication.
15 M. Piazzini, communication Turbomach AG.
16 Kabelwerke Brugg, communication.

Future systems

As the previous section indicates, methane cannot easily be stored. Converting it to methanol would reduce its volume and facilitate handling – can this aim be achieved with the help of microorganisms? Knowledge to date on this subject and projects for the future are presented by O. Ghisalba and F. Heinzer.
The final article deals with yet another topic of relevance for the future. Can man mimick photosynthesis in artificial systems? P. Cuendet and M. Grätzel discuss the efforts to overcome the problem of the rather low efficiency in photobiological processes by simplifying the energy storing process as well as the molecular system which transforms light energy into a usable chemical form.

Methanol from methane – a hypothetical microbial conversion compared with the chemical process

by Oreste Ghisalba and Franz Heinzer

Central Research Laboratories of Ciba-Geigy Ltd, CH–4002 Basel (Switzerland)

1. *Introduction*

The synthesis of methanol from organic feedstocks is in effect a conversion of stored solar energy into a liquid fuel. Methanol can be produced chemically via synthesis gas from a variety of feedstocks such as methane (natural gas, associated gas, biogas), biomass (wood, urban and agricultural refuse), coal, heavy fuel oils, shale oils and naphtha (fig. 1). The synthesis gas, a mixture of CO and H_2, is produced from these feedstocks either by steam reforming or by partial oxidation. At present, by far the largest amount of methanol (estimated world production in 1980: 12 million metric tons) is produced from natural gas[1,2]. The conversion efficiency for natural gas in terms of the heating value of methanol as a fraction of the total energy input, which consists of feedstock and processing energy, is about 60% or above[2,3]. In this paper we will consider whether the energy consuming indirect chemical conversion of methane (natural gas) into methanol might be replaced by an energy saving direct microbial oxidation process (fig. 1). The methane used for such a microbial conversion process could either be natural gas or biogas obtained by anaerobic digestion of biomass. To assess the feasibility of a microbial conversion we must first have a closer look at the well-known petrochemical conversion process.

2. *The chemical methanol synthesis based on methane*

The technical methanol synthesis based on methane is characterized by the 3 equations below[1,4,5]:

(1) steam reforming

$$CH_4 + H_2O \xrightarrow[\substack{700\text{–}900\ ^\circ C \\ 1\text{–}25\ bar}]{\substack{\text{Ni-catalyst} \\ (15\text{–}20\%\ \text{on}\ Al_2O_3\ \text{or}\ SiO_2)}} CO + 3\,H_2, \qquad \Delta H_{800} = +226.9\ kJ/mole$$

(2) shift reaction

$$CO_2 + H_2 \xrightarrow{\text{Ni-catalyst}} CO + H_2O, \qquad \Delta H_{25} = +38.5\ kJ/mole$$

(3) low pressure methanol synthesis

$$CO + 2\,H_2 \xrightarrow[\substack{230\text{–}280\ ^\circ C \\ 50\text{–}100\ bar}]{\text{Cu-catalyst}} CH_3OH, \qquad \Delta H_{25} = -90.9\ kJ/mole$$

The steam reforming of methane in this highly endothermic reaction (1) yields a synthesis gas too rich in hydrogen for the methanol synthesis (3). By adding 33% of CO_2 the CO/H_2 ratio can be shifted to the ideal value of 1:2 (2). If no CO_2 is available, excess hydrogen is purged from the methanol synthesis loop and burned as a fuel which can be used for the process[6]. Methanol is finally synthesized according to the exothermic step (3) from synthesis gas. The reactors constructed today follow exclusively the energy saving 'low pressure' concept (ICI, Lurgi, MGC process) at 230–280 °C and 50–100 bar[6,7]. The highly active catalysts used for this process are copper-based (together with metals like Zn, Al, Cr, Mn, V) and thus very sensitive to sulfur compounds. The sulfur content in the synthesis gas has to be below 0.1 ppm! The life-expectation of these catalysts is 3–4 years.

Low temperatures and ultra pure catalysts reduce the formation of unwanted by-products such as methane, dimethylether, higher alcohols and hydrocarbons to about 1%. Highly pure methanol (99.99%) is finally obtained by 2 successive distillations[6]. The table illus-

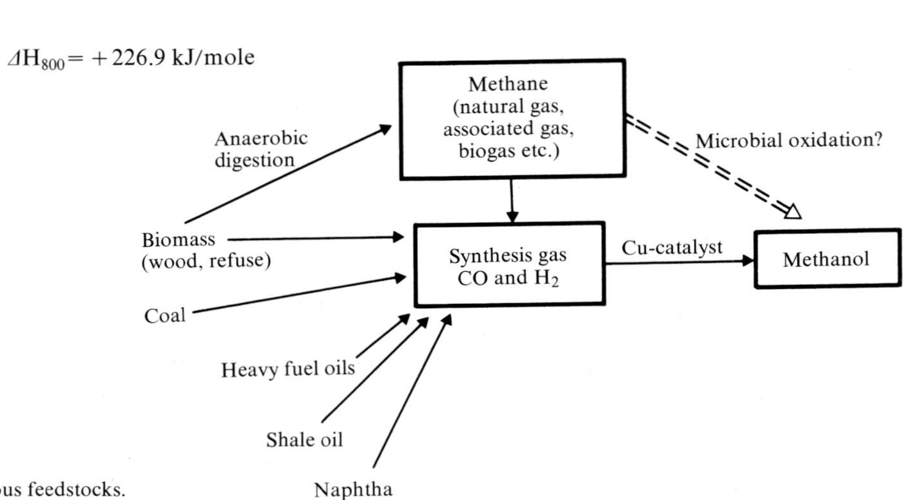

Figure 1. Methanol synthesis from various feedstocks.

Typical requirements per metric ton of methanol produced from methane by low pressure processes[7]

	ICI**	MGC**	Lurgi*
Feed and fuel (natural gas, GJ)	31.0	32.7	29.3
Electric power (kWh)	35	20	50
Feed water (m^3)	1.15	1.2	0.76
Cooling water (m^3)	70	100	45
Catalyst cost (US-dollars)	0.9	no data	1.02
Capacity of operating plants (tons/day)	50–1800	10–600	130–1360
Capacity of plants under construction (tons/day)		2000–4000	450–2000

ICI, Imperial Chemical Industries Ltd.; MGC, Mitsubishi Gas Chemical Co., Inc., * CO_2 available; ** no CO_2 available.

trates the energy and feed requirements for 3 typical modern low pressure processes.

3. *Wood as a feedstock for methanol synthesis*

Wood is a very interesting base for methanol production because it represents stored solar energy in the form of partially reduced carbon compounds and unlike petrochemical sources it is a renewable chemical feedstock. The interest in the economic feasibility of methanol production based on wood as a feedstock is worldwide[2,5]. So far there are no methanol plants operating on an industrial scale which use wood as a feedstock. Results from pilot plant studies in the United States and Canada[5] suggest that it is technically feasible but not as yet economically attractive to produce methanol from wood. Compared with natural gas (conversion efficiency above 60%) the energy yields for methanol from wood are poor. For a wood waste plant, the conversion efficiency is only 38% (heating value of methanol as a fraction of the total energy input into the plant)[5]. In addition the calculated investment costs for an industrial wood-based methanol plant (200,000 tons per year) are about 3 times higher than for a conventional natural gas facility[5]. However, in areas with large wood resources and high transport costs for petrochemical feedstocks (natural gas, petroleum) the production of methanol (via synthesis gas) from wood or refuse might become economically and ecologically attractive in the near future.

4. *Utilization and properties of methane and methanol*

Methane and methanol are important basic chemicals for the petrochemical industry and for the generation of energy (see fig. 2).

Sources of methane (world resources 70–230 · 10^{12} m^3)[8]: Natural gas[9] containing 65–95% methane, 1–6% ethane and 0–25% N_2, CO_2 H_2S, associated gas[9] containing 50–85% methane, 8–20% ethane, 3–12% propane and 0–10% CO_2 and biogas[10] containing 70% methane, 29% CO_2 and 1% H_2.

Storage and transport of methane: Methane can be liquified and stored at −162 °C under atmospheric pressure (estimated world production 1980: 43 · 10^9 m^3 liquified natural gas (LNG)[8]). Natural gas is transported either by pipelines or as LNG by refrig-

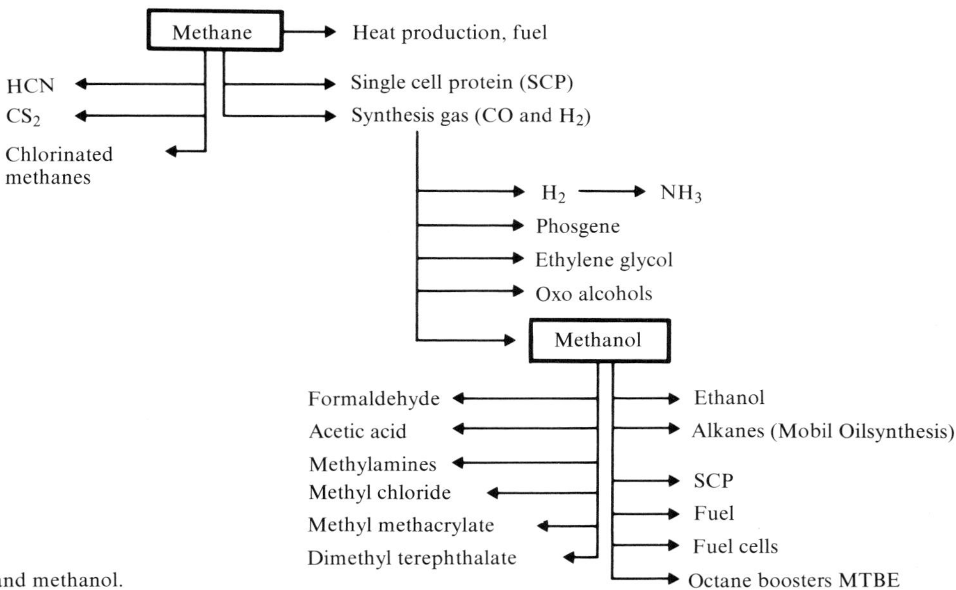

Figure 2. Utilization of methane and methanol.

erated tankers. LNG is condensed to 1/600 of the original volume of natural gas and is thus easier to store in large amounts. However, storage and transport of natural gas or LNG are both expensive and hazardous[8,11].

For political and commercial reasons (including the transport and storage problems mentioned above) large amounts of associated gas from production fields are still flared. For Saudi Arabia the reported amount of flared associated gas is $40 \cdot 10^9$ m^3 per year[8] and for all the OPEC states together a figure of 60% is given[12]. These annual losses in the Near East alone are equal to about 16% of the United States annual gas consumption[2]. This waste of a non-renewable resource should be stopped or at least reduced in the near future. Instead of undergoing the expensive and hazardous transport in pipelines or LNG-tankers, natural gas or associated gas could be converted into methanol and other derivatives on the spot. Methanol can be transported or stored as easily as gasoline and would be much cheaper and less risky to handle than natural gas or LNG. It is predictable that methanol will have a more important role as a basic petrochemical building block and as a source for energy generation in the future because of the multisources of synthesis gas[9]. Methanol is among the 20 most important organic chemicals[13]. The world production for 1980 is estimated to be 12 million tons but extrapolations predict a production of 68 million tons (or even more optimistic 200 million tons) of methanol for 1985[9]. Predicted uses for 1985 are: peak power generation (42%), established chemicals (25%), gasoline (21%), steel fabrication (6%), single cell protein (4%) and sewage treatment (2%)[9].

5. *Bioconversion or petrochemical synthesis?*

At present no *direct* chemical conversion of methane into methanol using less drastic conditions than those already described is known. All the processes in operation today synthesize methanol indirectly via synthesis gas. Studies on the direct catalytic oxidation of methane by a one-step reaction report only poor yields.

In contrast to ethanol, which can be produced from organic substrates such as sugars or cellulose by fermentation processes with yeasts or bacteria, no direct fermentation process for the production of methanol is known so far. The microbial production of methanol from organic substrates might involve the following steps (see also fig. 1):

1. Anaerobic digestion of biomass (manure, wood waste etc.) to produce biomethane (biogas). The theoretical fuel energy conversion efficiency for this process is reported to be above 90%[3].
2. Bioconversion of methane (natural gas, biogas) into methanol. The conversion efficiency of this step should be comparable or even better than the efficiency (above 60%) reached in the low pressure synthesis based on natural gas (theoretical maximum about 80%).

A semi-biological process could also be envisaged: low pressure methanol from biogas (step 1 followed by the chemical methanol synthesis described in section 2).

A considerable number of microorganisms (bacteria and yeasts), the methanotrophs or methylotrophs, are known to be able to grow on methane and/or methanol (and sometimes also on various other C_1 compounds such as methylamines, formamide, formate etc.) as the sole source of carbon and energy. Some years ago, Foo and Heden[3,14] proposed the use of such organisms for the biocatalytic production of methanol starting from methane. Since that proposition (1977/78) no practical results have been reported. Nevertheless, the idea is still very attractive. During the last few years a great deal of information on the physiology of the methylotrophs and especially on the mechanisms and enzymes of methane oxidation has been published.

6. *Methanotrophy – physiology of methane oxidation*

The microorganisms utilizing C_1 compounds are divided into 2 major groups, the obligate methylotrophs (growth only on C_1 compounds such as methane, methanol, methylamines etc.) and the facultative methylotrophs (growth on C_1 compounds or

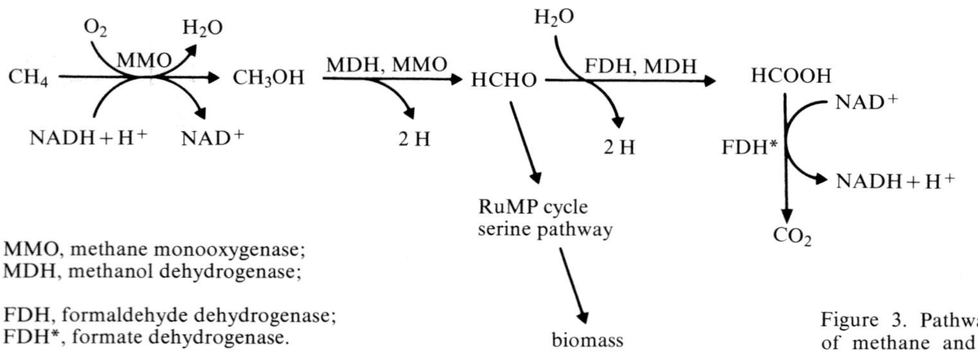

CH_4 → (MMO; O_2, $NADH + H^+$ in; H_2O, NAD^+ out) → CH_3OH → (MDH, MMO; 2 H) → HCHO → (FDH, MDH; H_2O in; 2 H) → HCOOH → (FDH*; NAD^+ → $NADH + H^+$) → CO_2

HCHO → RuMP cycle / serine pathway → biomass

MMO, methane monooxygenase;
MDH, methanol dehydrogenase;

FDH, formaldehyde dehydrogenase;
FDH*, formate dehydrogenase.

Figure 3. Pathway for the bacterial oxidation of methane and methanol.

other organic substrates such as acids, sugars etc.). In the following discussion we will concentrate on the methane-utilizing isolates (methanotrophs).

The obligate methanotrophs are generally classified into 5 genera: *Methylococcus, Methylomonas, Methylobacter, Methhylosinus* and *Methylocystis*[14–21]. All these organisms possess a complex internal arrangement of paired membranes either as bundles of vesicular discs (type I, in *Methylococcus, Methylomonas, Methylobacter*) or as layers around the periphery of the cells (type II, in *Methylosinus, Methylocystis*). These membranes are most probably involved in methane oxidation[18,20]. Only few facultative methanotrophs have been studied and described e.g. *Methylobacterium organophilum*[22] and *Methylobacterium* R 6[23].

The pathway of methane oxidation used by the methanotrophs is shown in figure 3 (for details see the reviews 14–21, pool of data from different organisms). The oxidation of methane to CO_2 is thought to proceed via a series of two-electron oxidation steps. Each step is sufficiently exergonic to permit, at least theoretically, the generation of 1 or more equivalents of ATP, but this has yet to be demonstrated. Formaldehyde occupies a central position in the oxidation of methane, for it is the branch point intermediate from where carbon is both assimilated into biomass (70%) via the ribulose monophosphate pathway (type I-organisms) or the serine pathway (type II-organisms) and dissimilated to CO_2 (30%) to provide energy for growth. The methane monooxygenase reaction (MMO) in *Methylococcus capsulatus, Methylomonas methanica (Pseudomonas methanica)* and *Methylosinus trichosporium* requires NADH as a cofactor (electron donor) whereas the cofactor for methanol dehydrogenase reaction (MDH) seems to be PQQ[28]. The formaldehyde dehydrogenase reaction (FDH) is mediated by different isoenzymes which are either dependent on NAD and glutathione *(Methylomonas methanica)* or NAD-independent. In *Methylococcus capsulatus* methanol oxidation and formaldehyde oxidation are both catalyzed by one single bifunctional methanol dehydrogenase. The formate dehydrogenase reaction (FDH*) is NAD-dependent in *Methylococcus capsulatus, Methylomonas methanica* and *Methylosinus trichosporium.*

For a biocatalytic production of methanol, MMO and MDH are the 2 relevant target enzymes at which we should have a closer look.

7. *Methane monooxygenase (MMO)*

Methane monooxygenase has been intensively studied in 3 methanotrophs: *Methylomonas methanica, Methylococcus capsulatus* and *Methylosinus trichosporium*[14–21]. The oxygen in the resulting methanol was shown to be exclusively derived from dioxygen. All the investigated MMOs involve cytochromes and show specific requirement for NADH as the electron donor. Ethane was also oxidized by the enzyme preparations.

The enzyme of *Methylococcus capsulatus* oxygenates a quite extraordinary range of compounds in addition to methane (co-oxidation) e.g. methanol, methane derivatives, n-alkanes, n-alkenes, ethers, alicyclic, aromatic and heterocyclic compounds. The enzyme thus shows a rare lack of substrate specificity[20]. The MMO of *Methylomonas methanica (Pseudomonas methanica)* was found to oxygenate the substrate analog bromomethane as well as ammonium chloride. The enzyme was inhibited by dithiothreitol, reduced glutathione and cyanide[20]. The MMO of *Methylosinus trichosporium* is possibly not obligatorily NAD-dependent. Ascorbate and methanol, in the presence of low concentrations of phosphate, were also effective electron donors[20]. However, these findings could not be reproduced by other researchers[19] and NADH is said to be the only effective electron donor. The MMO of *Methylosinus trichosporium* has a substrate specificity quite similar to that of *Methylococcus capsulatus* but the system of *Methylomonas methanica* is more restricted[19].

8. *Methanol dehydrogenase (MDH)*

In cell suspensions of *Methanomonas methanooxidans* (Brown and Strawinski) it was reported that 3 mM iodoacetate inhibited methanol oxidation but not methane oxidation and 75% of the consumed methane were accumulated as methanol[3,24]. However, these findings could not be reproduced by other researchers[25]. In fact it has been found that iodoacetate inhibits methane oxidation more than methanol oxidation[20,26]. Small amounts of methanol can be obtained from cell suspensions of *Methylomonas methanica* or *Methanomonas methanooxidans* at phosphate concentrations of 70–80 mM[3,26]. In cell free extracts of *Methylosinus trichosporium* the MDH activity was inhibited with 150 mM phosphate buffer[20,27] and also EDTA was found to have an inhibitory effect in the same system[3,14].

The MDHs of *Methylomonas methanica, Methylosinus trichosporium, Methylosinus sporium* and *Methylococcus capsulatus* can be coupled via phenazine methosulfate to oxygen or to artificial electron acceptors such as 2,6-dichlorophenol indophenol or cytochrome c[19,20]. Ammonia or methylamine act as activators and the enzyme has a broad substrate specificity, being able to oxidize primary alcohols and formaldehyde. For some methanol- but not methane-utilizing methylotrophs the cofactor of MDH has recently been identified as a new coenzyme PQQ[28].

9. *Research program for biocatalytic methane oxidation*

The following basic steps must be taken to establish a bioconversion process as proposed in section 5.

- Selection of highly efficient obligate and/or facultative methanotrophs including genetic improvement of strains (criteria: high MMO activity, reasonable generation time, increased methanol tolerance, eventually application of thermophilic strains).
- Studies on the regulation of methane and methanol oxidation. Activation of MMO and inhibition of MDH. Search for selective inhibitors of MDH which should not affect the methane oxidation.
- Search for artificial electron donors in vivo and in vitro to replace NADH in the methane monooxygenase reaction (eventually application of electroenzymological methods or co-oxidation systems). The energy metabolism is one of the crucial problems in this program.
- Selection of methanol accumulating mutants. Several types of mutants are possible: MDH^--mutants (facultative methanotrophs) or cytochrome c mutants. Cytochrome c is involved in electron transport during the methanol oxidation reaction.
- Studies on immobilization and stability of MMO or intact cells.
- Construction of a methanol producing bioreactor using immobilized cells or enzyme systems.

With our present knowledge it cannot yet be decided whether an obligate or a facultative methanotroph is more suitable for this work or whether the inhibition of the methanol oxidation is easier to achieve with inhibitors or by mutation. The knowledge acquired from the production of single cell proteins with methane and methanotrophs will be very helpful for the plant design.

10. *Outlook*

The major advantage of a biotechnological process for the conversion of methane into methanol compared with the petrochemical synthesis are the mild transformation conditions. Atmospheric pressure and low temperatures (maximum 60 °C for thermophilic strains) would be sufficient for a microbial process. Thus one would expect remarkable process energy savings, even in comparison with the modern low pressure chemical methanol synthesis. In addition the investment costs for a large scale methanol fermentation plant should be rather lower than for a conventional synthesis gas methanol plant. So far our comparison has dealt with large industrial facilities. But we should not forget that a microbial process has one major advantage over a petrochemical process, namely, microbial processes are the only choice for decentralized small and medium sized facilities. Even if a microbial conversion process might not be economically feasible on a large scale its ecological and economical advantages are obvious in the field of small technology. On the other hand a petrochemical methanol synthesis is not considered to be of economical interest on a scale below about 10,000 tons a year. A very attractive combination of 2 biological processes feasible on a small scale (for farms, villages etc.) would be methane production by anaerobic digestion of biomass followed by bioconversion into methanol. Thus storage and transportation problems associated with methane (biogas etc.) might be overcome. Such a combined biotechnological process allows the decentralized production of a liquid fuel based on renewable biomass.

* Acknowledgment. The authors thank J.A.L. Auden, J. Konecny, M. Küenzi and J. Nüesch for reviewing this manuscript.

1 F. Marschner, F.W. Möller and H.P. Gelbke: Methanol, in: Ullmanns Enzyklopädie der technischen Chemie, 4th edn, 1978, vol. 16, p. 621.
2 N.P. Cheremisinoff: Gasohol for Energy Production. Ann. Arbor Science Publ., Ann. Arbor, Mich., 1979.
3 E.L. Foo and G.C. Hedén, Is a biocatalytic production of methanol a practical proposition?, in: Microbial Energy Conversion, p. 267. Ed. H.G. Schlegel and J. Barnea. Pergamon Press, New York 1977.
4 H. Pichler and A. Hector, Synthesegas, in: Ullmanns Enzyklopädie der technischen Chemie, 3rd edn 1965, vol. 16, p. 599.
5 R.M. Rowell and A.E. Hokanson, Methanol from wood: A critical assessment, in: Progress in Biomass Conversion, vol. 1, p. 117. Ed. K.V. Sarkanen and D.A. Tillman. Academic Press, New York 1979.
6 J.A. Camps and D.M. Turnbull, Synthetic Gas production for Methanol: Current and Future Trends, in: Hydrogen: Production and Marketing, p. 123. Ed. W.N. Smith and J.G. Santangelo. ACS Symposium Series 116, American Chemical Society, Washington, DC, 1980.
7 Petrochemical Handbook 1979 Issue, Hydrocarb. Process., November 1979, p. 115.
8 Anon., Die grosse Zeit der Gasmänner. Der Spiegel Nr. 42, p. 79 (1979).
9 L.F. Hatch and S. Matar, From Hydrocarbons to petrochemicals, part 6 Petrochemicals from methane. Hydrocarb. Process., October 1977, p. 153.
10 J. Lawrie: Natural Gas and Methane Sources. Chapman and Hall, London 1961.
11 L.N. Davis: Gambling on 'frozen fire'. New Scient. *85,* 70 (1980).
12 L.F. Hatch and S. Matar, From Hydrocarbons to petrochemicals, part 1-Natural and associated gas. Hydrocarb. Process., May 1977, p. 191.
13 H.H. King, R. Williams and C.A. Stokes, Methanol - Fuel or Chemical? Hydrocarb. Process., June 1978, p. 141.
14 E.L. Foo, Microbial Production of Methanol. Process Biochem. *13,* 23 (1978).
15 D.W. Ribbons, J.E. Harrison and A.M. Wadzinski, Metabolism of single carbon compounds. A. Rev. Microbiol. *24,* 135 (1970).
16 J.R. Quayle, The Metabolism of One-Carbon Compounds by Micro-Organisms. Adv. Microbiol. Physiol. *7,* 119 (1972).
17 N. Kosaric and J.E. Zajic: Microbial Oxidation of Methane and Methanol. Adv. Biochem. Engng *3,* 89 (1974).
18 C. Anthony, The biochemistry of methylotrophic micro-organisms. Sci. Prog., Oxford *62,* 167 (1975).
19 J. Colby, H. Dalton and R. Whittenbury, Biological and biochemical aspects on microbial growth on C_1 compounds. A. Rev. Microbiol. *33,* 481 (1979).
20 R.S. Wolfe and I.J. Higgins, Microbial Biochemistry of Methane - A Study in Contrasts. Int. Rev. Biochem., Microbiol. Biochem. *21,* 267 (1979).
21 R. Whittenbury, K.C. Phillips and J.F. Wilkinson, Enrichment, Isolation and Some Properties of Methane-utilizing Bacteria. J. gen. Microbiol. *61,* 205 (1970).
22 T.E. Patt, G.C. Cole and R.S. Hanson, Methylobacterium, a New Genus of Facultatively Methylotrophic Bacteria. Int. J. Syst. Bact. *26,* 226 (1976).

23 R.N. Patel, C.T. Hou and A. Felix, Microbial Oxidation of Methane and Methanol: Isolation of Methane-utilizing Bacteria and Characterization of a Facultative Methane-Utilizing Isolate. J. Bact. *136*, 352 (1978).
24 L.R. Brown, R.J. Strawinski and C.S. McCleskey, The Isolation and Characterization of *Methanomonas methanooxidans* Brown and Strawinski. Can. J. Microbiol. *10*, 791 (1964).
25 J.F. Wilkinson, Physiological Studies of Bacteria Grown on Methane, in: Microbial Growth on C_1 Compounds, p.45. Ed. The Society of Fermentation Technology, Japan 1975.
26 I.J. Higgins and J.R. Quayle, Oxygenation of Methane by Methane-Grown *Pseudomonas methanica* and *Methanomonas methanooxidans*. Biochem. J. *118*, 201 (1970).
27 G.M. Tonge, D.E.F. Harrison, C.J. Knowles and I.J. Higgins, Properties and partial purification of the methane-oxidizing enzyme system from *Methylosinus trichosporium*. FEBS Lett. *58*, 293 (1975).
28 J.A. Duine, J. Frank and P.E.J. Verwiel, Structure and Activity of the Prosthetic Group of Methanol Dehydrogenase. Eur. J. Biochem. *108*, 187 (1980).

Artificial photosynthetic systems

by Pierre Cuendet and Michael Grätzel*

Institut de Chimie Physique, Ecole Polytechnique Fédérale, CH-1015 Lausanne (Switzerland)

Presently endeavors are being undertaken to design systems capable of converting solar radiation into fuels. These efforts are timely and mandatory in order to reduce our dependence on fossil energy reserves. Recently there has been an explosion of information on this subject[1–13] and a multitude of processes are under investigation which attempt to mimic photosynthesis. The present article gives a brief overview over these artificial systems and explains some basic principles of their operation.

Plant photosynthesis serves to convert light into energy-rich compounds such as carbohydrates. This biological device is, however, a rather poor energy converter if the amount of biomass produced by the incident solar flux is considered. Although the primary photoredox reactions that occur in the chloroplasts proceed with high quantum efficiency the overall conversion yield is approximately 5-6% and falls to 1-3% at best when averaged over the whole year[14]. Major losses are due to growth, adaptation and reproduction processes. Thus the photosynthetic machinery could work more efficiently had it been designed mainly for fuel production without any constraints due to evolutionary history[15].

Artificial systems try to overcome this shortcoming of the biological counterpart by simplifying both the energy storing process and the molecular units that accomplish this transformation. Three different approaches are presently being pursued. In the first or hybrid system the thylakoid membranes or individual photosystems are employed as light harvesting units. The objective is to exploit the high efficiency of the primary photosynthetic redox events without attempting to synthesize carbohydrates from CO_2. Instead, hydrogen generation from water is achieved through artificial redox relays and catalysts.

The second approach is to employ synthetic molecular assemblies such as micelles or membranes as reaction systems. These aggregates simulate the microenvironment present in biological systems and serve as a host for hydrophobic entities participating in the photoreactions.

Finally, artificial systems with very little resemblance to their biological counterpart are also under study. Prominent and promising at the same time are here colloidal semiconductor solutions.

None of the presently available model systems will achieve the conversion of carbon dioxide into sugar under illumination with visible light. However, they will perform other more simple endoergic transformations such as the photocleavage of water and the production of methanol from CO_2.

Hybrid systems

Before the recent development of bifunctional catalysts and the design of totally artificial systems, which will be described later, the photolysis of water by visible light was achieved only by the chloroplast and cyanobacteria machinery. The light-induced water decomposition

$$H_2O \xrightarrow{h\nu} H_2 + \tfrac{1}{2}O_2 \tag{1}$$

is an attractive means by which to convert solar photon energy into chemical potential. Cell-free hybrid systems, composed of isolated chloroplasts coupled to suitable redox catalysts have been shown to achieve this decomposition[16]. Figure 1 illustrates

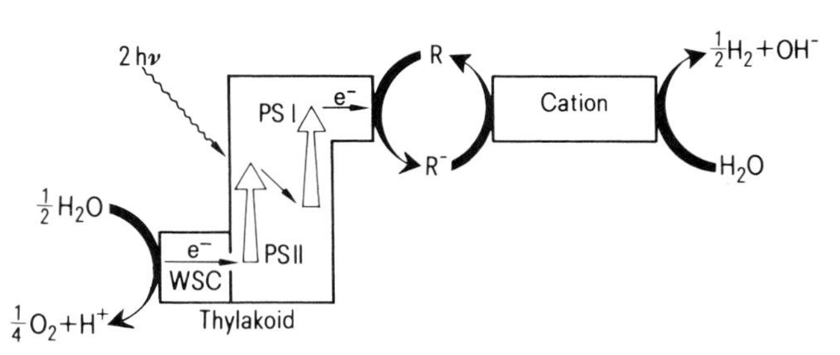

Figure 1. Schematic illustration of a cell-free hybrid system for the biophotolysis of water.

the mechanisms of this biophotolysis where the whole electron transport chain supported by the thylakoidal membrane is involved. Electrons are abstracted from water by the water splitting complex (WSC) at the oxidative end of the chain. The electron relay R is reduced by the terminal PSI acceptor and then transfers electrons to the redox catalyst which evolves hydrogen.

Since the first hydrogen production obtained in such systems with water as the sole electron donor[17] about 10 years ago, great improvements have been obtained and hydrogen yields of 20–50 μmoles per h and mg chlorophyll are now currently measured for several hours.

An oxygen scavenging system is, however, usually needed in these hybrid systems in order to remove oxygen and peroxide radicals which are formed under illumination and which inactivate the chloroplast components and diminish the energy conversion efficiency. Various means have been proposed to enhance the stability and the efficiency of the processes. Longer life-times have been recently obtained by using alginate-immobilized[18] or albumine cross-linked chloroplasts[19]. Oxygen-insensitive hydrogenases still possessing good catalytic activities have also been studied[20]. It is worthwhile to note that synthetic compounds can replace part of the system. Artificial catalysts, such as PtO_2[21], platinum or palladium asbestos[22], or ultrafine platinum colloids developed in our laboratory[18,23] have been used instead of hydrogenases and can evolve hydrogen at comparable rates. Various synthetic relays can also replace the natural electron carriers. Particularly, synthetic iron-sulfur and iron-selenium analogues have been developed to mimic ferredoxin[24].

The low efficiency of the hydrogen evolution compared to the electron transport capacity of the photosynthetic membrane has been shown in recent studies to be due to an important loss of reductive equivalents by reoxidation of methylviologen, a relay commonly employed, between PSII and PSI[22,25]. The electron transfer efficiency between the thylakoidal membrane and the redox catalyst has been shown in our laboratory to depend on the pH and the redox potential of the relay[23]. Our present work in this field is now devoted to the study of new compounds which back react much slower with the membrane.

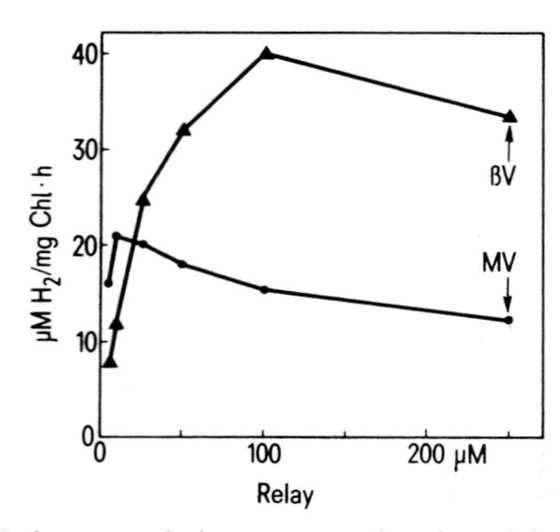

Figure 2. Hydrogen evolution rates as a function of the concentration of 2 electron relays in a hybrid system composed of broken spinach chloroplasts, platinum colloids and oxygen scavengers.

As an example, figure 2 shows the hydrogen evolution rates obtained in a system containing broken spinach chloroplasts, platinum colloids, O_2 scavengers and methylviologen (MV) or betaine-viologen (βV) as electron relays. This last compound, although thermodynamically less favorable because of its higher redox potential, mediates a two times higher hydrogen production rate. This is attributed to the negative charge of its reduced form which produces electrostatic repulsion from the membrane leading to smaller extent of back reaction with the electron transport chain than in the case of the positive analogue methyl viologen.

Great improvements, particularly in the stabilization of the water splitting complex and the pigment system will certainly be achieved in the future. This will lead to a better understanding of the natural photosynthetic energy converter and could help in the design of more efficient biomimetic photoredox devices for the photolysis of water.

Biomimetic systems

These devices make use of the effect of self-assembly of surfactant agents in aqueous solution to produce molecular assemblies mimicking the microenvironment present in the biological systems. Examples are given in figure 3 which illustrates structural features of micelles, microemulsions and vesicles. These aggregates are distinguished by a charged lipid water interphase that may be exploited to control kinetically the electron transfer events. In the light-induced redox reaction

$$S + R \underset{}{\overset{h\nu}{\rightleftharpoons}} S^+ + R^- \qquad (2)$$

the goal is to enhance the rate of the forward reaction and at the same time retard that of the backward electron transfer. A micellar system that achieves this goal is now discussed in detail.

Consider an artificial system where chlorophyll a (Chla) is used as a sensitizer and duroquinone (DQ) as an electron relay[26]. Such a system is reminiscent of the redox species involved in photosystem II. Both Chla and DQ are incorporated into an anionic micelle (figure 4). Light is used to excite Chla and to promote electron transfer to DQ. A radical ion pair is thereby produced within the aggregate. In an anionic micelle DQ^- is clearly destabilized with respect to the aqueous bulk solution and will therefore be ejected into the water. Conversely, $Chla^+$ is electrostatically stabilized by the micelle and remains associated with it. Once $Chla^+$ and DQ^- are separated, their diffu-

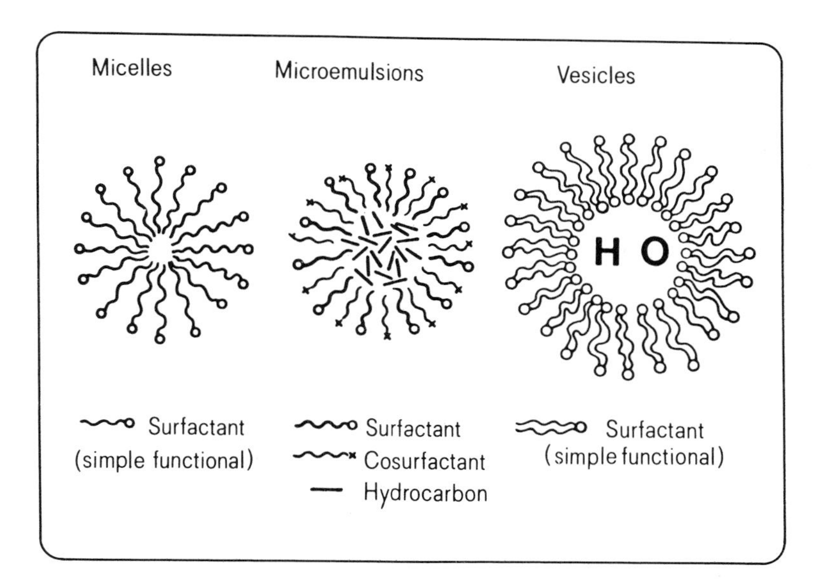

Figure 3. Structural features of colloidal assemblies employed in light-induced charge separation.

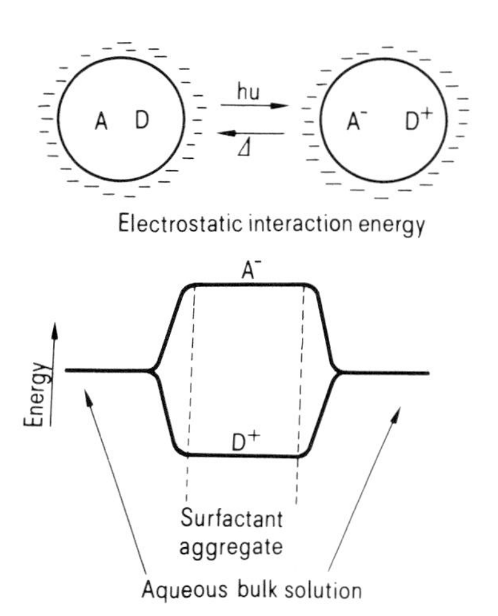

Figure 4. Schematic illustration of a light-induced charge transfer reaction in an anionic surfactant aggregate.

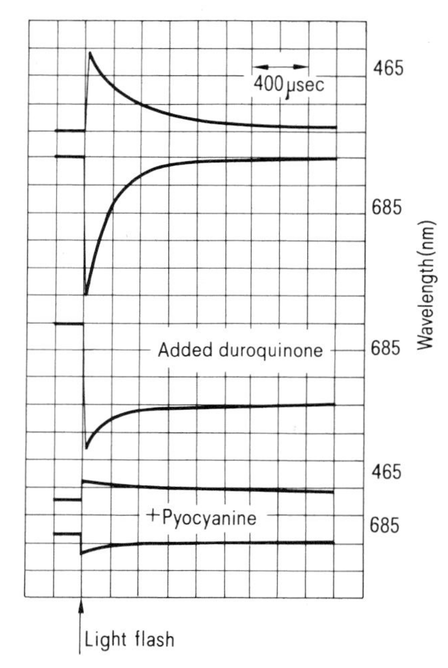

Figure 5. Oscilloscope traces obtained in the 694 nm laser photolysis of chlorophyll a (3×10^{-5} M) in sodium lauryl sulfate (0.1 M) micelles. From the top: decays of chlorophyll a triplet (trace 1) and bleaching of absorption (trace 2) in the absence of additives; traces 3 and 4: same decays in presence of 3×10^{-3} M duroquinone; trace 5: added 3×10^{-3} M duroquinone and 8×10^{-5} M pyocyanine.

sional re-encounter will be obstructed by the ultrathin barrier of the micellar double layer. The efficiency of the charge separation process will crucially depend on the relative rates of DQ^- ejection and intramicellar back transfer of electrons from DQ^- to $Chla^+$. Flash photolysis data obtained from this system are presented in figure 5. The upper 2 traces reflect the behaviour of Chla-triplet states in the absence of DQ. The time course of both the 685 and 465 nm absorption is drastically affected when DQ is added to the micelles. The long term bleaching can be attributed to the formation of $Chla^+$ cation radicals which through the field gradient in the micelle/water interface are prevented from recombining with DQ^-. If pyocyanine (PC^+) is added to the solution the charge separation is annihilated as PC^+ reoxidizes DQ^-. The neutral PC radical produced can enter the micelle and reduce in turn $Chla^+$.

Further progress in the development of molecular assemblies converting light into chemical energy was made by designing and synthesizing surfactants with suitable functionality. These are distinguished from simple surfactants by the fact that the micelle itself participates in the redox events. Noteworthy examples include:

a) micelles with acceptor relays as counterions (transition metal ion micelles[27,28]),

b) crownether surfactants[29],

c) amphiphilic redox relays[30,31] and redox chromophores[32].

The advantage of these aggregates is that electron donor and acceptor are held in close proximity. Thus, although the analytical amounts of these species in solution may be small, the formation of aggregates provides high local concentration which, in turn, insure that the photoredox reaction proceeds at a high rate. Very often these types of molecular assemblies display also cooperative effects assisting light-induced charge separation.

As an example, consider the detergent cupric lauryl sulfate ($Cu(LS)_2$) which is obtained from the commercially available sodium lauryl sulfate via ion exchange[27,28]. The counter ion atmosphere of these micelles is constituted by cupric ions. It has been shown that the reduction of cupric to cuprous ion by a sensitizer incorporated into these aggregates can occur at extremely high rates. The electron is transferred from the excited state of the sensitizer to the Cu^{2+} ions in less than 1 nsec. Thus, as is the case in photosynthesis, even singlet excited states can be used for this reaction. The Cu^+ ion escapes into the aqueous phase before back reaction can occur and charge separation can be achieved.

Apart from micelles, molecular assemblies such as microemulsions[33,34] and vesicles[35] deserve particular attention in the context of photoredox reactions in biomimetic aggregates. Calvin and co-workers have for the first time illustrated light-included electron transfer across the bilayer of liposomes[35]. $Ru(bipy)_3^{2+}$ was incorporated together with a sacrificial electron donor into the inner water core of a vesicle while a viologen was the ultimate electron acceptor in the bulk aqueous phase. The formation of the viologen radical was detected under illumination and from the kinetic analysis rate parameters obtained for the transmembrane electron transfer. This process requires a time of several μseconds and follows probably an electron exchange mechanism involving $Ru(bipy)_3^{2+}$ and $Ru(bipy)_3^{3+}$ on opposite sides of the membrane.

The same group has also performed elegant studies of electron transfer reaction in the water core of inverted micelles[34].

Artificial systems

Totally artificial systems are presently under intense investigation that show no apparent similarity with their natural counterpart. Figure 6 summarizes the principle of light harvesting and energy conversion of 3 systems typically employed here. A sensitizer/relay pair is used in the first device. Light induced electron transfer produces the radical ions S^+ and R^- which are subsequently employed to oxidize and reduce water respectively. The success of this system is mainly dependant on the use of active redox catalysts that can intervene extremely rapidly and, moreover, specifically in the hydrogen and oxygen formation from water. Several formidable problems have to be overcome here.

a) the catalysts have to intercept the thermal back reaction which occurs in the micro-to-millisecond time domain.

b) The water reduction catalyst 2 must compete with oxygen reduction by R^- which is expected to occur at a diffusion controlled rate. This sets the rate limit required for H_2 generation at several μseconds.

c) The intervention of the catalysts has to be specific in order to avoid short circuitry of the back reaction.

d) In order to achieve H_2/O_2 separation an oxygen carrier has to be present in solution that adsorbs the oxygen produced during photolysis.

A few years ago it would have seemed impossible to overcome all of these difficulties. At that time there was even no O_2 producing catalyst available. This was discovered in 1978 in our laboratory[36,37] in the form of noble metal oxides such as PtO_2, IrO_2 and RuO_2. The latter has been most widely investigated since then[38–40]. Platinum has been known for a long time to mediate water reduction by agents such as V^{2+}, Cr^{2+} and also reduce viologens. However, it required several years of research to develop a Pt catalyst that would

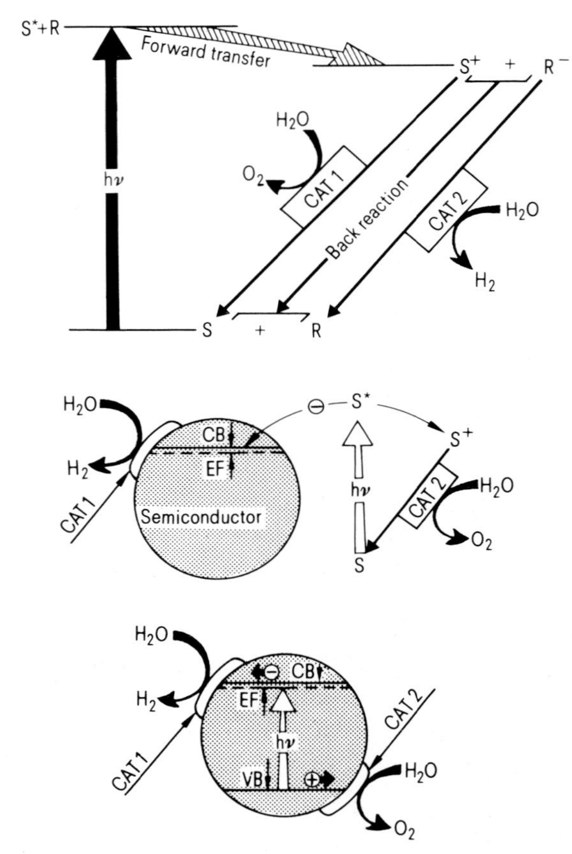

Figure 6. Light harvesting and catalytic units for light-induced water decomposition.

satisfy conditions (a)–(c) indicated above, i.e. produce H_2 in the microsecond time domain at reasonably low Pt concentrations. The water splitting catalyst which evolved finally[41,42] was colloidal TiO_2 loaded simultaneously with ultrafine deposits of Pt and RuO_2. Sustained water cleavage by visible light was observed in solutions containing $Ru(bipy)_3^{2+}$ as a sensitizer, methylviologen as an electron relay and this bifunctional catalyst, the quantum yield being around 6% (75 °C).

Figure 6b introduces a second light harvesting unit for light-induced water cleavage. It consists of a colloidal semiconductor loaded with a suitable catalyst that mediates hydrogen production from water. A dye S is adsorbed on the colloids which upon excitation injects an electron in the semiconductor particle. The electron is channelled to Pt sites where hydrogen evolution takes place. A 2nd catalyst converts S^+ back into S under simultaneous formation of oxygen from water. An experimental verification of this system was found in colloidal TiO_2 particles loaded with Pt and RuO_2 onto which a surfactant $Ru(bipy)_3^{2+}$ complex is adsorbed[43]. The maximum quantum efficiency of water splitting observed with such a system is ca. 7% so far.

The third device depicted in figure 6c requires neither sensitizer nor electron relay. A semiconductor particle is charged with both a catalyst for water oxidation and water reduction. Band gap excitation produces an electron/hole pair in the semiconductor particle. Both charge carriers diffuse to the aqueous interface where hydrogen and oxygen generation occurs. TiO_2 charged simultaneously with Pt and RuO_2 was the first material to be employed in such a system. In this case UV-irradiation is required to excite the semiconducting support. Hydrogen and oxygen are cogenerated rather efficiently, the quantum yield Ø (H_2) approaching 30%. The domain of photoactivity has recently been displaced into the visible by using n-CdS particles as carriers for Pt and RuO_2. Interestingly, the RuO_2 deposit prevents photocorrosion by valence band holes and instead allows water oxidation[44].

Finally, an analogy between colloidal semiconductors and photosynthetic bacteria should be pointed out. CdS particles when loaded with RuO_2 alone are capable of splitting hydrogen sulfide into sulfur and hydrogen when illuminated by visible light.

$$H_2S \xrightarrow{h\nu} H_2 + S \qquad (3)$$

The RuO_2 serves here as a hole transfer catalyst accelerating the rate of charge exchange between the valence band of CdS and the sulfide ions in solution. No Pt is required as water reduction by conduction band electrons occurs very rapidly. The quantum yield for H_2S splitting approaches 35%. A similar photoreaction can be performed by photosynthetic bacteria that also employ sulfides as an electron source.

The key advantage of colloidal semiconductors over other functional organizations is that light-induced charge separation and catalytic events leading to fuel production can be coupled without intervention of bulk diffusion. Thus a single particle can be treated with appropriate catalysts so that different regions function as anodes and cathodes. It appears that this wireless photoelectrolysis could be the simplest means of large scale solar energy harnessing and conversion.

Acknowledgment. Acknowledgment is made to the Swiss National Energy Foundation for support of this work.

1 V. Balzani, L. Moggi, M.F. Manfrin, F. Boletta and M. Gleria, Science *189*, 852 (1975).
2 M. Calvin, Photochem. Photobiol. *23*, 425 (1976).
3 G. Porter and M.D. Archer, Interdisc. Sci. Rev. *1*, 119 (1976).
4 J. Bolton, Science *202*, 705 (1978).
5 A. Harriman and J. Barber, in: Photosynthesis in relation to model systems. Ed. J. Barber. Elsevier, Amsterdam 1979.
6 E. Schumacher, Chimia *32*, 194 (1978).
7 M. Grätzel, Ber. Bunsenges. Phys. Chem. *84*, 981 (1980).
8 S.N. Paleocrassas, Solar Energy *16*, 45 (1974).
9 S. Claesson, ed., Photochemical Conversion and Storage of Solar Energy. Swedish National Energy Board Report, Stockholm 1977.
10 M. Tomkiwics and H. Fay, Appl. Phys. *18*, 1 (1979).
11 J. Bolton and D.O. Hall, A. Rev. Energy *4*, 353 (1979).
12 M. Grätzel, Discussions Faraday Society on Photoelectrochemistry. Oxford 1980.
13 K.I. Zamaraev and V.N. Parmon, Catal. Rev. Sci. Engl. *22*, 261 (1980).
14 D.O. Hall, Fuel *57*, 322 (1978).
15 J.J. Katz and M.R. Wasielewski, in: Biotechnology and Bioengineering Symposium No.8. Ed. C.D. Scott. Interscience, New York 1979.
16 K.K. Rao and D.O. Hall, in: Photosynthesis in relation to model systems. Ed. J. Barber. Elsevier, Amsterdam 1979.
17 J.R. Benemann, J.A. Berensen, N.O. Kaplan and M.D. Kamen, Proc. natl. Acad. Sci. USA *70*, 2317 (1973).
18 P.E. Gisby and D.O. Hall, Nature *287*, 251 (1980).
19 M.F. Cocquempot, V. Larreta Garde and D. Thomas, Biochimie *62*, 615 (1980).
20 K.K. Rao, P. Morris and D.O. Hall, in: Hydrogenases: their catalytic activity, structure and function. Ed. H.G. Schlegel and K. Schneider. Goltzke K.G., Güttingen 1978.
21 M.W.W. Adams, K.K. Rao and D.O. Hall, Photobiochem. Photobiophys. *1*, 33 (1979).
22 M.M. Rosen and A.I. Krasna, Photochem. Photobiol. *31*, 259 (1980).
23 P. Cuendet and M. Grätzel, Photobiochem. Photobiophys. *2*, 93 (1981).
24 M.W.W. Adams, K.K. Rao, D.O. Hall, G. Christou and C.D. Garner, Biochim. biophys. Acta *589*, 1 (1980).
25 A.A. Krasnovski, C. Van Ni, V.V. Nikandrov and G.P. Brin, Pl. Physiol. *66*, 925 (1980).
26 C. Wolff and M. Grätzel, Chem. Phys. Lett. *52*, 542 (1977).
27 Y. Moroi, A.M. Braun and M. Grätzel, J. Am. chem. Soc. *101*, 567 (1979).
28 Y. Moroi, P.P. Infelta and M. Grätzel, J. Am. chem. Soc. *101*, 573 (1979).
29 R. Humphry-Baker, M. Grätzel, P. Tundo and E. Pelizzetti, Angew. Chem., int. Ed. *18*, 630 (1979).
30 P.-A. Brugger and M. Grätzel, J. Am. chem. Soc. *102*, 2461 (1980).
31 P.-A. Brugger, P.P. Infelta, A.M. Braun and M. Grätzel, J. Am. chem. Soc. *103*, 320 (1981).

32 K. Kalyanasundaram, Chem. Soc. Rev. *7,* 453 (1978).
33 J. Kiwi and M. Grätzel, J. Am. chem. Soc. *100,* 6314 (1978).
34 I. Willner, W.E. Ford, J.W. Otvos and M. Calvin, Nature *280,* 823 (1979).
35 M. Calvin, Int. J. Energy Res. *3,* 73 (1979).
36 J. Kiwi and M. Grätzel, Angew. Chem., int. Ed. *17,* 860 (1978).
37 J. Kiwi and M. Grätzel, Angew. Chem., int. Ed. *18,* 624 (1979).
38 J. Kiwi and M. Grätzel, Chimia *33,* 289 (1979).
39 J.M. Lehn, J.-P. Sauvage and R. Ziessel, Nouv. J. Chim. *3,* 423 (1979).
40 K. Kalyanasundaram, O. Micic, E. Promauro and M. Grätzel, Helv. chim. Acta *62,* 2432 (1979).
41 K. Kalyanasundaram and M. Grätzel, Angew. Chem., int. Ed. *18,* 701 (1979).
42 J. Kiwi, E. Borgarello, E. Pelizzetti, M. Visca and M. Grätzel, Angew. Chem., int. Ed. *19,* 646 (1980).
43 E. Borgarello, J. Kiwi, E. Pelizzetti, M. Visca and M. Grätzel, J. Am. Chem. Soc. *103* (1981).
44 K. Kalyanasundaram, E. Borgarello and M. Grätzel, Helv. chim. Acta *64,* 362 (1981).

Concluding remarks

The articles published in this volume assess the current state of development of biological systems that are harnessing solar energy – and offer some outlooks for their use in the future. These systems may very well change our lives. Of course, it is not easy to make projections for the future when each projection must vary from country to country and with system to system. Certainly the options open to the Third World countries in answering their energy problems will be different from those in the industrial countries. Of great importance is how much a country's gross national product has to be spent for energy imports! Moreover, projections must account for economic and ecological alterations brought about by the implementation of such biological systems. The production of ethanol in Brasil, for example, has increased over the last years from $0.6 \cdot 10^6$ m^3 in 1975 to $3.7 \cdot 10^6$ m^3 in 1979; the forecast for 1987 is $14 \cdot 10^6$ m^3 (Trindade, 1980). Will such an increase in ethanol production not pose many new restrictions and limitations on food production or on the ecology of the land used for the sugar crop necessary to accomodate this high ethanol production?
Today, mankind does not seem to be optimistic enough about being able to solve the problems of energy supply without creating even more problems for future generations. However, when one looks back through history, one sees that man has always been engaged in a learning process that has demanded his adaptation to the changing conditions and pressures of the environment. Not always have the best long term solutions to his problems been selected, as we are well aware today. However, we should not lose confidence that appropriate solutions can be found to secure our future energy supply and that of our childrens' children.

S.C. Trindade, Energy crops – the case of Brasil, in: Energy from Biomass. Preprint of abstracts, p. 1:K 2. Int. Conference, Brighton 1980.